Springer
Proceedings in Physics 40

Springer Proceedings in Physics

Managing Editor: H. K. V. Lotsch

Volume 30	*Short-Wavelength Lasers and Their Applications* Editor: C. Yamanaka
Volume 31	*Quantum String Theory* Editors: N. Kawamoto and T. Kugo
Volume 32	*Universalities in Condensed Matter* Editors: R. Jullien, L. Peliti, R. Rammal, and N. Boccara
Volume 33	*Computer Simulation Studies in Condensed Matter Physics: Recent Developments* Editors: D. P. Landau, K. K. Mon, and H.-B. Schüttler
Volume 34	*Amorphous and Crystalline Silicon Carbide* Editors: G. L. Harris and C. Y.-W. Yang
Volume 35	*Polycrystalline Semiconductors* Editors: H. J. Möller, H. P. Strunk, and J. H. Werner
Volume 36	*Nonlinear Optics of Organics and Semiconductors* Editor: T. Kobayashi
Volume 37	*Dynamics of Disordered Materials* Editors: D. Richter, A. J. Dianoux, W. Petry, and J. Teixeira
Volume 38	*Electroluminescence* Editors: S. Shionoya and H. Kobayashi
Volume 39	*Disorder and Nonlinearity* Editors: A. R. Bishop, D. K. Campbell, and S. Pnevmatikos
Volume 40	*Static and Dynamic Properties of Liquids* Editors: M. Davidović and A. K. Soper

Volumes 1–29 are listed on the back inside cover

Static and Dynamic Properties of Liquids

Proceedings of the International Symposium
Dubrovnik, Yugoslavia, June 27 – July 2, 1988

Editors: M. Davidović and A. K. Soper

With 127 Figures

Springer-Verlag Berlin Heidelberg New York
London Paris Tokyo Hong Kong

Dr. Milorad Davidović

Laboratory of Solid State Physics, "Boris Kidrič" Institute of Nuclear Sciences
YU-11001 Belgrade, Yugoslavia

Dr. Alan K. Soper

Neutron Division, Rutherford Appleton Laboratory, Chilton, Didcot
Oxon, OX11 0QX, UK

ISBN-13:978-3-642-74909-4 e-ISBN-13:978-3-642-74907-0
DOI: 10.1007/978-3-642-74907-0

Softcover reprint of the hardcover 1st edition 1989

2154/3150-543210 – Printed on acid-free paper

Preface

The idea of a workshop devoted to the static and dynamic properties of liquids arose from two smaller meetings on the same topic held at the Interfaculty Reactor Institute, Delft, The Netherlands in 1985 and 1987, and from the LAM conference (International Conference on Liquid and Amorphous Metals) held at Garmisch-Partenkirchen, FRG, in 1986. The object of the Workshop was to bring together experimentalists, theoreticians and people working on computer simulations in order to learn about developments in the microscopic behaviour of liquids. Approximately 35 scientists from Austria, France, Israel, Italy, Sweden, The Netherlands, the United Kingdom, the Federal Republic of Germany and Yugoslavia gathered at the Hotel "Libertas", Dubrovnik, in Yugoslavia, June 27-30, 1988 to hear ten invited talks and nineteen contributed papers on a wide range of subjects, including generalized hydrodynamics, mode coupling theory, liquids at surfaces, extraction of effective potentials, and the glass transition. The most numerous were the papers dealing with computer simulation, and then those on supercooled liquids and the glass transition. The number of experimental results was relatively small, which points to difficulties in this kind of research (evaluation of experimental data).

The workshop closed with an expanded round-table discussion of the main themes of the meeting, with an emphasis on the likely trend of investigations in the future. This discussion is summarized at the end of this volume.

We would like to express special gratitude to professors and colleagues who helped in organizing the workshop: Prof. Karl-Eric Larsson and Prof. Ulf Dahlborg, from the Royal Institute of Technology in Stockholm; Prof. Stephen Lovesey, from the Rutherford Appleton Laboratory, who initiated the main themes of the workshop. We are greatly indebted to Dr. Peter Verkerk from the Interfaculty Reactor Institute, Delft, and his colleagues, who spared no effort in defining the themes of the workshop and in organizing the first meeting of the Scientific Advisory Committee of the workshop. We thank very much Dr. Wolfram Schommers from Karlsruhe for his substantial help in making arrangements with Springer-Verlag, as well as Dr. H. Lotsch from Springer-Verlag for pleasant cooperation during the preparation of this book. We very much appreciate the help of people from the Laboratory of Solid State Physics and Radiation Chemistry, the "Boris Kidrič" Institute of Nuclear Sciences, Vinca, who were engaged in the organization of the workshop.

The papers included in this book have been refereed by experts, and represent, we believe, a realistic account of current scientific interest in liquids and the glassy state. They demonstrate that much progress has been made in recent years.

Belgrade
Chilton, Oxon.
March 1989

M. Davidović
A.K. Soper

Contents

Part I

Generalized Hydrodynamics and Mode Coupling Theories

Analysis of Neutron Scattering and Computer Simulation Studies on Noble-Gas Fluids

L.A. de Graaf

Interfaculty Reactor Institute, Delft University of Technology, Mekelweg 15, NL-2629 JB Delft, The Netherlands

Inelastic neutron scattering is a powerful technique to study the atomic dynamics in liquids and gases on nanometer and picosecond scales. The dynamic structure factor $S(k,\omega)$, which is directly connected to the neutron scattering cross-section, reflects density fluctuations with wavelength $\lambda = 2\pi/k$ and frequency ω. Except for very small k values, it is a fairly featureless function in ω. Therefore, in order to extract useful information from $S(k,\omega)$, it is necessary to have excellent data, reliable correction procedures, and a sophisticated theory, or model, with which to describe, or compare, the experimental results. Over the last five years or so, new high-quality neutron data have become available on the noble-gas fluids argon, neon and krypton, over a large range of densities. These have led to further activities on the theoretical side, especially on the basis of the Revised Enskog Theory for hard-sphere systems. Moreover, extensive computer Molecular-Dynamics simulations have been performed to get a better understanding of the influences of approximations in the theory, and of potential-shape effects. In this review the intercomparison of experiments, theory and computer simulations will be discussed, with examples mainly taken from the work of the author and his colleagues.

1. Introduction

For the study of the propagation and dissipation of spontaneous fluctuations in dense fluids, the main techniques of investigation are inelastic neutron scattering (INS) experiments and computer molecular-dynamics (CMD) simulations. In both, the relevant correlation functions, or their Fourier transforms, are probed with wavenumbers $k = 2\pi/\lambda > 1\ \text{nm}^{-1}$ and frequencies $\omega > 1\ \text{ps}^{-1}$. The neutron scattering cross section is related to the weighted sum of two functions, $S_s(k,\omega)$ and $S(k,\omega)$, by the expression [1]

$$\frac{d^2\sigma}{d\Omega d\omega} = \frac{k_f}{k_0}\frac{1}{4\pi}\left\{\sigma_c S(k,\omega) + \sigma_i S_s(k,\omega)\right\}. \tag{1}$$

k_0 and k_f are the momenta of incident and scattered neutrons in units of $(2\pi)^{-1}h$, and $k = |\mathbf{k}| = \mathbf{k}_0 - \mathbf{k}_f$. $S_s(k,\omega)$, the single-particle, or incoherent, dynamic structure factor, describes the dynamics of single particles in the fluid, while $S(k,\omega)$, the (coherent) dynamic structure factor, describes their relative motion in terms of the particle density. σ_c and σ_i are the bound atom coherent and incoherent scattering cross sections, respectively, which have been measured for all the elements and for many isotopes. Most samples scatter both coherently and incoherently. If one wants to study the coherent part, the incoherent part has to be corrected for, which can be done by using realistic models (if available). If the incoherence is caused by spinflip scattering, a separation can be achieved by neutron polarization analysis, as was done recently for liquid sodium [2]. Single-particle motions, and mode-coupling effects on the diffusion behaviour,as obtained from incoherent scattering from liquid sodium and dense hydrogen gas, are discussed by MORKEL [3] and VERKERK [4] in these proceedings. In the present paper only recent results of coherent scattering experiments from dense fluid Ne [5], ^{36}Ar [6], and Kr [7] will be considered. Earlier work on classical fluids, including liquid metals, has been reviewed before [8- 11], for recent results on Pb and Bi see [12].

In CMD simulations essentially exact results can be obtained for time-dependent correlation functions (also those unobservable in real experiments), for a given interatomic force law. Solving the equations of motion for a system of several hundred particles is in

Springer Proceedings in Physics, Vol. 40
Editors: M. Davidović and A.K. Soper
Static and Dynamic Properties of Liquids

most cases sufficient to simulate the behaviour of a macroscopic system. After the earliest work on hard spheres by ALDER and WAINWRIGHT [13] and on a system interacting with a more realistic, effective two-particle, potential by RAHMAN [14], this field has rapidly developed [8, 15]. Readers interested in these techniques are referred to the recent 'how-to-do-it' book by ALLEN and TILDESLEY [16]. For the simulation of noble-gas fluids, and especially 'computer-argon', the Lennard- Jones 12-6 potential has been the favourite choice as an effective two-particle approximation of true two- and multi-particle interactions. Different aspects of the collective density fluctuations are probed by the density- correlation (or intermediate-scattering) function

$$F(k,t) = \frac{1}{N} \left\langle \sum_{j=1}^{N} e^{ik \cdot r_j(t)} \sum_{l=1}^{N} e^{-ik \cdot r_l(o)} \right\rangle \quad (2)$$

and by the longitudinal velocity correlation function $C(k,t) = - k^{-2}(d^2/dt^2)F(k,t)$, which are the Fourier transforms of $S(k,\omega)$ and $C(k,\omega) = \omega^2 S(k,\omega)/k^2$, respectively. In (2) $r_j(t)$ is the position of particle j at time t. Comparisons between simulations and real experiments are important in order to ascertain that the former contain the essential features of the real systems.

The theories to describe the atomic dynamics as observed in INS and CMD have been developed either as a generalization of linear hydrodynamics or at the kinetic level. In generalized hydrodynamics both k- and ω-dependent transport coefficients are introduced. The kinetic theories are based on Enskog's modification of the Boltzmann equation for hard spheres, which includes particle size effects and collisional transfer of momentum and energy. For hard spheres the situation is easier than for particles interacting with a continuous potential, because the binary collisions can be described exactly. Theories have been developed that try to include the effects of collisions no longer being instantaneous — nor strictly binary — in more realistic systems, a.o. by the introduction of mode-coupling effects [17]. For a complete overview, and extensive reference lists to the original papers , one should consult the excellent books of BOON and YIP [18] and HANSEN and MCDONALD [15]. The advanced theories have certainly improved our understanding of the dynamics of dense fluids compared to earlier more phenomenological models. However, it is unfortunate that they require elaborate computations, and mostly ad-hoc assumptions at some point, which make them hard to use, not only in the comparison with INS and CMD results, but also as input for corrections of INS data (Sect.2).

The outline of this paper is as follows: In Sect.2 INS experiments, and the necessity for reliable correction procedures, are discussed. Sect.3 is devoted to the interpretation of neutron scattering data on ^{36}Ar and Ne, based on fits to an equation derived from the Revised Enskog Theory (RET) [19]. Here, generalized 'extended-heat' and 'extended-sound' eigenmodes, and their behaviour with k and as a function of density, are discussed. Also the recent method of EGELSTAFF and GLAESER [20] for derivation of a memory function from INS data, and applied by them to Kr and ^{36}Ar results, will be discussed. Differences between the eigenmode behaviour of the experimental data and the hard-sphere system predictions could be caused by potential-shape differences or by approximations in the theoretical calculations. Here CMD simulations can be used to advantage, and results for hard-spheres [21, 22], and for full- [23,24] and repulsive [25] Lennard- Jones systems are analyzed in Sect.4, and compared with the INS results and with further theoretical calculations [22, 24, 26]. Conclusions are given in Sect.5.

2. Neutron Scattering Experiments

Both higher neutron fluxes and improvements to data correction procedures have been needed to improve on the results of SKOLD *et.al.* [27] on liquid argon near its triple point. For a long time these data were used as one of the important test cases for newly- developed models and theories [15, 18]. Recently, new experimental data of high statistical accuracy were reported for Kr, Ar and Ne, with correction procedures aimed at the reduction of systematic errors to a level comparable to that caused by statistical uncertainties.

EGELSTAFF *et al.* [7] examined krypton at 12 dense gas states, ranging from the critical- to 0.8 times the triple- point density, along the 297 K isotherm ($T^* = k_B T/\varepsilon$ = 1.71, with ε the potential depth). For these experiments use was made of the IN-4 rotating- crystal spectrometer at ILL, Grenoble, with an incident wavelength λ= 0.254 nm, and covering a wavenumber-transfer range of 4.0 < k < 33 nm^{-1}. Of the large quantity of data, only the fully corrected S(k,ω) functions for n = 10.6 and 13.8 nm^{-3} ($n^* = n\sigma^3$ = 0.49 and 0.64, respectively, with an atomic diameter σ of 0.359 nm) have been published so far.

The rotating- crystal spectrometer RKS I in Delft, with λ = 0.408 nm, was used by VERKERK *et al.* [28-30] to obtain data on fluid ^{36}Ar at 120 K (T^* = 0.97), and at pressures of 20, 270 and 850 bar (0.8 - 1.0 triple- point density; n^*= 0.710, 0.787, and 0.872 with σ = 0.343 nm), for 4.2 < k < 22.2 nm^{-1}. VAN WELL *et.al.* [6] used the IN-4 with λ = 0.2544 nm to obtain data with high statistical accuracy on ^{36}Ar at 120 K, at p = 20.1, 115, 270, and 394 bar (n^* = 0.710, 0.747, 0.787, and 0.812), for 4.2 < k < 39.2 nm^{-1}. The data correction procedures applied to these results will be discussed in some detail later in this section. An experiment at larger k values, 38 - 72 nm^{-1}, to study the transition to free- gas behaviour, was performed at the RKS II spectrometer in Delft by VAN WELL and DE GRAAF [31] on ^{36}Ar at p = 20, 270, and 400 bar, with an incident wavelength of 0.1436 nm. In order to study the dynamics at lower densities in the k = 4 - 22 nm^{-1} region, VERKERK and WOUTERS [32] did experiments with the RKS I on ^{36}Ar at 4 pressures between 160 and 840 bar along the 212 K isotherm (T^*= 1.71, like for Kr [7]; n^*= 0.289, 0.387, 0.513, and 0.646).

Liquid neon at 35 K (T^*= 0.97) has been studied with the RKS I by VAN WELL and DE GRAAF [5] for k values between 3 and 22 nm^{-1} (below the position of the first diffraction peak at k = 23.5 nm^{-1}). Results were obtained for n^*= 0.725 and 0.752, which made a direct comparison with the argon data as to the validity of the corresponding-states principle possible.

Inelastic neutron scattering experiments on liquids and dense gases are difficult and time- consuming. EGELSTAFF has, based on his long experience in the field, given very clear guidelines for this kind of experiments [11]. Forced by the necessity to improve on existing data reduction programs, in order not to spoil the high statistical accuracy of the data on ^{36}Ar obtained at IN-4, VERKERK and VAN WELL have developed a system of computer programs to calculate the dynamic structure factor from thermal neutron time-of-flight experiments [33]. Intermediate results are stored for inspection and for input in other programs, while the variances of the datapoints are carried through the whole procedure, giving as a final result the fully corrected symmetrized S(k,ω) at a rectangular (k,ω)-grid, together with error bars [5, 6, 31].

One of the most important corrections is that for multiple scattering and duty-cycle overlap. Based on the Monte Carlo program developed by COPLEY [34], an updated version, MSCAT85, was used [35]. In Fig.1 simulation results are shown. The correction is applied iteratively, because S(k,ω) is not known sufficiently accurately in advance. Especially at small scattering angles, where the static structure factor S(k)= $\int S(k,\omega)d\omega$ is small, the multiple scattering contribution in the spectra is very important. The time-of-flight data were corrected for finite spectrometer resolution using a fast method developed by VERKERK [36], and are normalized absolutely using the elastic part of (also fully-corrected) vanadium spectra. Care should be taken to use vanadium in the same geometry as the sample container. The background ("empty-container") correction can be improved by filling the container with low-density ^3He gas with equal transmission as the sample. The quality of the S(k,ω) data was checked by comparing energy-gain and energy-loss data, and by the first and second frequency moments. Quantum corrections to the second frequency moment for neon are discussed in [5].

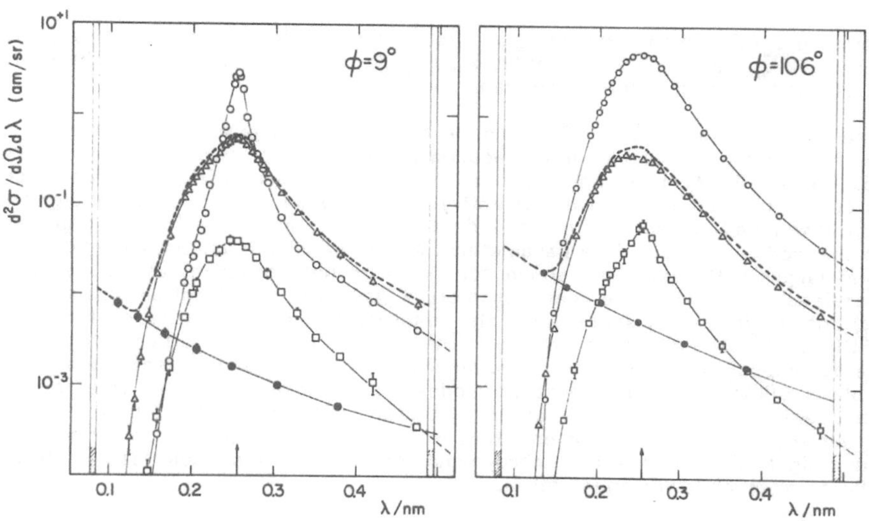

Fig.1. Spectra for ^{36}Ar measurement at 115 bar, at two scattering angles, as simulated by MSCAT85 [35]. The arrow indicates the incoming neutron wavelength and the shaded vertical bars the dead time of the TOF analyzer. The curves are (with s = sample, and c = container) depicting the following contributions: open circles, s; triangles, ss; squares, cs ≈ sc; closed circles, all overlap scattering (s + ss + 2cs); dashed line, all multiple scattering and overlap. 1 am/sr = 1 barn/Å·sr. From [6]

3. Interpretation of INS Results

The dynamic structure factors obtained for ^{36}Ar and Ne could very well be described in terms of three extended hydrodynamic modes, that for small k go over continuously into the Landau-Placzek triplet of one Rayleigh (heat) and two Brillouin (sound) peaks. This description is based on the generalized Enskog kinetic theory (RET) for hard spheres as developed by DE SCHEPPER and COHEN [19,37]. The generalized Enskog operator L(**k**), that determines the time evolution of all the correlation functions in the fluid, is a sum of three terms: a free-streaming term, a collision term, and a mean-field term:

$$L(\mathbf{k}) = -i\mathbf{k}\cdot\mathbf{v} + ng(\sigma)\Lambda_{\mathbf{k}} + nA_{\mathbf{k}}. \tag{3}$$

Here **v** is the velocity, g(σ) the radial distribution function of two spheres at contact, $\Lambda_{\mathbf{k}}$ a binary collision operator, and $A_{\mathbf{k}}$ the mean-field operator. L(**k**) is applied through a spectral decomposition in eigenmodes $z_j(k)$, and then is [19,37]

$$S(k,\omega) = \frac{1}{\pi} S(k)\,\mathrm{Re}\left\langle \frac{1}{i\omega - L(\mathbf{k})} \right\rangle = \frac{1}{\pi} S(k)\,\mathrm{Re} \sum_{j=-\infty}^{\infty} \frac{A_j(k)}{i\omega + z_j(k)} \quad , \tag{4}$$

where $A_j(k)$ and $z_j(k)$ are either real, or complex conjugate pairs.

To determine the $z_j(k)$ explicitly an ∞ x ∞ matrix of a complete set of orthonormal polynomials in **v** must be approximated by a smaller one that can be inverted. Reasonably convergent results were obtained with an 11x11 submatrix (M=11) for the first six eigenmodes, of which the most important are the " generalized- hydrodynamic" (kinetic) extended- heat (j=0) and - sound (j=+1, -1) modes [37]. Then (4) can, for most k values, be written as a sum of three "Lorentzians":

$$\frac{S(k,\omega)}{S(k)} = \frac{A_o}{\pi}\frac{z_o}{\omega^2 + z_o^2} + \frac{A_s}{\pi}\left(\frac{z_s + (\omega+\omega_s)\tan\Phi}{(\omega+\omega_s)^2 + z_s^2} + \frac{z_s - (\omega-\omega_s)\tan\Phi}{(\omega-\omega_s)^2 + z_s^2}\right), \tag{5}$$

with heat damping $z_o(k)$, sound damping $z_s(k) = \mathrm{Re}\, z_1$, sound frequency $\omega_s(k) = \mathrm{Im}\ z_1$, and $\tan \Phi(k) = (\mathrm{Im}\ A_1)/A_s$. For some values of k, z_1, z_{-1}, A_1, and A_{-1} turned out to be real, leading to a $S(k,\omega)$ as sum of three Lorentzians about $\omega = 0$.

Fits of the experimental data on ^{36}Ar and Ne to (4) and (5) gave excellent results. The number of six free parameters can be reduced by using one or two known frequency moments [5]. Except for some k regions, three "eigenmodes" are sufficient to describe the neutron data. An extension to five (effective) eigenmodes, and its physical significance, will be discussed in Sect. 4. In Fig. 2 neutron results are given for Ne, together with a fit to the model with four free parameters. At the smaller k values,shoulders are visible, but at larger k the heat and sound contributions both broaden, and $S(k,\omega)$ becomes featureless. A fit with five free parameters gave virtually identical results. In Fig. 3 $S(k,\omega)$ and the frequency spectrum of the longitudinal current correlation function, $C(k,\omega) = \omega^2 S(k,\omega)/k^2$, are displayed for one of the Ar experiments, together with the fit to (5), and a decomposition in heat and sound modes. It is clear that the heat mode plays the major role in $S(k,\omega)$, whereas the sound modes are dominant in $C(k,\omega)$. This applies to the whole k region covered in the experiment. It is also clear from Fig. 3 that in order to measure the second frequency moment, that is,the area of $C(k,\omega)$, at this relatively small k value,a higher incident neutron energy would be needed. This would lead to either worse resolution or much lower count rates.

Although there is qualitative agreement for the behaviour of the parameters A_j and z_j between experiment and theoretical calculations, the quantitative differences for $z_o(k)$, but especially $\omega_s(k)$, are appreciable. These can to a large extent be accounted for by differences in potential, which will be discussed in Sect. 4. The extended-sound dispersion curve has very intriguing behaviour as a function of k and density, as shown in Fig. 4. The "sound-propagation gap" becomes narrower with increasing density, and ultimately disappears in the 850 bar experiment [28,39]. The positive anomalous sound dis-

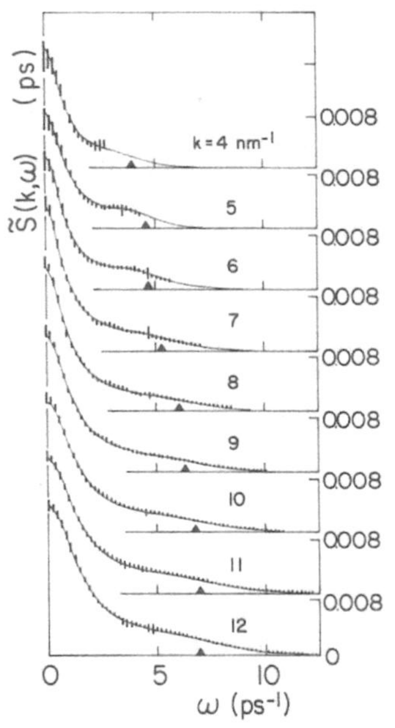

Fig.2. Fully corrected $S(k,\omega)$ for Ne at 35 K and 140 bar (error bars), and the fits to a sum of three Lorentzians (solid line). Sound frequencies ω_s are indicated by triangles. From [5]

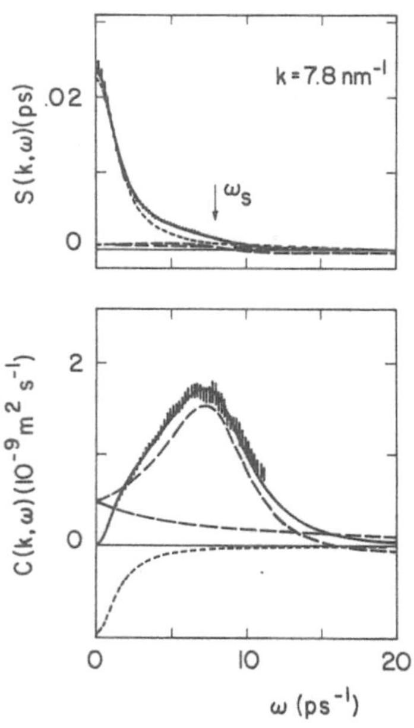

Fig.3. $S(k,\omega)$ and $C(k,\omega)$ for Ar at 120 K, 115 bar, $n^* = 0.747$ (error bars) [6]. The full lines are the best fits to (5). Short dashes: heat-mode , long dashes: sound-mode contributions. From [38]

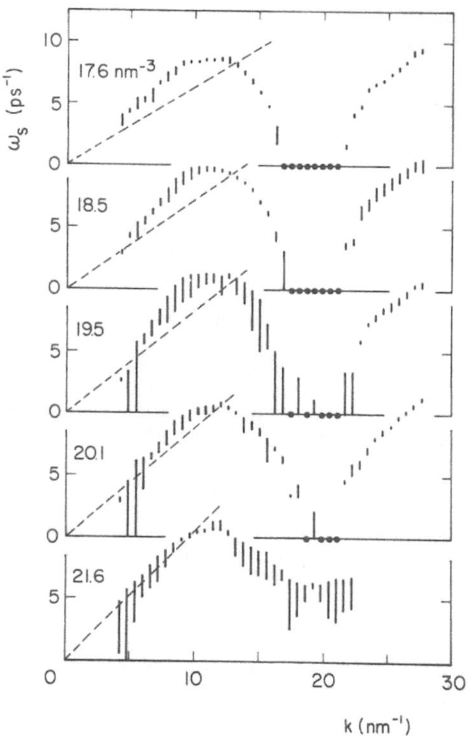

Fig.4. Sound dispersion curves for argon at 120 K and different number densities. Vertical bars connect values from fits with either one or two frequency moments fixed. Dashed lines: $\omega_s = ck$. From [38]

persion at small k, *i.e.* its upward curvature from the hydrodynamic behaviour $\omega_s = ck$, with c the adiabatic speed of sound, can, both as a function of k and of density, very well be described with mode coupling theory: $\omega_s = ck + a_s k^{5/2} + O(k^{11/4})$ [40,41]. Here a_s is a sum of thermodynamic quantities that arise from the coupling of two hydrodynamic modes (heat, sound, shear) to the sound mode.

VAN WELL and DE GRAAF [5] have compared the fits to the neon data in terms of the decay of the eigenmodes, with a generalized hydrodynamics description using k- and ω-dependent transport coefficients. If the transport coefficients are only k dependent, the two descriptions can be translated into each other [5,42]. With the eigenmode description the parameters are less correlated, and can be determined better. Therefore, this method is considered to be preferred.The fit parameters of neon were compared with those of argon at corresponding states. Except for a small difference in ω_s, the values correspond very closely [5]. Argon and liquid Rb differ more, but the similarities, especially in the sound dispersion, are striking [38]. A discussion about the significance of the various parameters was held [42,43]. Arguments in favour of a visco-elastic theory [42] , were not accepted because of much worse agreement with the experimental data [43]. In Sect. 4 this comparison will be reconsidered. EGELSTAFF and GLAESER [20] have analyzed Kr, Rb, and Ar data with a new method by which the memory function of the longitudinal current correlation function can be determined at six specific frequencies. They found a marked variation with k, but only slight ω dependence. Evidence for propagating modes at higher k was given. Further work to establish the meaning of the derived quantities, and *e.g.* their relation to the eigenmodes, seems needed.

4. Interpretation of CMD Simulations

The Fourier transforms of $S(k,\omega)$ and $C(k,\omega)$, the intermediate scattering function $F(k,t)$ and the longitudinal current correlation function $C(k,t) = -k^{-2}(d^2/dt^2)F(k,t)$, can be determined directly from molecular dynamics simulations. Simulations of a dense hard-sphere (HS) fluid [22] could very well be fitted to a sum of three exponentials:

$$F(k,t) = \sum_{j=0,\pm 1} A_j(k)\exp\left(-z_j(k)t\right); \quad C(k,t) = \sum_{j=0,\pm 1} A_j(k)\, z_j^2(k)\exp\left(-z_j(k)t\right), \tag{6}$$

with parameters A_j and z_j in good agreement with an explicit RET calculation, thus validating the approximations made in RET. The heat-mode behaviour is in reasonable agreement with that found in argon, the sound-mode behaviour, and especially the sound dispersion, is quite different. So the heat mode is less sensitive to the potential shape than the sound mode. This effect was also found in full (LJ)- and repulsive (RLJ)-Lennard-Jones systems [23].

For a RLJ system at $T^*=0.97$, simulations at 11 different densities from $n^*= 0.20$ to 0.924 were performed by BRUIN *et al.* [25]. A repulsive potential was chosen to avoid phase separation. Results for F(k,t) and C(k,t) are shown in Fig.5. The error bars were determined by averaging six simulation runs with different initial configurations. Here F(k,t) is dominated by the heat mode, C(k,t) by the sound mode, which can be compared directly with Fig.3. Whereas the sound contribution oscillates visibly at $k^*= k\sigma = 3.84$ and 10.27, its frequency is much smaller at $k\sigma = 7.07$. The sound propagation gap was found for densities $0.37 < n^* < 0.88$, while at $n^* = 0.20$ ω_s behaves as in a free gas. For the two highest densities, simulations with 8788 particles were done to improve on the gap determination. The gap disappears before undercooling sets in, similar to Ar and full LJ, where the gap ceases at $n^* > 0.79$. This is quite different from RET calculations where a gap occurs at $n^* > 0.73$ and is still existent at $n^*= 0.99$ [44], which is in the undercooled region.

ALLEY and ALDER [21] performed CMD simulations for a HS fluid at several densities, and determined the 3x3 matrix $F_{ij}(k,t)$ with i,j = n, u, T (the microscopic density-, longitudinal velocity-, and temperature fluctuations, respectively). At high densities they found that their data could be described in terms of $F_{ij}(k,0)$ and a 3x3 matrix $H^{(3)}(k)$

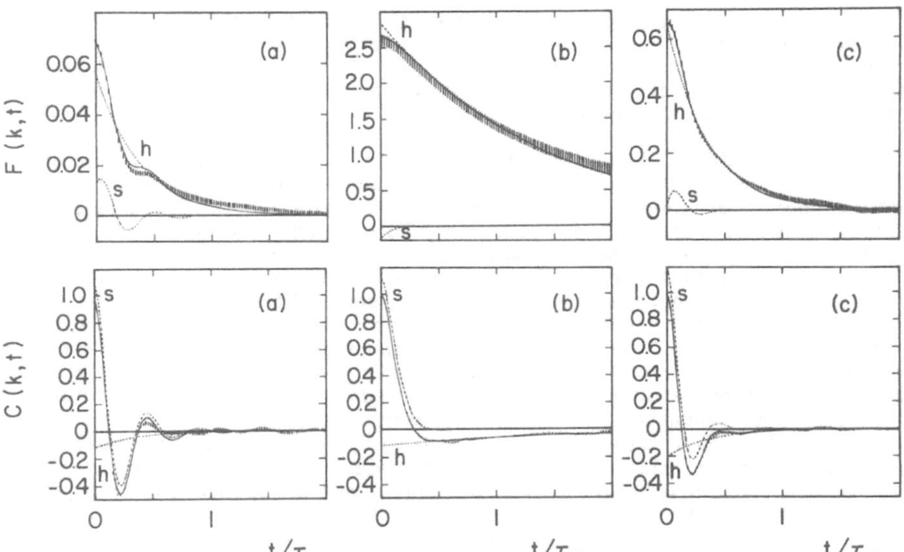

Fig.5. CMD results (error bars) for F(k,t) and C(k,t) for 864 RLJ particles at $T^*= 0.97$ and $n^*= 0.924$, for $k\sigma = 3.84$ (a), 7.07 (b), and 10.27 (c) as functions of t/τ_σ, with $\tau_\sigma = (\beta m)^{1/2}\sigma/2$. The best-fitted sums of three exponentials are shown as full curves, the contributions of the heat and the two sound modes as dotted and dashed curves. The heat mode is dominant in F(k,t), the sound modes in C(k,t). From [25]

which depends on k, but not on t. In such a generalized hydrodynamic description all nine correlation functions $F_{ij}(k,t)$ are given by a sum of three exponentials related to the eigenmodes (eigenvalues and eigenfunctions) of $H^{(3)}(k)$. These results could also be understood on the basis of RET, as shown by KAMGAR-PARSI, COHEN and DE SCHEPPER [26]. At low and intermediate densities, however, this description fails and instead one needs a 5x5 matrix $H^{(5)}(k)$, that describes the 25 correlation functions between n, u, T, the longitudinal stress-tensor fluctuation σ, and the longitudinal heat-flux fluctuation q. In [26] much better approximations were used than before to obtain the eigenmodes of L(**k**), and although the basic conclusions of previous calculations did not change, a more detailed picture emerged. A description in true hydrodynamic eigenmodes was found to be less stable than one in effective modes, *i.e.* the eigenmodes of an effective 3x3 matrix that effectively replaces the ∞x ∞ matrix.

A detailed study was made by DE SCHEPPER *et al.* [24] of a LJ fluid at T^*= 1.71 and n^*= 0.845. Using CMD, four correlation functions $F_{ij}(k,t)$ with i,j = nn, uu, ne, and ee were determined (e is a combination of T and n), as given in Fig.6. These could not all be fitted by a sum of three exponentials, and instead five were needed, related to the eigenmodes of $H^{(5)}(k)$. Of the 25 F_{ij}, only three are independent. Thus, also F_{nn} = F(k,t) is in principle a sum of 5 exponentials. In practice, however, it can, for all k with 0 < kσ < 15, be represented by 3 out of 5 exponentials. The character of these eigenmodes then changes continuously with k, and the 3 extended hydrodynamic modes can involve combinations of the three hydrodynamic quantities n, u, and e, and of the two nonhydrodynamic quantities σ and q. In Fig.7 the extended hydrodynamic mode eigenvalues are shown. It is emphasized that, although the CMD results for all $F_{ij}(k,t)$ can only consistently be described by 5 exponentials, F(k,t) - or equivalently S(k,ω) - alone can still be described in terms of three (extended) hydrodynamic eigenmodes of the 5x5 matrix. This is consistent with earlier fits of F(k,t) with three exponentials, as carried out for Ar, Ne, and LJ- and RLJ-fluids (Sects. 3 and 4).

The representation of F(k,t) with three exponentials made a comparison possible with the visco-elastic theory [8,42]. In [24] is clearly shown that for a LJ fluid the coupling of the microscopic quantities n, u, and σ with T and q is not weak enough for the visco-elastic description, that completely neglects temperature fluctuations, to hold.

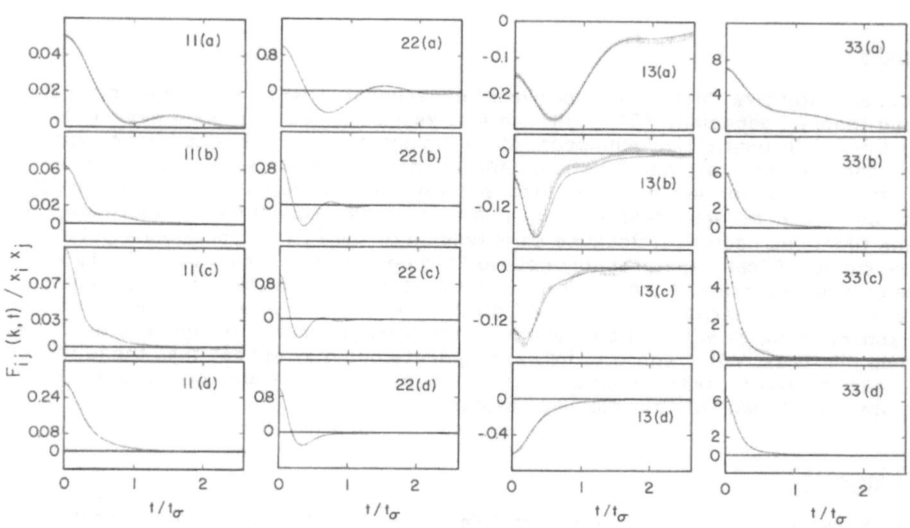

Fig.6. CMD results (vertical bars) for the reduced correlation functions $F_{ij}(k,t)/x_i x_j$ as functions of t/t_σ with i,j = 11, 22, 13, and 33 (or nn, uu, ne, and ee), for a LJ fluid at T^*= 1.71 and n^*=0.845; kσ = 1.24 (a), 2.49 (b), and 4.98 (c). The curves are best simultaneous fits to a sum of five exponentials. From [24]

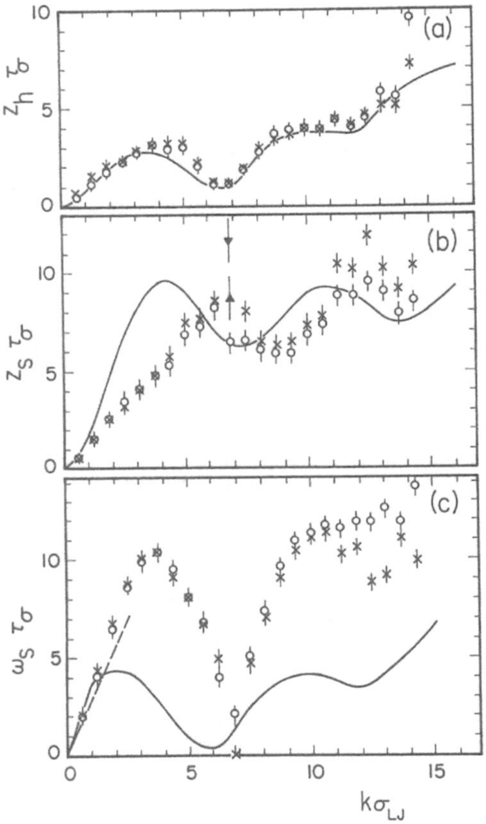

Fig.7. Reduced heat mode eigenvalue $z_h\tau_\sigma$, sound damping $z_s\tau_\sigma$, and sound dispersion $\omega_s\tau_\sigma$ for a LJ fluid at $T^* = 1.71$ and $n^* = 0.845$ versus $k\sigma$. Open circles: from 5x5 matrix; crosses: from 3-Lorentzians fit to F(k,t) and C(k,t). When $\omega_s = 0$, z_s is double valued. The dashed curve in (c) is the hydrodynamic sound dispersion $ck\tau_\sigma$. The solid curves are RET calculations for a hard sphere fluid at $n^* = 0.80$ [26]. From [24]

5. Conclusions

The sketched intensive interaction between experiment (INS), computer simulation (CMD), and theory (especially RET) over the last years has led to a good description of S(k,ω) and F(k,t) in terms of eigenmodes. More insight has been obtained in the various eigenmode contributions to these functions, and to their behaviour as a function of k. The dynamic structure factor seems to be less sensitive to the details in the microscopic dynamics, as it can be well described by a sum of three Lorentzians, whereas other correlation functions need five. Because an extension of the RET to continuous potentials seems extremely difficult, and probably not worthwhile pursuing, one would like to see a link made to other theories. A more detailed comparison between the noble-gas fluids, or their LJ computer counterparts, and liquid (alkali) metals is of interest. Theories based on hard-sphere systems will still play a role in the study of more complicated fluids, as for example mixtures. With regard to INS experiments one may expect that these will be carried out in the future, with conditions strongly based on the predictions of both theory and computer simulations of "sample-like" systems.

Acknowledgments

I want to thank Drs. P. Verkerk, A.A. van Well, I.M. de Schepper, and C. Bruin, and Professor E.G.D. Cohen for their fruitful cooperation and many helpful discussions.

References

1. S.W. Lovesey: Theory of Neutron Scattering from Condensed Matter (Clarendon, Oxford 1984)
2. W. Gläser: In Amorphous and Liquid Materials, ed. by E. Lüscher, G. Fritsch and G. Jacucci, NATO ASI Series E118 (Martinus Nijhoff Publ., Dordrecht 1987) p.258
3. C. Morkel: These Proceedings
4. P. Verkerk: These Proceedings
5. A.A. van Well, L.A. de Graaf: Phys. Rev. A32, 2396 (1985)
6. A.A. van Well, P. Verkerk, L.A. de Graaf, J.-B. Suck, J.R.D. Copley: Phys. Rev. A31, 3391 (1985)
7. P.A. Egelstaff, W. Gläser, D. Litchinsky, E. Schneider, J.-B. Suck: Phys. Rev. A27, 1106(1983)
8. J.R.D. Copley, S.W. Lovesey: Rep. Progr. Phys. 38, 461 (1975)
9. J.R.D. Copley, S.W. Lovesey: Inst. Phys. Conf. Ser. 30, 575 (1977)
10. J.R.D. Copley: In Physics of Modern Materials, Vol. 2 (IAEA, Vienna 1980) p.613
11. P.A. Egelstaff: In Neutron Scattering, ed. by K. Sköld and D.L. Price, Academic Press Ser.- Methods of Experimental Physics, Part B, Ch. 14 (Academic Press, New York 1988)
12. O. Söderström, U. Dahlborg, and M. Davidovic: Phys. Rev. A27, 470 (1983); U. Dahlborg and L.G. Olsson: J. Phys. F13, 555 (1983); K.-E. Larsson: these Proceedings
13. B.J. Alder and T.E. Wainwright: J. Chem. Phys. 31, 459 (1959)
14. A. Rahman: Phys. Rev. A136, 405 (1964)
15. J.P. Hansen and I.R. McDonald: Theory of Simple Liquids, 2nd.ed., (Academic Press, New York 1986)
16. M.P. Allen and D.J. Tildesley: Computer Simulation of Liquids (Clarendon Press, Oxford 1987)
17. A. Sjölander: In Amorphous and Liquid Materials, ed. by E. Lüscher, G Fritsch and G. Jacucci, NATO ASI Series E118 (Martinus Nijhoff Publ., Dordrecht 1987) p.239 discusses theoretical progress, with some emphasis on the work of the Gothenburg, Munich, and MIT groups
18. J.P. Boon and S. Yip: Molecular Hydrodynamics (Academic Press, New York 1982)
19. I.M. de Schepper and E.G.D. Cohen: Phys. Rev. A22, 287 (1980); J. Stat. Phys. 27, 223 (1982)
20. P.A. Egelstaff and W. Gläser: Phys. Rev. A31, 3802 (1985); W. Gläser and P.A. Egelstaff: Phys. Rev. A34, 2121 (1986)
21. W.E. Alley and B.J. Alder: Phys. Rev. A27, 3158 (1983); see also W.E. Alley, B.J. Alder, and S. Yip: Phys. Rev. A27, 3174 (1983)
22. C. Bruin, J.P.J. Michels, J.C. van Rijs, L.A. de Graaf, and I.M. de Schepper: Phys. Lett. 110A, 40 (1985)
23. I.M. de Schepper, J.C. van Rijs, A.A. van Well, P. Verkerk, L.A. de Graaf, and C. Bruin: Phys. Rev. A29, 1602 (1984)
24. I.M. de Schepper, E.G.D. Cohen, C. Bruin, J.C. van Rijs, W. Montfrooij, and L.A. de Graaf: Phys. Rev. A38, July (1988)
25. C. Bruin, J.C. van Rijs, L.A. de Graaf, and I.M. de Schepper: Phys. Rev. A34, 3196 (1986)
26. B. Kamgar-Parsi, E.G.D. Cohen, and I.M. de Schepper: Phys. Rev. A35, 4781 (1987)
27. K. Sköld, J.M. Rowe, G. Ostrowski, and P.D. Randolph: Phys. Rev. A6, 1107 (1972)
28. P. Verkerk: PhD Thesis, Delft University of Technology (1985)
29. P. Verkerk: J. de Physique C9, 17 (1985)
30. P. Verkerk and A.A. van Well: Physica B136, 168 (1986)
31. A.A. van Well and L.A. de Graaf: Phys. Rev. A32, 2384 (1985)
32. P. Verkerk and J.A.A. Wouters: to be published
33. P. Verkerk and A.A. van Well: Nucl. Instr. Meth. in Phys. Res. 228, 438 (1985)
34. J.R.D. Copley: Comput. Phys. Commun. 7, 289 (1974), 20, 459 (1980), 21, 431 (1981)
35. J.R.D. Copley, P. Verkerk, A.A. van Well, and H. Fredrikze: Comput. Phys. Commun. 40, 337 (1986)
36. P. Verkerk: Comput. Phys. Commun. 25, 325 (1982)
37. E.G.D. Cohen, I.M. de Schepper, and M.J. Zuilhof: Physica 127B+C, 282 (1984), and references given there
38. L.A. de Graaf, A.A. van Well, P. Verkerk, and I.M. de Schepper, In Proc. 2nd Int'l. Conf. on Phonon Physics, ed. by J. Kollar, N. Kroo, N. Menyhard, and T. Siklos, World Scientific, Singapore, 1985, p. 378

39. P. Verkerk, A.A. van Well, and I.M. de Schepper: J. Phys. C: Solid State Phys. 20, L979 (1987)
40. M.H. Ernst and J.R. Dorfman: J. Stat.Phys. 12, 311 (1975)
41. I.M. de Schepper, P. Verkerk, A.A. van Well, and L.A. de Graaf: Phys. Lett. 104, 29 (1984)
42. S.W. Lovesey: Phys. Rev. Lett. 53, 401 (1984); Z. Phys. B 58, 79 (1985)
43. I.M. de Schepper, P. Verkerk, A.A. van Well, L.A. de Graaf, and E.G.D. Cohen: Phys. Rev. Lett. 53, 402 (1984); Phys. Rev. Lett. 54, 158 (1985)
44. M.J. Zuilhof, E.G.D. Cohen, and I.M. de Schepper: Phys. Lett. 103 A, 120 (1984)

A Liquid Metal as an Extensive Test Case of Static and Dynamic Properties

K.-E. Larsson, W. Gudowski, and M. Dzugutov
Royal Institute of Technology, S-10044 Stockholm, Sweden

1. Background

The motion of atoms or molecules in liquids or in fluids in general has been the aim of study for a long time. It was, however, with the introduction of the inelastic slow neutron scattering [NS] technique in the middle of the 1950s that an experimental tool was invented which permitted a detailed study of the time and space evolution of such motions in a time and space domain of about 0.1 - 10 ps and 0.1 - 10 Å, respectively. Around 1960, computers had reached such a degree of development that molecular dynamics [MD] could be simulated by simultaneously solving Newton's equations of motion for a large number of particles in a box, typically 500 to 800 particles in the majority of early cases. These new techniques turned out to be complementary because they covered a similar range in space and time or equivalently Q and ω, respectively, in the Fourier transformed space, typically in a range $0.1 < Q < 10$ Å^{-1} and $0 < h\omega < 40$ meV. In order to perform MD simulation the force acting between the particles—or equivalently a pair potential V(r)—must be assumed known.

For many years in the 1960s and 1970s various phenomenological models were created and tested against such NS and MD data. Two lines of development in particular emerged as important: the generalized hydrodynamical approach [1] extending the usefulness of the hydrodynamical framework of equations - such as the Navier-Stokes equations - from the continuum domain to atomic distances, and the kinetic theory approach [2-4] going the opposite way, starting with a detailed description of atomic motions and extending its region of application to the continuum domain. The generalized hydrodynamical theory makes use of the Kubo formalism setting up series of memory function equations in a continued fraction form. Several of these memory functions may be identified as generalized transport coefficients such as the self-diffusion coefficient $\hat{D}(\mathbf{Q},\omega)$ and various viscosity coefficients $\hat{v}_l(\mathbf{Q},\omega)$ - the longitudinal viscosity, and $\hat{v}_{shear}(\mathbf{Q},\omega)$ - the shear viscosity.

The problem is to be able to derive such detailed information from observation of a natural many-body system such as a liquid. As is well known, most neutron scatterers are coherent ones such as most liquid metals; at best some are mixed coherent and incoherent scatterers such as liquid argon. Hydrogen-containing molecules are, to a certain degree, mainly incoherent scatterers but they exist in molecular form. Liquid metals and noble gases are the best examples of atomic fluids. Both early and later NS and MD studies on liquid rubidium [5,6] and liquid argon [7] are well known. The present work deals with another metal, namely liquid lead, a 100% coherent scatterer. The NS study [8] gives the scattering function $S(\mathbf{Q},\omega)$ from which the intermediate scattering function $F(\mathbf{Q},t)$ may be obtained. Writing the first two consecutive memory function equations for $F(\mathbf{Q},t)$ and solving them numerically gives $K(\mathbf{Q},t)$ and $M(\mathbf{Q},t)$, K being the first-order memory function and M the second-order memory function of $F(\mathbf{Q},t)$. It is known [9] that $M(\mathbf{Q},t)$ is equivalent to the generalized longitudinal kinematical viscosity coefficient $\hat{v}_l(\mathbf{Q},\omega)$:

$$M(\mathbf{Q},t) = Q^2 v_l(\mathbf{Q},t) \ . \tag{1}$$

Also $M(\mathbf{Q},t)$ is very closely related to the first-order memory function $K_l(\mathbf{Q},t)$ of the longitudinal current correlation function $J_l(\mathbf{Q},t)$ appearing in the generalized hydrodynamics:

$$K_l(\mathbf{Q},t) = M(\mathbf{Q},t) + \frac{< \omega^2 >}{< \omega^0 >} \ , \tag{2}$$

where $< \omega^0 >$ and $< \omega^2 >$ are zeroth and second frequency moments of $S(\mathbf{Q},\omega)$. Obviously the Fourier transform $\hat{v}_l(\mathbf{Q},\omega)$ taken at $\omega = 0$ gives the interesting quantity $\hat{v}_l(\mathbf{Q},0)$, the wave vector dependent kinematical viscosity. But nothing can be derived from the NS study regarding the important **self** motion. The only way to arrive at such information as $S_s(\mathbf{Q},\omega)$ or $F_s(\mathbf{Q},t)$ from which the memory functions $K_s(\mathbf{Q},t)$ and $M_s(\mathbf{Q},t)$ could be derived is by MD simulation. Similarly if one wants detailed information on the shear viscosity $\hat{v}_{shear}(\mathbf{Q},\omega)$ one must start out from the transverse current correlation function. It cannot be measured directly at present. But knowing the pair potential for the liquid it may be obtained by MD simulation.

Thus if one wants a very extensive set of information on the dynamics of a liquid one needs the following basic data:

a. An accurate NS determination of $S(\mathbf{Q},\omega)$.
b. A very good description of the pair potential $V(r)$.

Often $V(r)$ is derived from measurement of a static structure factor $S(\mathbf{Q})$. This is obtained by neutron or X-ray diffraction.

The idea of the coupling between NS and MD data is that if MD data well describe $S(\mathbf{Q})$ - necessary to obtain $V(r)$ - and $S(\mathbf{Q},\omega)$ then the many-body system described by MD using $V(r)$ is considered a quasi-real liquid. It may be assumed that real observation exists on, for instance, not only $S(\mathbf{Q},\omega)$ but also $S_s(\mathbf{Q},\omega)$ and $J_T(\mathbf{Q},\omega)$. In the following, such a relatively complete study of liquid lead is outlined.

2. Basic Relations

The interesting physical quantities that may be derived from $S(\mathbf{Q},\omega)$ or rather from its Fourier transform $F(\mathbf{Q},t)$ are the memory functions $K(\mathbf{Q},t)$ and $M(\mathbf{Q},t)$ defined from

$$\frac{\partial F}{\partial t} = -\int_0^t K(\mathbf{Q},t-t')F(\mathbf{Q},t')dt' , \tag{3}$$

$$\frac{\partial K}{\partial t} = -\int_0^t M(\mathbf{Q},t-t')K(\mathbf{Q},t')dt' . \tag{4}$$

Furthermore, there exists a very simple relation between $F(\mathbf{Q},t)$ and the longitudinal current correlation $J_l(\mathbf{Q},t)$

$$J_l(\mathbf{Q},t) = -1/Q^2 \frac{\partial^2 F(\mathbf{Q},t)}{\partial t^2} . \tag{5}$$

From generalizing the linearized hydrodynamical equations it may be shown that (if temperature fluctuations are neglected) the momentum conservation equation takes the form

$$\frac{\partial J_l(\mathbf{Q},t)}{\partial t} = -\int_0^t K_l(\mathbf{Q},t-t')J_l(\mathbf{Q},t')dt', \tag{6}$$

and in general the relation (2) above may be shown to hold with

$$K_l(\mathbf{Q},t) = Q^2 v_l(\mathbf{Q},t) + \frac{< \omega^2 >}{< \omega^0 >} \tag{7}$$

or

$$M(\mathbf{Q},t) = Q^2 v_l(\mathbf{Q},t) . \tag{8}$$

Taking the Laplace transform of (3) and putting $s = i\omega$ leads to an expression for $S(\mathbf{Q},\omega)$. If we concentrate our interest on $\omega = 0$, i.e. the time integral of the quantity $K(\mathbf{Q},t)$ which is $\hat{K}(\mathbf{Q},\omega = 0)$, we find

$$S(\mathbf{Q},0) = (1/\pi)\ F(\mathbf{Q},0)/\hat{K}(\mathbf{Q},0) . \tag{9}$$

It is easy to show that one may write

$$\hat{K}(\mathbf{Q},0) = \frac{< \omega^2 >}{S(\mathbf{Q})} \int_0^\infty n_K(\mathbf{Q},t)dt . \tag{10}$$

Similarly, taking the Laplace transform of (4) and putting $s = i\omega$ leads to a second expression for $S(\mathbf{Q},\omega)$. Putting $\omega = 0$ gives

$$S(\mathbf{Q},0) = (1/\pi)\ F(\mathbf{Q},0)\ \hat{M}(\mathbf{Q},0)/K(\mathbf{Q},0) . \tag{11}$$

Here

$$\hat{M}(\mathbf{Q},0) = \int_0^\infty M(\mathbf{Q},t)dt = \left(\frac{< \omega^4 >}{< \omega^2 >} - \frac{< \omega^2 >}{< \omega^0 >}\right) \int_0^\infty n_M(\mathbf{Q},t)dt . \tag{12}$$

Combining (9) and (11) gives

$$(\hat{K}(\mathbf{Q},0))^{-1} = \hat{M}(\mathbf{Q},0)/K(\mathbf{Q},0) . \tag{13}$$

Remembering that $\hat{M}(\mathbf{Q},0) = Q^2\hat{\nu}_l(\mathbf{Q})$ and $K(\mathbf{Q},0) = Q^2V_0^2/S(\mathbf{Q})$ where $V_0^2 = k_BT/M$ and M is the particle mass and T the absolute temperature one finds

$$\int_0^\infty K(\mathbf{Q},t)dt = V_0^2/(S(\mathbf{Q})\hat{\nu}_l(\mathbf{Q})) = nk_BT/(S(\mathbf{Q})nM\hat{\nu}_l(\mathbf{Q})) = (\chi_T(\mathbf{Q})\eta_l(\mathbf{Q}))^{-1} . \tag{14}$$

Here n is number density , $\chi_T(\mathbf{Q}) = S(\mathbf{Q})/(nk_BT)$ is generalized isothermal compressibility and $\eta_l(\mathbf{Q})$ is generalized longitudinal viscosity. It is remembered that in the hydrodynamic limit $\hat{\nu}_l(0,0) = 4/3\ \nu_{shear} + \nu_B$, where ν_B is bulk viscosity.

Turning to self-motion, similar relations may be derived from the memory function equations as were derived from (3) and (4). Particularly we find

$$S_s(\mathbf{Q},0) = (1/\pi)F_s(\mathbf{Q},0)/\hat{K}_s(\mathbf{Q},0) , \tag{15}$$

where $K_s(\mathbf{Q},t)$ can be interpreted as

$$K_s(\mathbf{Q},t) = Q^2D(\mathbf{Q},t) ; \tag{16}$$

$D(\mathbf{Q},t)$ is a generalized diffusion coefficient.

We may regard in analogy with (10)

$$\hat{K}_s(\mathbf{Q},0) = Q^2V_0^2 \int_0^\infty n_K^s(\mathbf{Q},t)dt = Q^2V_0^2 \int_0^\infty (<V^x(\mathbf{Q},0)V^x(\mathbf{Q},t)>)/V_x^2dt \tag{17}$$

corresponding to

$$\hat{D}(\mathbf{Q}) = V_0^2 \int_0^\infty n_K^s(\mathbf{Q},t)dt . \tag{18}$$

Here $< V_x(\mathbf{Q},0)V_x(\mathbf{Q},t) >$ is a Q-dependent velocity autocorrelation function, the integration of which leads to the Q-dependent self-diffusion coefficient. We also observe that this coefficient follows from $S_s(\mathbf{Q},0)$ directly, (15):

$$\hat{D}(\mathbf{Q}) = (\pi Q^2S_s(\mathbf{Q},0))^{-1} . \tag{19}$$

Using a relation similar to (11) one finds

$$S_s(\mathbf{Q},0) = (1/\pi)\ F_s(\mathbf{Q},0)\hat{M}_s(\mathbf{Q},0)/K_s(\mathbf{Q},0) . \tag{20}$$

This defines the generalized self-friction coefficient $\hat{M}_s(\mathbf{Q})$ or equivalently the self part of the kinematical longitudinal viscosity $\hat{\nu}_l^s(\mathbf{Q})$:

$$Q^2\hat{\nu}_l^s(\mathbf{Q}) = \hat{M}_s(\mathbf{Q},0) = \pi Q^2V_0^2S_s(\mathbf{Q},0) . \tag{21}$$

Comparing (15) and (20) one finds

$$\hat{D}(\mathbf{Q},0)\hat{M}_s(\mathbf{Q},0) = V_0^2 . \tag{22}$$

This is a generalization of the Einstein relation $D\gamma = V_0^2$, where $\gamma = \hat{M}_s(0,0)$ and $D = \hat{D}(0,0)$. It is further to be observed that $M_s(\mathbf{Q},t)$ is the memory function of the generalized velocity autocorrelation function defined in (17). This function was also the aim of the modern kinetic theories - mode coupling theories.

Finally the transversal current correlation function $J_T(\mathbf{Q},t)$ may be obtained from MD simulation. In the hydrodynamic limit at small Q-values, the well-known solution for J_T is

$$J_T(\mathbf{Q},t) = J_T(\mathbf{Q},0)e^{-Q^2 v_{shear} t} . \tag{23}$$

In the shorter wavelength region the simple differential equation leading to (23), namely

$$\frac{\partial J_T}{\partial t} = -Q^2 v_{shear} J_T, \tag{24}$$

is to be replaced by its generalized form

$$\frac{\partial J_T}{\partial t} = -\int_0^t Q^2 v_{shear}(\mathbf{Q},t-t')J_T(\mathbf{Q},t')dt'. \tag{25}$$

In analogy with the result for self-motion, (15), one finds for $\hat{v}_{shear}(\mathbf{Q},0)$

$$\hat{v}_{shear}(\mathbf{Q}) = J_T(\mathbf{Q},0)/(Q^2 \int_0^\infty J_T(\mathbf{Q},t)dt) . \tag{26}$$

3. Results

The experimental basis for the present data analysis is two NS series on liquid lead giving $S(\mathbf{Q},\omega)$ at 623K and 1173K reported earlier [8]. Useful data in this context extend from 1.8 to about 7 $Å^{-1}$. Furthermore, a neutron diffraction study of $S(\mathbf{Q})$ [10] at temperatures 623, 643, 863 and 1163K formed the basis for derivation of a pair potential for liquid lead by a MD fitting procedure using a system of 16 384 particles [11]. As described above, the derived $V(r)$ was first used to simulate $F(\mathbf{Q},t)$ which when Fourier transformed was compared to the NS measured $S(\mathbf{Q},\omega)$. As can be seen in Fig. 1 the simulated $S(\mathbf{Q},\omega)$ describes the measured one with reasonable accuracy.

From this it is concluded that either the NS or MD data may be used to continue the analysis. To obtain the Fourier transform $F(\mathbf{Q},t)$ of $S(\mathbf{Q},\omega)$ from NS the latter is represented by a sum of three gaussians centered at the origin $\omega = 0$. This gives the analytical form for $F(\mathbf{Q},t)$. The same analytical form also fits the MD determined $F(\mathbf{Q},t)$. From (3) and (4) one then computes the memory function $K(\mathbf{Q},t)$ and $M(\mathbf{Q},t) = Q^2 v_l(\mathbf{Q},t)$. A few examples of results are given in Fig. 2. Here, only the time-dependent parts $n_K(\mathbf{Q},t)$ and $n_M(\mathbf{Q},t)$ are given (normalized to 1 at t= 0). It is seen that both types of memory functions show the typical behavior of a first rapid decrease followed by a slower and somewhat structured tail.

The time integral of $v_l(\mathbf{Q},t)$ at once gives the generalized viscosity coefficient $v_l(\mathbf{Q})$ (= $\hat{v}_l(\mathbf{Q},0)$). In Fig. 3 the obtained kinematical viscosity is given as derived both from NS and MD data. It is seen that the two data sets give results in good agreement. Whereas useful NS data stops at Q = 1.8 $Å^{-1}$, the MD data deliver useful results down to Q = 0.5 $Å^{-1}$. Below this value the data become uncertain due to slow decay of $F(\mathbf{Q},t)$. It is observed that NS and MD data agree well within limits of error. It is also seen that a closer approach to the hydrodynamic value could be reached for smaller Q-values in the present MD simulation if more extensive data were available.

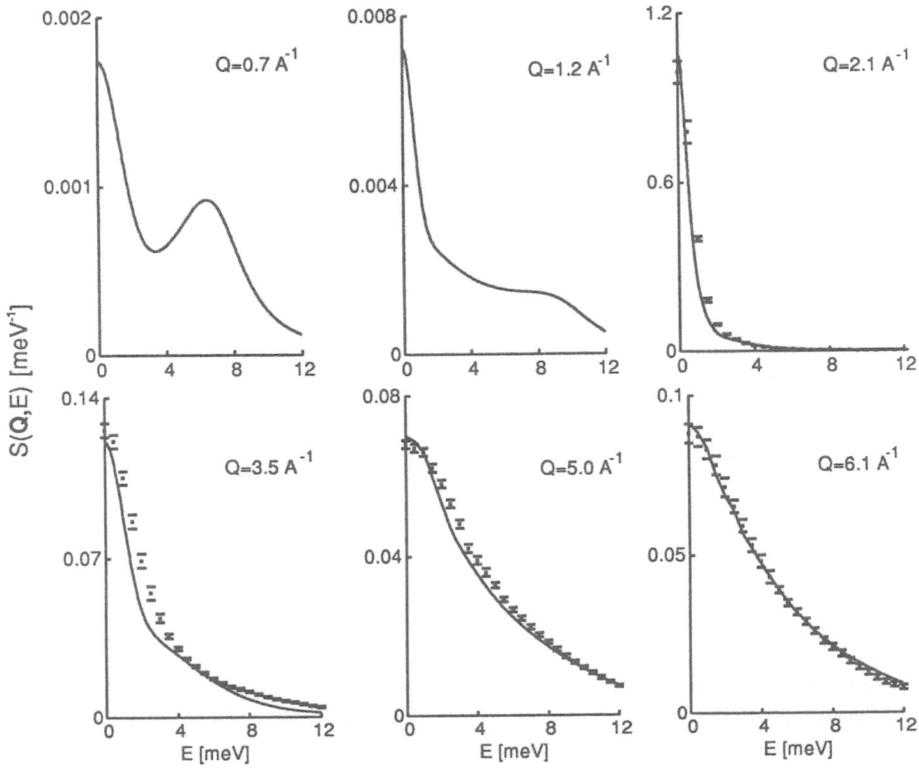

Fig. 1 NS-determined $S(\mathbf{Q},\omega)$ (o) and MD simulation (——) compared for a few Q-values.

Another quantity of interest is the time integral of $K(\mathbf{Q},t)$. According to (14) this is the inverse of the product $\chi_T(\mathbf{Q})\eta_l(\mathbf{Q})$. In Fig. 4 the MD data are compared to the NS data. It is again seen that a satisfactory agreement exists between the data sets and that the MD data tend toward the hydrodynamic limit. It is also found that the inverse of the product $\chi_T(\mathbf{Q})\eta_l(\mathbf{Q})$ well describes the observed variation of $\hat{K}(\mathbf{Q},0)$.

The next step in this series of investigations involves an MD simulation of the self part $F_s(\mathbf{Q},t)$ for a particle system corresponding to liquid lead at 623K. Applying the memory function (3) and (4) with index "s" added for "self" one derives the memory functions $K_s(\mathbf{Q},t) = Q^2D(\mathbf{Q},t)$ and $M_s(\mathbf{Q},t)$. As discussed in connection with (15) - (22) above, $D(\mathbf{Q},t) = V_0^2 n_K^s(\mathbf{Q},t)$, where $n_K^s(\mathbf{Q},t)$ is a generalized form of the velocity autocorrelation function. Furthermore $M_s(\mathbf{Q},t)$ is the memory function for the velocity autocorrelation function. Equivalently it may be said that $M_s(\mathbf{Q},t)$ defines the self part of the longitudinal kinematical viscosity $\nu_l^s(\mathbf{Q},t)$, the time integral of which is $\nu_l^s(\mathbf{Q})$ $(= \hat{\nu}_l^s(\mathbf{Q},0))$.

In Fig. 5 a few examples of $n_K^s(\mathbf{Q},t)$ and its memory function $n_M^s(\mathbf{Q},t)$ are displayed. The velocity correlation $n_K^s(\mathbf{Q},t)$ shows the behavior at small Q-values well known from studies on other substances. At larger Q-values it narrows and the negative swing becomes insignificant. Due to statistical inaccuracy, details of the memory function $n_M^s(\mathbf{Q},t)$ are lost. It is, however,

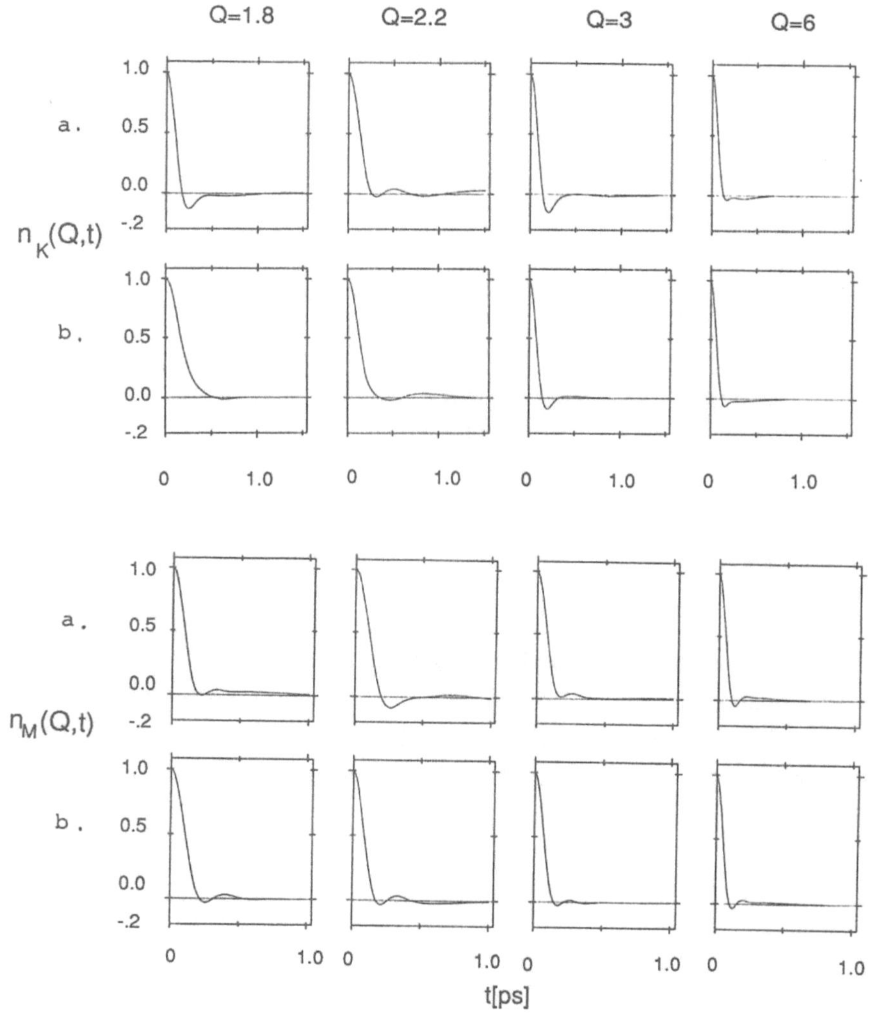

Fig. 2 $n_K(\mathbf{Q},t)$ and $n_M(\mathbf{Q},t)$ for a few Q-values derived from the NS (or MD) data for lead at a. 623 and b. 1173K.

clear that one observes the rapidly decaying part followed by a tail the shape of which remains unknown with present accuracy. This tail decreases quickly in amplitude when Q has passed Q_o (the value of Q at the peak of the structure factor).

The generalized diffusion coefficient $D(\mathbf{Q}) = \hat{D}(\mathbf{Q},0)$ may be obtained in three ways, namely from (18),(19) and (22). They should of course give the same result but may differ due to inaccuracies in the computational procedures. In Fig. 6 the three sets of values are displayed. Not surprising the $D(\mathbf{Q})$-values from (22) involving the integration of $M_s(\mathbf{Q},t)$ have a value that differs from the two by some 5-9%. This may be produced by the integration of the uncertain tail of $M_s(\mathbf{Q},t)$ (compare Fig. 5). The macroscopic value of D at Q = 0

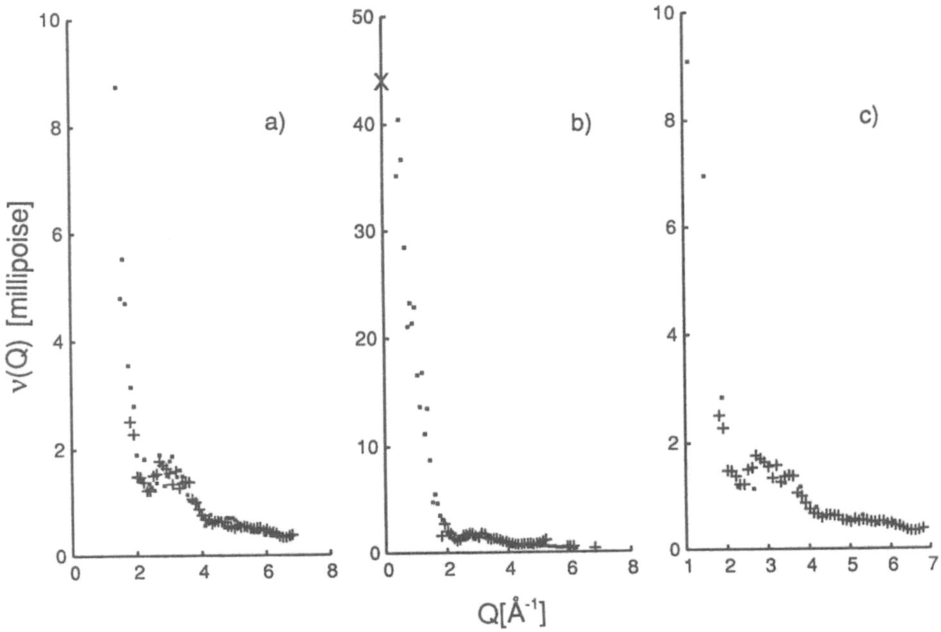

Fig. 3 Derived longitudinal kinematic viscosities ν_l from NS (+) and from MD data (▪) for lead at 623 and 1173K. a) and c) are given on an enlarged scale to display the degree of agreement between NS and MD data. The cross (X) on the ν_l axis marks the position of the experimentally determined macroscopic value.

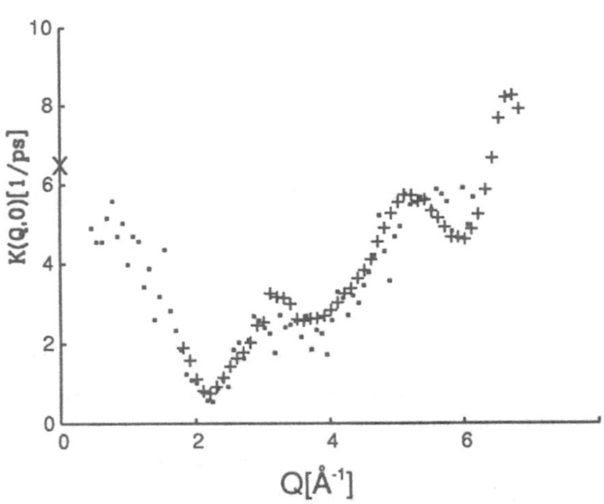

Fig. 4 The time integral $\acute{K}(\mathbf{Q},0)$ for NS (+) and MD (▪) data for Pb at 623K. The hydrodynamic limit (Q=0) for the quantity $(\chi_T\eta_l)^{-1}$ is shown by the cross on the $\acute{K}(\mathbf{Q},0)$-axis.

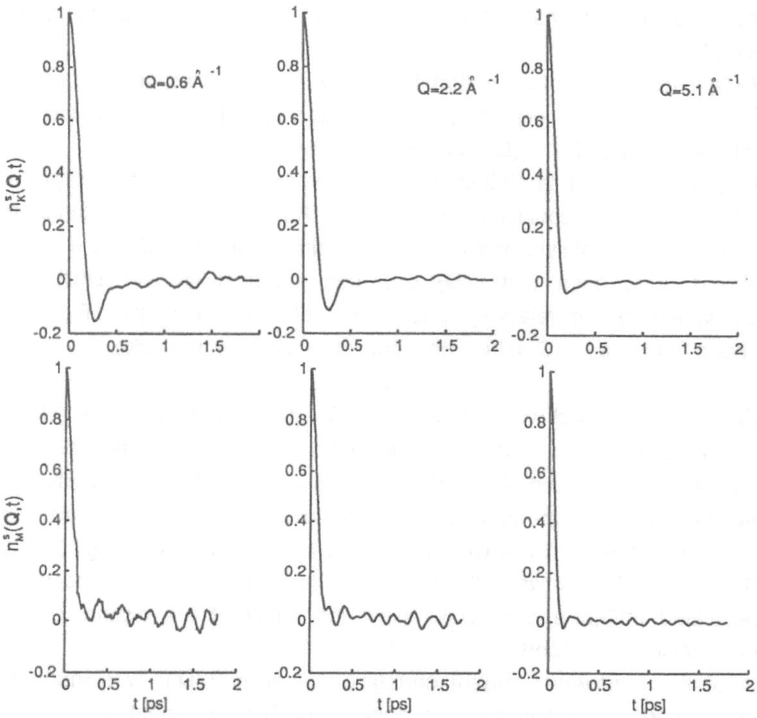

Fig. 5 The generalized velocity autocorrelation function $n^s_K(\mathbf{Q},t)$ and its memory function $n^s_M(\mathbf{Q},t)$ both normalized to 1 at t=0 for liquid lead at 623K.

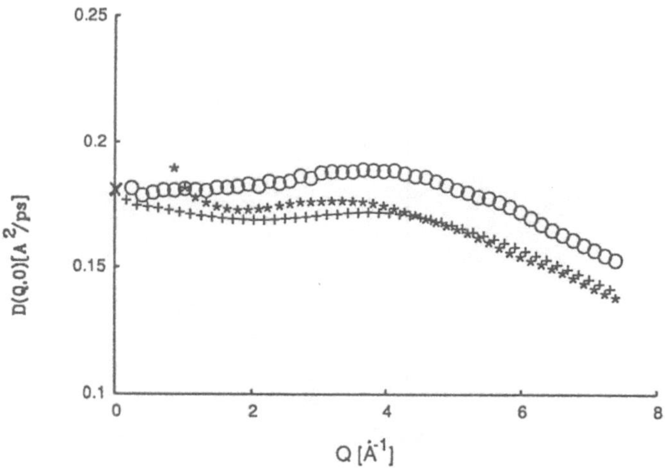

Fig. 6 The generalized self-diffusion coefficient D(**Q**) from integration of $K_s(\mathbf{Q},t)$ (+), integration of $M_s(\mathbf{Q},t)$ (o) and from $S_s(\mathbf{Q},0)$ (*). The cross (X) on the D-axis marks the hydrodynamic value.

determined from the mean square displacement in the MD simulation is marked by the cross; it is 0.184 $Å^2$/ps (= $1.84*10^{-5}$ cm^2/s).

The product $D(\mathbf{Q})M_s(\mathbf{Q})$ should according to (22) be just V_0^2. In the present case at T = 623K V_0^2 = 2.5 $(Å/ps)^2$. From Fig. 7 it is seen that its observed value is some 5-9% too low, which again indicates that time integral of $M(\mathbf{Q},t)$ is low by this amount.

The time integral of $K_s(\mathbf{Q})$ equivalent to $Q^2D(\mathbf{Q})$ may now be compared to the integral of $K(\mathbf{Q},t)$. One may consider $K(\mathbf{Q},t)$ as consisting of $K_s(\mathbf{Q},t)$ + $K_d(\mathbf{Q},t)$, where "d" stands for distinct part. In Fig. 8 these two quantities, derived from MD simulation of $F_s(\mathbf{Q},t)$ and $F(\mathbf{Q},t)$, respectively, are displayed. It is seen that $K(\mathbf{Q})$ oscillates round $K_s(\mathbf{Q})$ = $Q^2D(\mathbf{Q})$ from Q ~ 1.8 $Å^{-1}$. This is equivalent to the relatively small effect of the distinct part $F_d(\mathbf{Q},t)$ in $F(\mathbf{Q},t)$. On the other hand for smaller Q-values the collective phenomena takes over and dominates the scene.

A behavior of a similar nature is found when $\hat{M}(\mathbf{Q},0) = Q^2(\mathbf{Q})v_l(\mathbf{Q})$ is compared to $\hat{M}_s(\mathbf{Q},0) = Q^2v_l^s(\mathbf{Q})$. In Fig. 9a the two quantities v_l and v_l^s are displayed and in 9b the ratio between $v_l(\mathbf{Q})$ and $v_l^s(\mathbf{Q})$ is given. It is again found that $v_l(\mathbf{Q})$ oscillates round $v_l^s(\mathbf{Q})$, the oscillations being to a large degree determined by the second frequency moment $< \omega_l^2 >$ (compare (12)) of the longitudinal current correlation. This is well manifested by comparison of Fig. 9b and any plot of $< \omega_l^2 >$ [12]. The deviation between the two curves in 9 b) reflects differences and inaccuracies in the calculated $M(\mathbf{Q},t)$ and $M_s(\mathbf{Q},t)$. Again the effect of the distinct part of $M(\mathbf{Q},t)$ is seen to decrease quickly in importance for $Q>Q_o$.

Finally, a simulation of the transversal current correlation function $J_T(\mathbf{Q},t)$ was made for lead at 623K. As this is a very lengthy and costly calculation we had to restrict ourselves to an accuracy of ~6%. A few examples of $J_T(\mathbf{Q},t)$ are shown in Fig. 10. The oscillations in$J_T(\mathbf{Q},t)$ for Q>0.3 $Å^{-1}$ are strongly disturbed by noise, which complicates the interpretation. To obtain the shear viscosity $v_{shear}(\mathbf{Q})$ one has to integrate $J_T(\mathbf{Q},t)$ over time (26). This is not

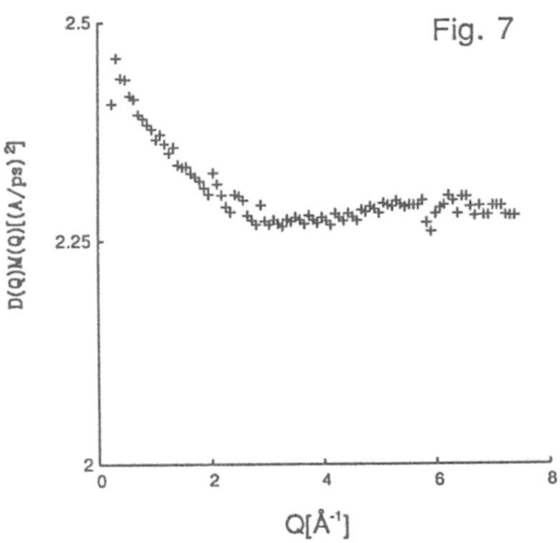

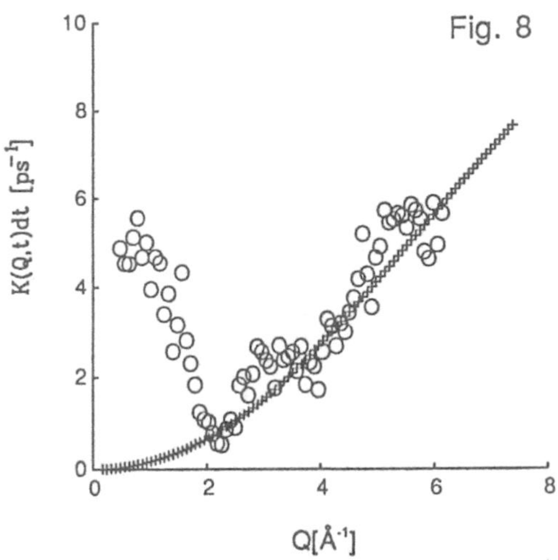

Fig. 7 The product $D(\mathbf{Q})M_s(\mathbf{Q})$ as a function of Q for lead at 623K.

Fig. 8 Time integrals of $K(\mathbf{Q},t)$ (o) and $K_s(\mathbf{Q},t)$ (+) for lead at 623K.

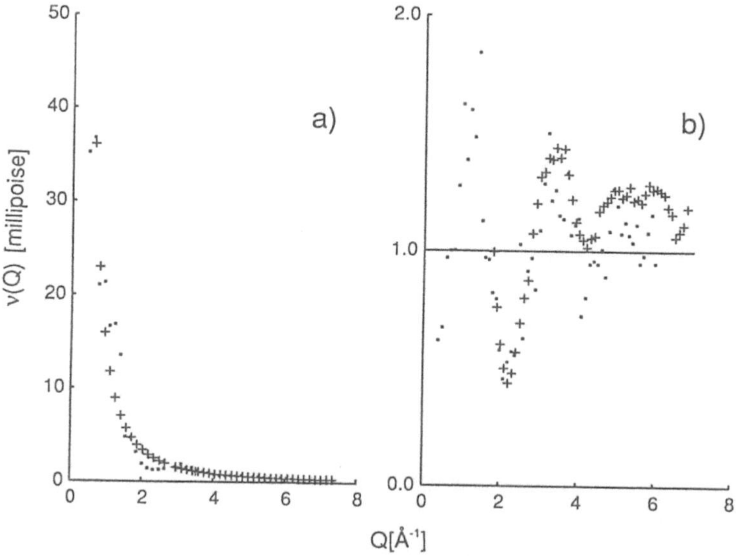

Fig. 9 a) Plot of $\nu_l(\mathbf{Q})$ (■) and $\nu_l^s(\mathbf{Q})$ (+) for liquid lead at 623K.
b) Ratio between $\nu_l(\mathbf{Q})$ and $\nu_l^s(\mathbf{Q})$. $\nu_l(\mathbf{Q})$ from experiment (+) and $\nu_l^s(\mathbf{Q})$ from MD (■).

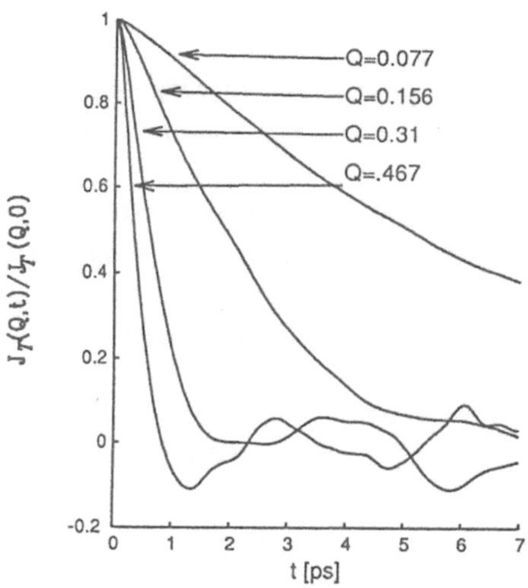

Fig. 10 Transversal current correlation function, $J_T(\mathbf{Q},t)$, for lead at 623K.

practical for Q = 0.0777 and 0.1553 $Å^{-1}$ for which J_T has not decayed. Instead, for these cases (23) was used, which is justified considering the curve shape and the long wavelengths (λ = 80 and 40 Å) involved. For the rest of the Q-scale the integration method (26) is used. As shown in Fig. 11, the quantity $\nu_{shear}(\mathbf{Q}) = \hat{\nu}_{shear}(\mathbf{Q},0)$ approaches the macroscopically

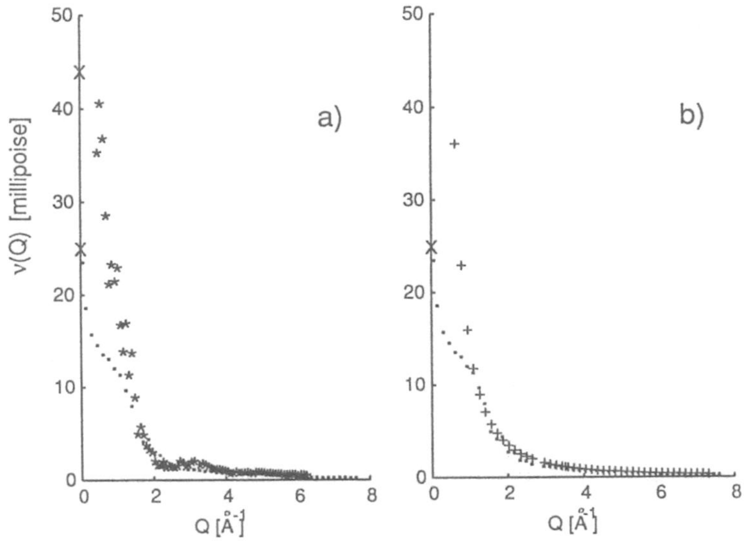

Fig. 11 Generalized shear viscosity $\nu_{shear}(Q)$ (■) compared to generalized total longitudinal, a) $\nu_l(Q)$ (*) and b) self-longitudinal $\nu_l^s(Q)$ (+) viscosities for lead at 623K. The crosses (X) on the viscosity axis mark the experimental macroscopic value for longitudinal and shear viscosities, respectively.

measured hydrodynamical value 25 Å^2/ps (= 2.5 cp) for the smallest Q = 0.0777Å^{-1} for which λ~80 Å. It then decays gradually, showing a change in slope at Q ~ 1 Å^{-1}, and finally at Q ~ 1.8 Å^{-1} takes on a value not far from $\nu_l(Q)$ or $\nu_l^s(Q)$.

It is even more revealing to plot the ratios between the longitudinal and shear viscosities. It is noted that in the small Q limit $\nu_l(Q) = 4/3\nu_{shear}(Q)+\nu_B(Q)$ and $\nu_l^s(Q)$ is not determined. However, plotting the ratios $\nu_l(Q)/\nu_{shear}(Q)$ and $\nu_l^s(Q)/\nu_{shear}(Q)$ as done in Fig. 12, one finds that both ratios fluctuate round the value 4/3, $\nu_l(Q)/\nu_{shear}(Q)$ in a regular oscillatory way as determined by $< \omega_l^2 >$ and $\nu_l^s(Q)/\nu_{shear}(Q)$ being fairly constant and equal to 4/3 for Q > 2 Å^{-1}, possibly showing small insignificant oscillations.

If one assumes that the hydrodynamic expression for $\nu_l(Q) = 4/3\nu_{shear}(Q)+\nu_B(Q)$ remains valid at all Q-values it follows that

$$\nu_l(Q)/\nu_{shear}(Q) = 4/3 + \nu_B(Q)/\nu_{shear}(Q) .$$

The present observation thus gives the result that $\nu_B(Q)$ oscillates round $\nu_{shear}(Q)$ and decreases to zero at larger Q-values.

Furthermore as $\nu_l^s(Q)/\nu_{shear}(Q)$ remains about constant = 4/3 from Q ~ 2 Å^{-1} it is concluded that $\nu_B(Q)$ is associated mainly or totally with the distinct part of $\nu_l(Q)$. As the pair interactions are included in the distinct part the conclusion is that the effect on $\nu_l(Q)$ from pair interactions as described by $F_d(Q,t)$ becomes negligible at Q-values >> Q_o (here when Q_o = 2.2 Å^{-1} already at Q-values of about 7 Å^{-1}). On the other hand it is well known that the time

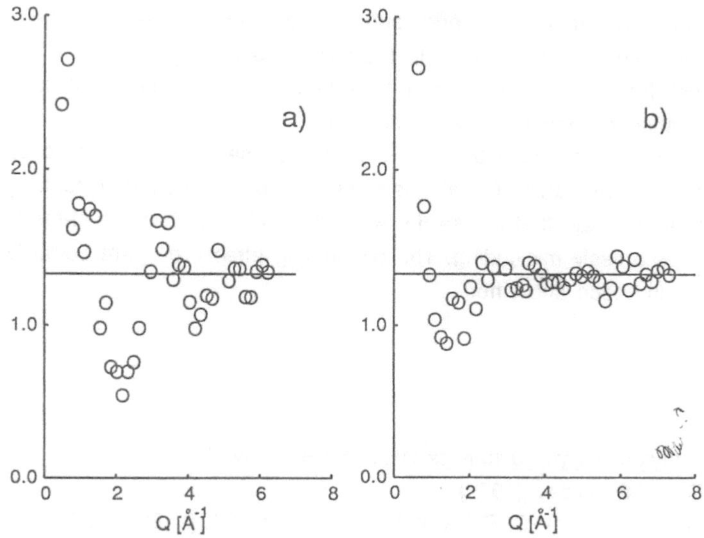

Fig. 12 Ratios of generalized longitudinal viscosities a) $v_l(\mathbf{Q})$ and b) $v_l^s(\mathbf{Q})$ to generalized shear viscosity for lead at 623K.

integral of $F_d(\mathbf{Q},t)$ makes up a contribution almost as large as the time integral of $F_s(\mathbf{Q},t)$ for small Q-values (this effect is well illustrated by the time integrals of first-order memory functions $K(\mathbf{Q},t)$ and $K_s(\mathbf{Q},t)$ in Fig. 8, and by Figs. 11 and 12, which show how the contribution from $v_B(\mathbf{Q})$ increases when $Q < 1.0$ Å^{-1}).

It is to be noted that in this larger Q-region, collective effects described by the long time tail in the memory functions $M(\mathbf{Q},t)$ and $M_s(\mathbf{Q},t)$ become negligible. The time integrals in this Q-region are determined to 90-95% by the first rapidly decaying part of $M(\mathbf{Q},t)$ and $M_s(\mathbf{Q},t)$ (compare Figs. 2 and 5). So therefore the parts describing the binary collisions plus associated rapid readjustments within the sphere of nearest neighbors caused by this collision - in effect a many-body collision - almost totally describe the viscosities at short wavelengths (of order of atomic distance). The possible collectivity active here is of very short range in space and time (space range comparable to radius of nearest neighbor sphere, time range of order 0.1 ps). This is the range where the normal mode-coupling models cannot be applied and this range covers most of the easily observable neutron scattering range (Q-values from the peak of the structure factor and up).

4. Conclusions

A combination of NS-data and MD-data based on an earlier derived pair potential led to the following results for liquid lead:

a. Simulation of $F(\mathbf{Q},t)$ and $F_s(\mathbf{Q},t)$ gave generalized longitudinal viscosities and diffusion coefficients.
b. Simulation of transversal current correlation gave generalized shear viscosity.

c. Comparison of obtained longitudinal and shear viscosities shows a general behavior of the bulk viscosity and relates it to the distinct part of F(**Q**,t).

d. The hydrodynamical framework works as a hypothesis up to large Q-values if these generalized transport coefficients are used in it.

e. It is shown that, for Q-values above about 1.8 $Å^{-1}$ in lead self-motion tends to be dominating and, for Q-values above Q_o, effects of self-motion are very dominant.

f. For this reason, it can be said that to use the word "hydrodynamic" in the Q-range above Q ~ 1.8 $Å^{-1}$ is grossly misleading. The dominating effects in ν_l are certainly of short range order in space and time.

References

1. J.P. Boon and S. Yip: Molecular Hydrodynamics (McGraw-Hill,New York 1980)
2. G.F. Mazenko: Phys. Rev. A**6**, p.2545 (1972)
3. L. Sjögren and A. Sjölander: J. Phys. C: Solid State Physics Vol. 12, p.4369 (1979),
   L. Sjögren: J. Phys. C: Solid State Physics Vol. 13, p.705 (1980),
   L. Sjögren: Phys. Rev. A**22**, p.2860 (1980)
4. W. Götze and M. Lücke: Phys. Rev. A**11**, p.2173 1975,
   J. Bosse, W. Götze and M. Lücke: Phys. Rev. A**17**, p.434 (1978),
   J. Bosse, W. Götze and M. Lücke: Phys. Rev. A**17**, p.447 (1978),
   J. Bosse, W. Götze and M. Lücke: Phys. Rev. A**18**, p.1176 (1978)
5. J.R.D. Copley and J.M. Rowe: Phys. Rev. Lett. **32**, p.49 (1974),
   J.R.D. Copley and J.M. Rowe: Phys. Rev. A **9**, p.1656 (1974)
6. A. Rahman: Phys. Rev. A**9**, p.1667 (1974)
7. L.A. de Graaf: "Analysis of Neutron Scattering and Computer Simulation Studies on Noble Gas Fluids", Europhysics Liquid State Workshop on Static and Dynamic Properties of Liquids, June 27 - July 1, Dubrovnik, Yugoslavia (1988)
8. O. Söderstrom: Phys. Rev. A**23**, p.785 (1981),
   O. Söderstrom, U. Dahlborg and M. Davidovic: Phys. Rev. A**27**, p.470 (1983)
9. O. Söderstrom, U. Dahlborg and W. Gudowski: J. Phys. F **15**, **L**23 (1985)
10. U. Dahlborg, O. Söderstrom, W.Gudowski, K.E. Larsson and M. Davidovic: J. Phys. F **15**, p.2053 (1985)
    U. Dahlborg and L.G. Olsson: Phys. Rev. A **25**, p.2712 (1982)
11. M. Dzugutov, K.E. Larsson and I. Ebbsjö: Phys. Rev. A**38**, p.3609 (1988)
12. K.E. Larsson and W. Gudowski: Phys. Rev. A**33**, p.1975 (1986)

The Transition from the Hydrodynamic Regime to the Kinetic Regime in Helium/Xenon Gas Mixtures as Measured by Light Scattering

G.H. Wegdam, A. Bot, and H.M. Schaink

Laboratory for Physical Chemistry, University of Amsterdam,
Nieuwe Achtergracht 127, NL-1018 WS Amsterdam, The Netherlands

The dynamics of a He+Xe gas mixture is to a large extent determined by the disparity in mass of the components ($m_{Xe}/m_{He} = 33$), which causes a large difference in the thermal velocity of the particles ($v_{Xe}/v_{He} = 1/6$). In the limit of low xenon concentration it resembles a system with Brownian dynamics, while at the opposite end of the concentration range it tends to the Lorentz gas limit.

Due to the disparity in mass between helium and xenon, a strong correlation is to be expected between concentration fluctuations and fluctuations in the pressure and entropy. This correlation will affect the lineshape of the Rayleigh-Brillouin spectrum, especially when the product of the wavevector k and the mean free path l approaches $kl=O(1)$. We will show here that for a range of compositions of the He+Xe mixture the transition from the hydrodynamic regime ($kl<<1$) to the kinetic regime ($kl=O(1)$) is accompanied by a dramatic change in the dispersion of the sound waves. We will do this by measuring the Rayleigh-Brillouin spectrum as a function of the reduced wavevector kl. The manner in which the features of the spectra deviate from the thermodynamic limit ($kl<<1$) are compared with the predictions of the hydrodynamic theory. The data presented in this way form a more accurate and sensitive test for the proposed models and dynamic descriptions than a likeness between the overall shape of the light scattering spectrum from calculations and experiment.

When a light wave passes through a medium, it is scattered by the fluctuations of the dielectric constant of the medium. The light scattering spectrum $I(k,\omega)$ is proportional to the temporal Laplace transform of the time-correlation function of the dielectric fluctuations with a wavevector k [1]:

$$I(k,t) = \sum_m A_m(k)\, e^{-z_m(k)t} \quad . \tag{1}$$

Here $z_m(k)$ represents the m-th eigenvalue of the operator which governs the time evolution of the fluctuations, and $A_m(k)$ the wavevector-dependent amplitude of the m-th eigenmode. In the limit of small wavevectors the light scattering spectrum consists of three lines: one of these, the Rayleigh-line is unshifted in frequency, while the two Brillouin lines are symmetrically shifted in frequency with respect to the Rayleigh line. The integrated intensity of the scattered light is proportional to the square of the polarizability of the particles that interact with the light wave. The polarizability of the xenon particles is about 20 times larger than the polarizability of the helium particles [2]. Therefore the light scattering spectrum of a He+Xe mixture mainly reflects the dynamics of the fluctuations of the xenon density. For large wavevectors this lineshape passes over into a Gaussian whose shape is given by the Doppler broadening of a Maxwellian velocity distribution of the xenon particles [3].

For sufficiently small wavevectors the evolution operator of the fluctuations can be described by the hydrodynamic matrix which is obtained in the process of solving the linearized conservation laws of the density, momentum density, energy density and the concentration. A detailed expression of the hydrodynamic matrix of a binary mixture is given elsewhere [4]. Here we just state that the elements of the hydrodynamic matrix are terms of order ik and k^2. As a result the eigenvalues $z_m(k)$ can be written as a power-law in ik, of which the term of order k^2 has a positive sign [4]. All elements of the hydrodynamic matrix which stem from the conservation law of concentration are of the order of

k^2. Therefore comparing the one-component fluid with the binary mixture, we may conclude that the constraint of a conserved concentration causes an increased damping of the eigenmodes (see eq. 1).

We have, for several compositions of disparate mass gas mixtures, calculated the wavevector-dependence of the eigenvalues of the hydrodynamic matrix [5]. The thermodynamic functions needed for the calculation of the eigenvalues $z_m(k)$ are calculated using an equation of state up to the third virial coefficient, while for the calculation of the transport coefficients the Chapman-Enskog expressions are used in conjunction with empirical relations for the collision-integrals [6]. The results of the hydrodynamic calculations will be presented as a function of the reduced wavevector kl_{He}, where l_{He} stands for the mean free path of a helium particle. Of course, the results of the calculations are correct only for low values of kl_{He}. However they serve as a qualitative indication of the effects to be expected when kl_{He} approaches values of order 1.

The calculations show that the damping of the eigenmodes increases as the molecular mass-ratio of the components increases and as the mole fraction of the heavier component decreases. In particular for a He+Xe gas mixture with $x_{Xe}=0.45$ the wavevector-dependence of the sound propagation frequency Im $z_s(k)$ has a point of inflexion [5]. And for He+Xe mixtures with $x_{Xe}<0.17$ the calculations show sound propagation gaps (see Fig. 1) [5], similar to those which have been measured in liquid argon by DE SCHEPPER et al. [7]. Moreover, calculations of the amplitudes of the Rayleigh-Brillouin triplet show that the light scattering spectrum becomes featureless as the "gap-region" is approached [5]. The damping of the sound mode is to a large extent due to the correlation between fluctuations in the concentration and the pressure. All the phenomena mentioned above occur for wavevectors which can be accessed by light scattering experiments, provided a proper choice of the density is made.

The experimental results we want to present here concern wavevector-dependent and density-dependent experiments on various He+Xe mixtures. The density-dependent experiments were performed in a 90° scattering geometry, while in the wavevector-dependent experiments the scattering angle varied from 44° to 133°. The mixtures were studied at room-temperature, while the pressure ranged from 2 to 40 bar. The results of the wavevector-dependent experiments have been published elsewhere [8].

At a pressure of 40 bar we have found for a large composition range of the He+Xe mixture that the wavevector-dependence of the Brillouin-shift is in accordance with the small k-limit: $\Delta\omega(k)$ = Im $z_s(k) = c_s k$ [8]. Here c_s represents the adiabatic sound velocity. However, as the pressure is lowered, accompanied by a broadening of the Rayleigh and Brillouin-lines, a clear deviation from the linear dispersion relation can be observed [8,9]. In the wavevector-regime where the linear law is no longer valid, the sound damping also deviates from its quadratic k-dependence for small wavevectors [4]. In Fig. 2 we show a non-linear dispersion relation for the He+Xe mixture with $x_{Xe}=0.58$ [8] and at a pressure of 9.9 bar.

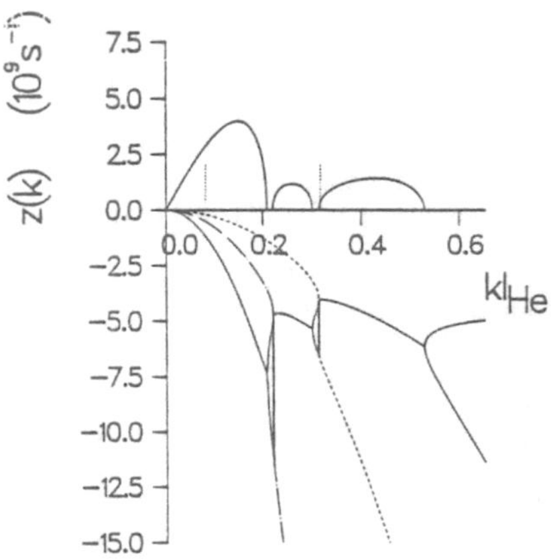

Fig. 1: Eigenvalues z(k) of the hydrodynamic matrix for a He+Xe mixture with x_{Xe} = 0.1, p = 10 bar, T = 293K, as a function of the reduced wavevector kl_{He}; the positive values refer to Im z; the negative values to Re z; the drawn lines represent the sound mode eigenvalue, the dashed lines the eigenvalues of the diffusive eigenmodes. The dotted bars indicate the wavevector region which is accessible by light scattering.

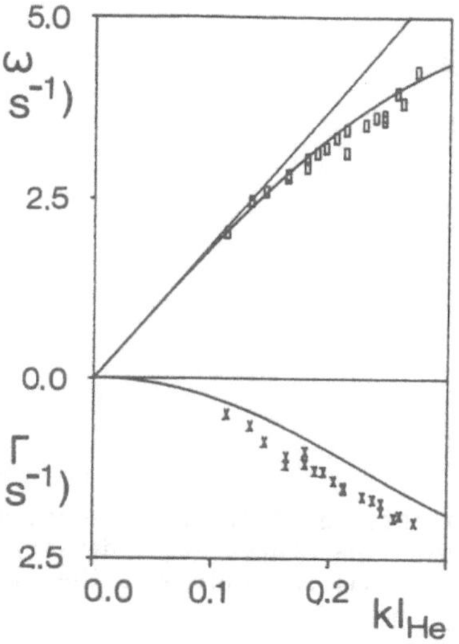

Fig. 2: The Brillouin-shift $\omega(k)$ and the width of the Brillouin lines $\Gamma(k)$ for the He+Xe mixture with $x_{Xe} = 0.58$, $p = 9.9$ bar, $T = 293K$, as a function of the reduced wavevector kl_{He} (l_{He}: mean free path of a He particle); the straight line represents the small wavevector limit of the sound dispersion; the drawn line represents the calculated eigenvalue $z_s(k)$.

The experimental points are compared with the calculated wavevector-dependence of the sound-mode eigenvalue $z_s(k)$ (solid lines). In these calculations no adaptable parameters are used. In Fig. 2 it can be seen that qualitative agreement between experiment and theory is obtained.

Having established experimentally a deviation from the small wavevector limit, we next studied the kl_{He}-dependence of $I(k,\omega)$ over a larger kl_{He}-domain by means of density-dependent light scattering experiments. These experiments were performed on He+Xe mixtures with xenon mole fractions x_{Xe}=0.79, 0.61, 0.45 and 0.22. In the case of a He+Xe mixture with x_{Xe}=0.22, we found that the light-scattering spectrum becomes featureless when kl_{He} exceeds 0.15. However, the function $\omega^2 I(k,\omega)$ has well-defined peaks for all values of kl_{He}, even in the region where sound propagation gaps are predicted to exist. It should be noted that $\omega^2 I(k,\omega)$ is closely related to the longitudinal current-current autocorrelation function [3]. For the peak positions $\omega_m(k)$ of $\omega^2 I(k,\omega)$ analytic expressions are known for both the small and large wavevector limits [10] . In the small wavevector limit $\omega_m(k) = c_s k$. At the other extreme, for large wavevectors, $\omega^2 I(k,\omega)$ has peaks which are located at the Gaussian limit of $\omega_m(k) = v_{Xe} k\sqrt{2}$. Therefore, we study the reduced wavevector dependence of $\omega^2 I(k,\omega)$ and compare the results of the experiments with the results of the hydrodynamic calculations for the small kl_{He} and with the Gaussian limit for large kl_{He}.

In Fig. 3 we show our results for $\omega_m^*(k) = \omega_m(k)/c_s k$ for various mixtures, plotted as a function of kl_{He}. For all mixtures considered here the hydrodynamic theory predicts the peak positions of $\omega^2 I(k,\omega)$ well up to kl_{He}=0.2. The measured deviations from the value of the small k-limit are small for mixtures with a large xenon mole fraction and become larger as x_{Xe} decreases. However, for all mixtures $\omega_m^*(k)$ deviates significantly from the calculated dispersion of the sound mode when kl_{He} exceeds 0.3. For these high kl_{He}-values the dispersion relation of $\omega_m^*(k)$ tends towards the value of the Gaussian limit (horizontal bars).

An intuitive model [5] based on the description in terms of two independent subsystems is helpful in illustrating the observed and calculated phenomena. Consider a system which consists of independent helium and xenon excitations. To the excitations one may assign a kinetic energy-density $E_i = n_i m_i v_i^2$ (i=Xe,He) and a momentum-density $p_i = n_i m_i v_i$. Here n_i represents the number-density of species 1. If the velocity v_i is taken to be of the order of magnitude of the average thermal velocity of particle i, it can be shown that the energy-densities balance each other when x_{Xe}=0.5. Similarly one finds that the momentum-densities balance at x_{Xe}=0.15. Clearly, according to this model there exists a composition range for which the modes are dominated by the Xe-excitations ($0.5<x_{Xe}<1.0$), a

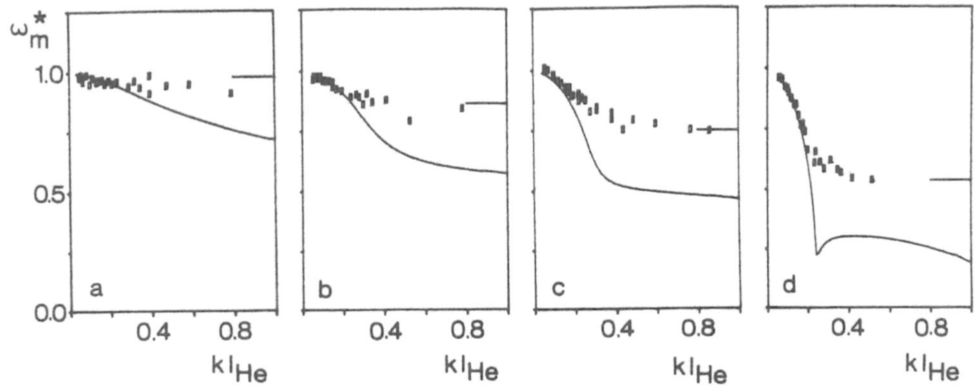

Fig. 3: The dispersion of $\omega_m^*(k) = \omega_m(k)/c_s k$ obtained from density-dependent experiments, as a function of kl_{He}; the drawn line represents a density-dependent hydrodynamic calculation; the horizontal bar indicates the value of the large wavevector limit;

a: $x_{Xe} = 0.79$ T = 293 K
b: $x_{Xe} = 0.61$ scattering angle $\Theta = 90°$
c: $x_{Xe} = 0.45$
d: $x_{Xe} = 0.22$

range for which He-excitations dominate ($0.0<x_{Xe}<0.15$) and an intermediate range. It is in the intermediate range that strong damping of the sound mode occurs.

We have shown that in the He+Xe gas mixture, the transition from the hydrodynamic regime to the kinetic regime is accompanied by damping of the sound mode. The hydrodynamic theory predicts the peak positions of $\omega^2 I(k,\omega)$ well up to $kl_{He} \cong 0.2$.

REFERENCES

1. B.J. Berne and R. Pecora: Dynamic Light Scattering (John Wiley & Sons Inc., New York, 1976)
2. R.C. Burns, C. Graham and A.R.M. Weller: Mol. Phys., 59, 41 (1986)
3. J.P. Boon and S. Yip: Molecular Hydrodynamics (McGraw-Hill, New York, 1980)
4. C. Cohen, J.W.H. Sutherland and J.M. Deutch: Phys. Chem. Liquids 2, 213 (1971)
5. G.H. Wegdam and H.M. Schaink: Mol. Phys., 65, 531 (1988)
6. J. Kestin, H.E. Khalifa and W.A. Wakeham: Physica, 90A, 215 (1978)
   J. Kestin, S.T. Ro and W.A. Wakeham: Physica, 58, 165 (1972)
7. I.M. de Schepper, P. Verkerk, A.A. van Well and L.A. de Graaf: Phys. Rev. Lett., 50, 974 (1983)
8. H.M. Schaink and G.H. Wegdam: Physica, 141A, 211 (1987)
   The compositions given in ref. 3 are in error.
9. B.Y. Baharudin, P.E. Schoen and D.A. Jackson: Phys. Lett. A, 42, 77 (1972)
10. A.A. van Well, P. Verkerk, L.A. de Graaf, J.-B. Suck and J.R.D. Copley: Phys. Rev. A, 31, 3391 (1985)

Transition to Hydrodynamic Behaviour and Mode Coupling Effects in Dense Hydrogen Gas

P. Verkerk, J. Westerweel, U. Bafile, I.M. de Schepper, and L.A. de Graaf

Interfacultair Reactor Instituut, Technische Universiteit Delft,
NL-2629 JB Delft, The Netherlands

The incoherent inelastic neutron scattering has been measured in dense hydrogen gas at 119 K and 768 bar for wave vectors $k \geq 2.5$ nm^{-1}. The experimental data clearly deviate from simple hydrodynamics. For $k \leq 5$ nm^{-1} we find agreement with mode-coupling theory by fitting the self-diffusion coefficient D and the viscosity η. If we include also data with $k > 5$ nm^{-1} it is no longer possible to fit the mode-coupling model to the experimental dynamic structure factor with single values for D and for η. The value for η from this experiment at $k \leq 5$ nm^{-1} agrees with data found in the literature, but the value we find for D (0.0581 nm^2 ps^{-1}) differs significantly from the result of a recent NMR experiment (D = 0.0623 nm^2 ps^{-1}).

1. Introduction

Recent neutron scattering experiments have indicated the existence of mode-coupling effects in liquid argon [1], dense hydrogen gas [2] and molten sodium [3]. The argon data as well as the sodium data agree with mode-coupling theory (MCT) for $k \leq 10$ nm^{-1}. The hydrogen data are qualitatively in agreement with MCT, but deviate slightly. A significant deviation would imply that hydrogen behaves differently from argon and sodium. Therefore the hydrogen experiment was repeated with the Delft time-of-flight spectrometer optimized to increase the accuracy of the data at small scattering angles. Knowledge of the self-diffusion coefficient D and the viscosity η is useful for the analysis of the neutron data. Since D in H_2 had not been measured before, an NMR experiment was performed at the Van der Waals Laboratory, Amsterdam, on hydrogen at a nearly identical thermodynamic state to determine the value of D [4].

Static and Dynamic Properties of Liquids

2. Theory

We give a summary of the relevant theoretical results. According to MCT [5] the self-part $S_s(k,\omega)$ of the dynamic structure factor is for small k given by

$$S_s(k,\omega) = \frac{1}{\pi}\,\frac{D\,k^2}{\omega^2 + (D\,k^2)^2} + \frac{1}{\pi}\,\frac{1}{D\,k\,k^*}\;\mathrm{Re}\,G\left(\frac{i\omega + Dk^2}{\delta\,D\,k^2}\right)$$

$$+\; O\{\omega^2 / k^{3/2}\}, \tag{1}$$

with $k^* = 16\pi\beta MnD^2$, $\beta = 1/k_BT$, k_B Boltzmann's constant, M the particle mass, n the number density, $\delta = D/(D + \nu)$, $\nu = \eta/(Mn)$ the kinematic viscosity, η the shear viscosity and

$$G(z) = \arctan\,(z-1)^{-1/2} - (z-2)\cdot(z-1)^{1/2} / z^2 \;. \tag{2}$$

The first term in (1) results from Fick's law and is called the simple-diffusion model. From (1) and (2) follows

$$k^2\,S_s(k,\omega=0) = \frac{1}{\pi\,D}\,[1 + a\,k + O(k^{3/2})\,]\;, \tag{3}$$

$$\omega_H^s(k)/k^2 = D\,[1 - b\,k + O(k^{3/2})\,]\;, \tag{4}$$

with $\omega_H^s(k)$ the half-width at half maximum of $S_s(k,\omega)$, $a = G(\delta^{-1})/k^*$, $b = H(\delta)/k^*$ and

$$H(\delta) = 1.4531\;\delta^{3/2}[1 - 0.7276\;\delta - 0.1523\;\delta^2 - O(\delta^3)\;]\;. \tag{5}$$

3. Experiment

The new measurements on hydrogen were performed with the time-of-flight spectrometer at the Delft 2 MW reactor. The angular resolution was reduced to 2.8° at all scattering angles. The incident wavelength was 0.200 nm and the k-range at zero energy transfer 2.2 - 24 nm^{-1}. The improved energy resolution was approximately 1.3 meV. The sample container was made of 5052 aluminium alloy capillary with an inner diameter of 0.75 mm and is described by VERKERK and PRUISKEN [6].

During the measurements the temperature of the hydrogen sample was 119.2 ± 0.5 K with a variation of less than 1 ‰ and the average pressure was 768 bar with a spread of 16 bar, leading to a particle number density $n = 21.2 \pm 0.2\ nm^{-3}$ (using PVT-data compiled by YOUNGLOVE [7]). The ortho-para composition was checked for changes by monitoring the transmission of the sample for neutrons; 100 hours after setting the temperature no more changes could be detected. The data were corrected in the usual way [8] for background, multiple scattering and time-of-flight resolution [9]. The data at scattering angles < 11° were however corrected for resolution by means of the much more elaborate but superior maximum entropy method [10], because the width of the spectra at these angles is small compared to the energy resolution. The $S(k,\omega)$ data at constant k were corrected for inelastic scattering due to rotational transitions of the hydrogen molecule and for coherent scattering, using results of calculations by BAFILE [11].

LUIGJES and PRINS [4] measured D with an absolute accuracy of 2 % by means of NMR at the Van der Waals Laboratory, Amsterdam, in H_2 at 119.21 K and 749.5 bar ($n = 21.0\ nm^{-3}$, [7]). Assuming $D\ n^2$ to be constant as function of n we obtain $D = 0.0623\ nm^2\ ps^{-1}$ for $n = 21.2\ nm^{-3}$. From TOULOUKIAN and HO [12], who give η with an accuracy of 10 % for compressed hydrogen, we calculate $\nu = 0.172 \pm 0.017\ nm^2\ ps^{-1}$.

4. Analysis

First we fitted the mode-coupling model (1) to the experimental data at a number of k values separately with D and η as fitting parameters. Since MCT is a hydrodynamic theory it is not expected to be valid at large ω and the fitting was limited to $\omega \leq 2\ D'\ k^2$ using $D' = 0.0623\ nm^2\ ps^{-1}$, the value from LUIGJES and PRINS [4]. We find that D and η do not change significantly as functions of k for $k \leq 5\ nm^{-1}$, but as k increases beyond $5\ nm^{-1}$ D starts to decrease. Therefore we redetermined D and η fitting (1) simultaneously to all experimental data for $k \leq 5\ nm^{-1}$ and $\omega < 2\ D'\ k^2$. We obtain in this way $D = 0.0581\ nm^2\ ps^{-1}$ and $\nu = 0.169\ nm^2\ ps^{-1}$. The standard deviations due to the statistical errors of the experimental data are $0.0002\ nm^2\ ps^{-1}$ for D and $0.005\ nm^2\ ps^{-1}$ for ν. The mode-coupling

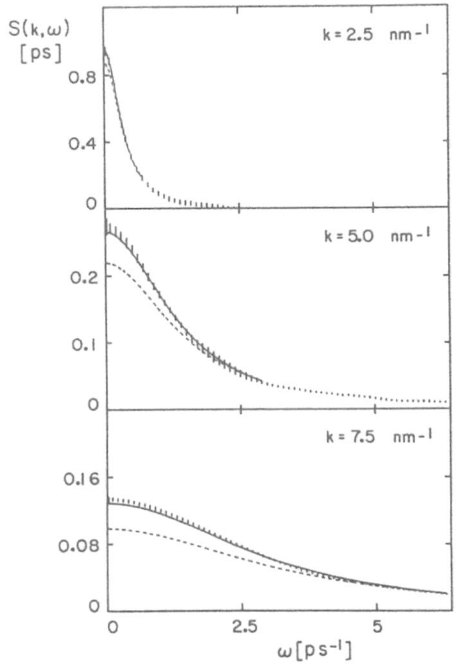

Fig. 1. Experimental $S(k,\omega)$ (error bars) compared with the mode-coupling model (solid line) and with the simple-diffusion model (dashed line). The value for D used in both models is 0.0581 nm^2 / ps and ν = 0.169 nm^2 / ps.

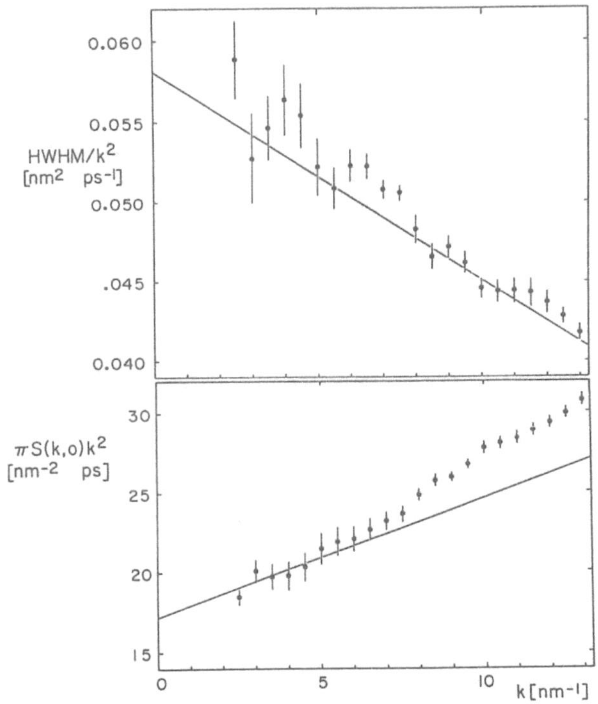

Fig. 2. Experimental data for $S(k,\omega=0)$ and $\omega_H^s(k)$ (dots with error bars) compared with MCT (solid line). D = 0.0581 nm^2 / ps, ν = 0.169 nm^2/ps.

model fitted in this way is shown in Fig. 1 for k = 2.5 and 5.0 nm^{-1} (the boundaries of the fit region) and for k = 7.5 nm^{-1} together with the experimental data and the simple-diffusion model. In Fig. 2 the experimental values for $S(k,\omega=0)$ and $\omega_H^s(k)$ are compared with MCT, using the aforementioned values for D and ν.

5. Discussion

We find that the single-particle motion in hydrogen at 119.2 K, 768 bar and k > 2.5 nm^{-1} cannot be described by simple diffusion. However, a very good description for 2.5 < k < 5 nm^{-1} is obtained when one uses the MCT with the self-diffusion coefficient D and the kinematic viscosity ν as adjustable parameters. The value we find in this way for ν is consistent with the value for η given by TOULOUKIAN and HO [12], but the value we find for D is 7 % lower than the value D_{NMR} measured by LUIGJES and PRINS with NMR [4]. It is tempting to conjecture that the difference between D and D_{NMR} is related to the difference between the bare (D^*) and the full or macroscopic ($\tilde{D}$) diffusion coefficients, respectively, discussed in mode-coupling theories [13]. For, then one also has that $D^* < \tilde{D}$ while D^* is the diffusion coefficient relevant for single-particle motions on a very small (microscopic) scale.

If so, the difference between D and D_{NMR} will gradually vanish when $k \to 0$. Thus, it might be interesting to see whether neutron scattering experiments at increasingly smaller k would show that the difference between D and D_{NMR} decreases.

We conclude that the present experimental data for $S_s(k,\omega)$ with $2.5 \leq k \leq 5\ nm^{-1}$ can be understood on the basis of the mode-coupling theory but that, most probably, the observed value of the self-diffusion coefficient D has not yet reached its full hydrodynamic value D_{NMR}.

It is interesting to note that MORKEL et al. [3] have found that $S_s(k,\omega)$ of liquid sodium at high temperatures and a wavenumber range comparable with ours follows the mode-coupling prediction (1), when the full hydrodynamic values of both D and ν are used. It appears therefore that hydrogen and sodium behave differently on comparable microscopic length scales, since the diffusion coefficient of hydrogen has not yet reached its full hydrodynamic value at $k = 2.5\ nm^{-1}$.

References

1. I.M. de Schepper, P. Verkerk, A.A. van Well, L.A. de Graaf: Phys. Lett. 104A, 26 (1984)
2. P. Verkerk, J.H. Builtjes, I.M. de Schepper: Phys. Rev. A 31, 173 (1985)
3. C. Morkel, C. Gronemeyer, W. Gläser, J. Bosse: Phys. Rev. Lett. 58, 1873 (1987)

4. H. Luigjes, K. O. Prins: to be published
5. I.M. de Schepper, M.H. Ernst: Physica A 98, 189 (1979)
6. P. Verkerk , A.M.M. Pruisken: Nucl. Instr. Meth. 160, 439 (1979)
7. B. A. Younglove: J. Phys. Chem. Ref. Data 11, Suppl.1 (1982)
8. P. Verkerk , A. A. van Well: Nucl. Instr. Meth. Phys. Res. 228, 438 (1985)
9. P. Verkerk: Comput. Phys. Commun. 25, 325 (1982)
10. J. Skilling , S. F. Gull: In Maximum-Entropy and Bayesian Methods in Inverse Problems, ed. by C. Ray Smith and W. T. Grandy, Jr. (Reidel, Dordrecht, 1985) p.83; S. Steenstrup: Aust. J. Phys. 38, 319 (1985)
11. U. Bafile: Low-Temperature Neutron Scattering Cross Section of Hydrogen, IRI-report 132-88-03 (IRI, Delft, 1988)
12. Y. S. Touloukian , C. Y. Ho: Properties of Nonmetallic Fluid Elements (McGraw-Hill, New York, 1981) p.97
13. D. Bedeaux, P. Mazur: Physica 73, 431 (1974); Y. Pomeau, P. Résibois: Phys. Repts. 19C, 64 (1975)

Generalized Viscosity and Generalized Diffusion: MD and Neutron Scattering Data for Some Liquids

W. Gudowski, K.-E. Larsson, and M. Dzugutov

Royal Institute of Technology, Department of Reactor Physics, S-10044 Stockholm, Sweden

Extensive studies on the liquid metals **Pb** at 623K and **Bi** at 578K were performed based on neutron scattering data and molecular dynamics simulations. The kinematic viscosity for these liquids in time and frequency spaces was calculated based on the memory function approach and then compared with MD results. The precision of the memory function calculations was evaluated. The generalized diffusion was derived and calculated from MD simulations for the self-part of the intermediate scattering function and its memory function.

1. Introduction

Recently many attempts have been done to perform an in-depth analysis of neutron scattering data from liquid metals. The kinematic viscosity $\nu_l(\mathbf{Q})$ for some liquid metals was calculated and compared to MD results [1]. The way of calculating memory functions raised the question about the errors introduced in the calculation process involving multiple differentiation of the intermediate scattering functions [2]. However, careful analysis of the memory function equations in Laplace space shows a new exciting possibility of calculating the generalized viscosity and generalized transport coefficients with a satisfactory precision limited only by the precision of experimental and MD data. It proves also that the results based on the integration of memory functions in time space [1,3] are of very good quality and can be used for testing theoretical predictions.

2. Equations

The generalized longitudinal viscosity $\nu_l(\mathbf{Q})$ can be defined as [1]

$$\nu_l(\mathbf{Q}) = \hat{M}(\mathbf{Q},0)/Q^2 \tag{1}$$

where $\hat{M}(\mathbf{Q},\omega)$ is the second-order memory function of the scattering function.
As shown in [4] the memory function formalism enables one to calculate the generalized longitudinal viscosity also directly from the scattering function $S(\mathbf{Q},\omega)$ and the first-order memory function $\hat{K}(\mathbf{Q},\omega)$

$$\nu_l(\mathbf{Q}) = \pi V_0^2 S(\mathbf{Q},0)/S(\mathbf{Q})^2 \,, \tag{2}$$

$$\nu_l(\mathbf{Q}) = V_0^2/(S(\mathbf{Q})^*\hat{K}(\mathbf{Q},0)) \,. \tag{3}$$

Using the same procedure for the self-part of the scattering function for which $S(\mathbf{Q})=1$, and setting the generalized diffusion coefficient as

Static and Dynamic Properties of Liquids

$$\dot{D}(\mathbf{Q},\omega) = \dot{K}_s(\mathbf{Q},\omega)/Q^2 \,, \tag{4}$$

one can write

$$\dot{D}(\mathbf{Q},0) = (\pi Q^2 S_s(\mathbf{Q},0))^{-1} \,, \tag{5}$$

$$\dot{D}(\mathbf{Q},0) = V_0^2/(\dot{M}_s(\mathbf{Q},0)) \,. \tag{6}$$

So the generalized diffusion coefficient can be calculated directly from the time integral of the intermediate self-scattering function.

3. Results and Discussion

Figure 1 shows the generalized viscosity $\eta_l(\mathbf{Q})$ for liquid bismuth at 578K derived from (2). The measurements extend only to Q= 0.5 $Å^{-1}$ and the sudden rise of $\eta_l(\mathbf{Q})$ to about 60.0 millipoise indicates that the macroscopic value of the viscosity of about 100 millipoise published in [5] is very reasonable. The large discrepancy with the value published at [1] based on the normalization of the hard-sphere results of Alley et al. [6] means that the hard-sphere model works very well for the low temperature liquid lead (623K) but does not for bismuth. The comparison of the viscosity data obtained from the integration of the $v_l(\mathbf{Q},t)$ memory function and directly from S(Q,0) in Fig. 1b shows that the calculation procedure for the memory

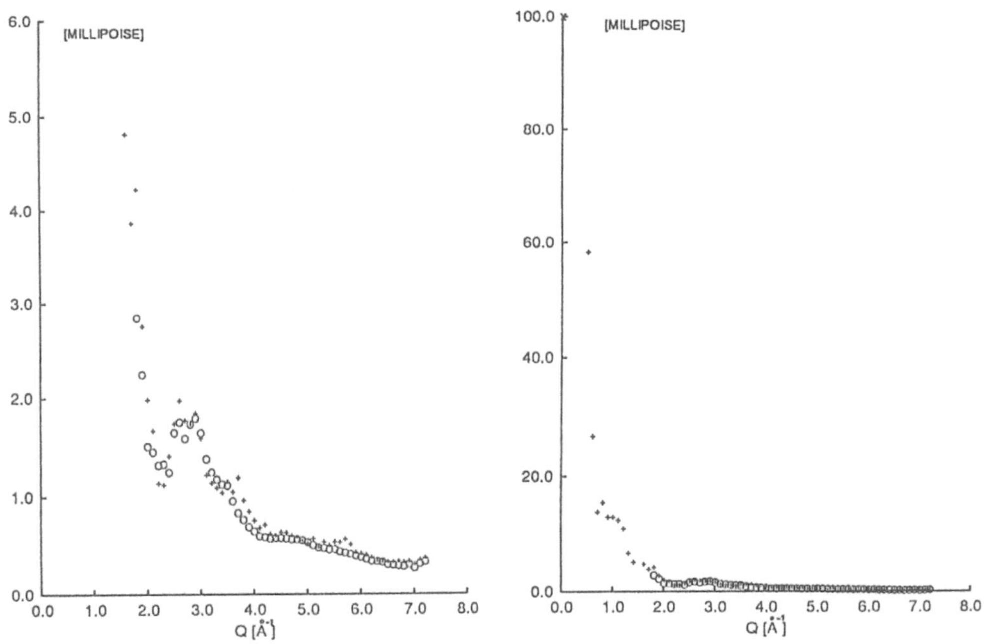

Fig. 1: Generalized viscosity $\eta_l(\mathbf{Q})$ for liquid bismuth at 578K. Data calculated directly from S(Q,0) (+), data from integration of $\eta_l(\mathbf{Q},t)$ (o). X marks an experimental macroscopic value.

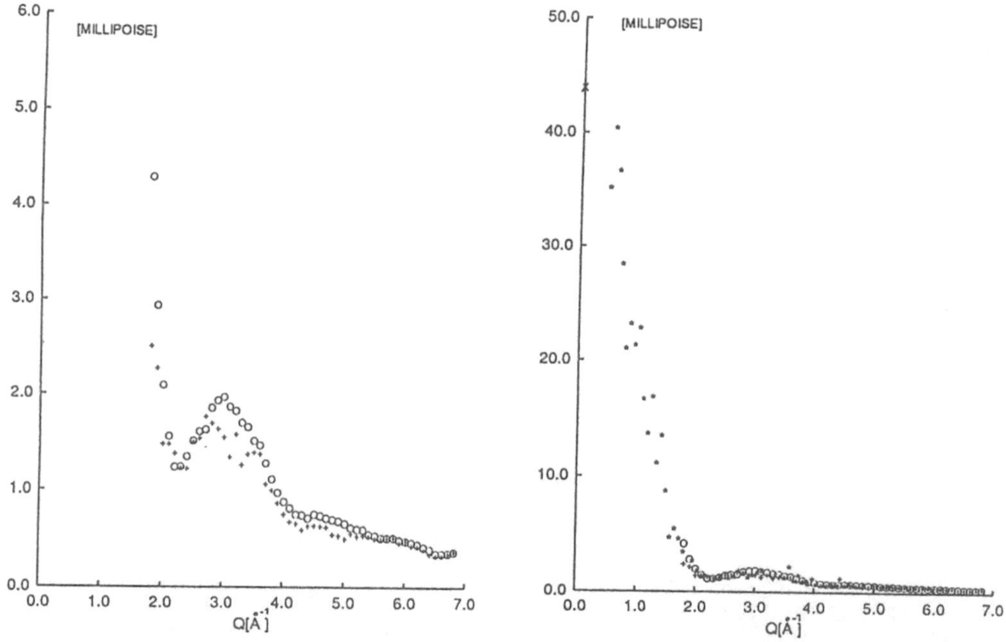

Fig. 2: Generalized viscosity $\eta_l(Q)$ for liquid lead at 623K. Data calculated directly from S(Q,0) (+), data from integration of $\eta_l(Q,t)$ (o), MD data (*). X marks an experimental macroscopic value.

functions does not introduce any significant errors and the data are definitely not of a qualitative nature as claimed by Egelstaff [2]. The same conclusions can be drawn from Fig. 2 where the data for liquid lead at 623K are presented. Figure 2b presents also the molecular dynamics results together with experimental ones. It can be observed that the MD results give a correct value for the macroscopic viscosity of about 45 millipoise [5].

Finally Figs. 3 and 4 show the generalized diffusion for lead at 623K and 1173K, respectively, calculated from MD data. The plot for Pb at 623K shows that there are practically no differences between the data calculated from $S_s(Q,0)$ and from the time integral of the $K_s(Q,t)$ memory function except for the very low **Q** area where the cut-off error for the intermediate scattering function occurs. For the high temperature lead there is a certain systematic error of the order of 8% due to the inaccurate MD data for intermediate scattering functions. One observes a drastic change in the evolution of D(Q) curves between the two temperatures. This very interesting phenomenon will be investigated in the nearest future.

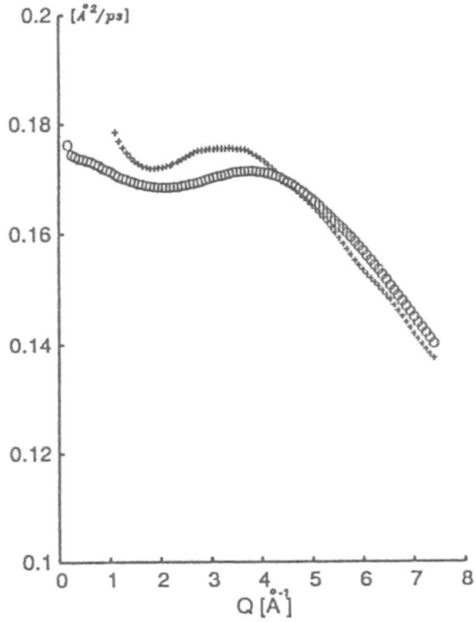

Fig. 3: Generalized diffusion D(**Q**) for liquid lead at 623K. Data calculated directly from the MD intermediate scattering function (+), (o) - data from integration of $K_S(\mathbf{Q},t)$ memory function.

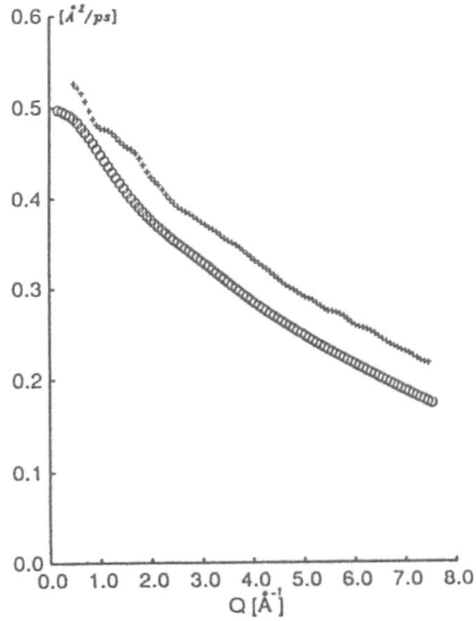

Fig. 4: Generalized diffusion D(**Q**) for liquid lead at 1173K. Data calculated directly from the MD intermediate scattering function (+), (o) - data from integration of $K_S(\mathbf{Q},t)$ memory function.

REFERENCES

1. K.E. Larsson and W.Gudowski: Phys. Rev. A 33, 3 (1986)
2. P.A. Egelstaff: Zeit. fur Physik. Chemie, Neue Folge, Vol. 1 (1987)
3. O. Söderström, U. Dahlborg and W. Gudowski: J. Phys. F 15, L23 (1985)
4. K.E. Larsson, W.Gudowski and M.Dzugutov:"A Liquid Metal as an Extensive Test Case of Static and Dynamic Properties", Europhysics Liquid State Workshop on Static and Dynamic Properties of Liquids, June 27 - July 1, Dubrovnik, Yugoslavia (1988)
5. M. Flinn: Ph. D. dissertation, Catholic University of America (1969)
6. W.E Alley and B.J. Alder: Phys. Rev. A 27, 3174 (1983)

Hydrodynamic Mode Coupling in a Simple Liquid

*C. Morkel, C. Gronemeyer, and W. Gläser**

Physik-Department TU München, D-8046 Garching, Fed. Rep. of Germany
*Institut Laue-Langevin, F-38042 Grenoble, France

A refined analysis of high resolution incoherent neutron scattering data from liquid sodium at high temperatures is presented. The well-known Egelstaff formula $z(\omega) = \lim_{Q\to 0} \frac{\omega^2}{Q^2} S_s(Q,\omega)$ for the extrapolation of the frequency distribution $z(\omega)$ from small Q incoherent scattering data $S_s(Q,\omega)$ is extended properly to the hydrodynamic region ($Q \to 0$, $\omega \to 0$). The new analysis reveals a low frequency $\omega^{1/2}$ cusp in $z(\omega)$, indicating a $t^{-3/2}$ decay of the autocorrelation function $\langle \vec{v}(t)\cdot\vec{v}(0)\rangle$ at long times as predicted by mode coupling theory (MCT).

By a linear extrapolation of the experimental data to $Q = 0$, a direct determination of the diffusion constant D and the shear viscosity η from the neutron scattering data is possible. The results are discussed in terms of mode coupling theories for single particle behavior and give strong confirmation for mode coupling effects in the hydrodynamic region of simple high temperature liquids.

I. Introduction

Since the theoretical development of mode coupling concepts /1-6/ and corresponding molecular dynamics simulations /7-9/ there has been a lack of experimental work in this field. But in recent years some successful experimental efforts have been made, to describe liquid state dynamics in terms of mode coupling theories. Low Q incoherent neutron scattering, especially, has provoked interest, as mode coupling effects manifest themselves clearly in the hydrodynamic regime of single-particle motion /10-14/.

II. Mode Coupling in Single-Particle Motion

To study mode coupling effects in single particle dynamics the incoherent scattering law $S_s(Q,\omega)$ or the generalized diffusion coefficient $D(Q,\omega)$ is of central interest with $D(Q,\omega) = D'(Q,\omega) + i\, D''(Q,\omega)$, the generalized complex diffusivity:

$$S_s(Q,\omega) = \frac{1}{\pi} \frac{D''(Q,\omega)\, Q^2}{[\omega + D'(Q,\omega)Q^2]^2 + [D''(Q,\omega)Q^2]^2} . \qquad (1)$$

Several features of $S_s(Q,\omega)$ or $D(Q,\omega)$ are used to describe mode coupling effects. So the reduced peak height $\Sigma(Q) = \pi DQ^2\, S_s(Q,0)$ and the reduced half width $\gamma(Q) = \omega_{1/2}(Q)/DQ^2$ have been used /6,10-13/, where $[\pi DQ^2]^{-1}$ and DQ^2 are peak

Springer Proceedings in Physics, Vol. 40
Editors: M. Davidović and A.K. Soper

Static and Dynamic Properties of Liquids

height and half width of the corresponding Lorentzian scattering law, which is approached at $Q = 0$ by $D(Q,\omega) = iD$, where D is the diffusion constant. Both quantities $\Sigma(Q)$ and $\gamma(Q)$ are known to vary linearly at low Q in the mode coupling approximation /10-13/:

$$\Sigma(Q) = 1 + bQ + \ldots . \qquad b = (2D/\omega_0)^{1/2}$$
$$\gamma(Q) = 1 - aQ + \ldots . \qquad a = (D/2\omega_0)^{1/2} \tag{2}$$

with $\omega_0 = 72\ (\pi\ n\ m\ D/k_BT)^2(D + D_T)^3$ /4/, where n·m is the mass density and D_T is the kinematic viscosity of the liquid at temperature T.

$\Sigma(Q)$ and $\gamma(Q)$ for liquid sodium at high temperature ($T = 803$ K, $D = 2.2 \cdot 10^{-4}$ cm²/sec, $D_T = 2.75 \cdot 10^{-3}$ cm²/sec corresponding to $\eta = 2.28 \cdot 10^{-3}$ g/(cm·sec), $\omega_0 = 5.1 \cdot 10^{13}$ sec^{-1} /16/) are shown in Figs. 1 and 2, where the straight lines are the linear mode coupling approximations (2) with $a = 0.14$ Å and $b = 0.28$ Å /12/. A detailed description of the experiment is found elsewhere /13, 15/. At larger Q values ($Q \gtrsim 1$ Å^{-1}) a deviation from linear behavior occurs especially for $\Sigma(Q)$, which indicates higher-order terms in (2).

A further interesting quantity in mode coupling theory for single-particle motion is the imaginary part $D''(Q,\omega)$ of the generalized diffusivity $D(Q,\omega)$. The static absorptive part $D''(Q,0)$ can be understood as a Q-dependent diffusion coefficient. From (1) $D''(Q,0)$ is seen to be directly related to $S_s(Q,0)$ via

$$D''(Q,0) = \frac{1}{\pi\ Q^2 S_s(Q,0)} \quad , \tag{3}$$

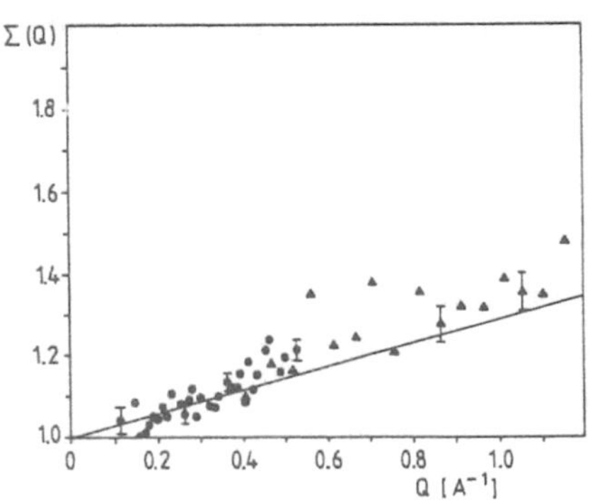

Fig.1 Reduced peak height vs Q (• IN5-, ▲ IN6 experiment /12,13/, straight line: linear MCT)

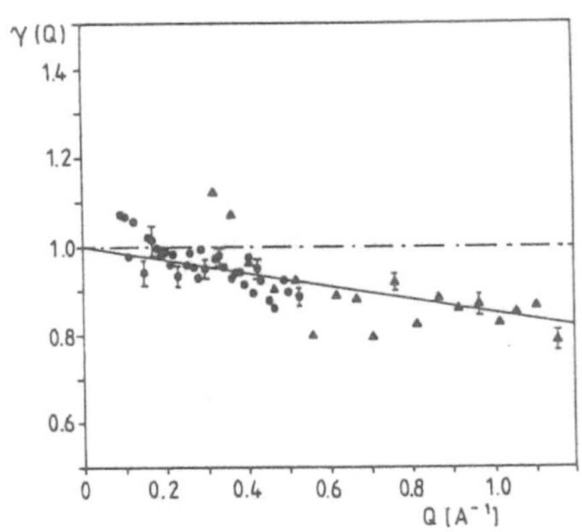

Fig.2 Reduced half width vs Q (symbols as Fig.1, dash-dotted line: Brownian limit)

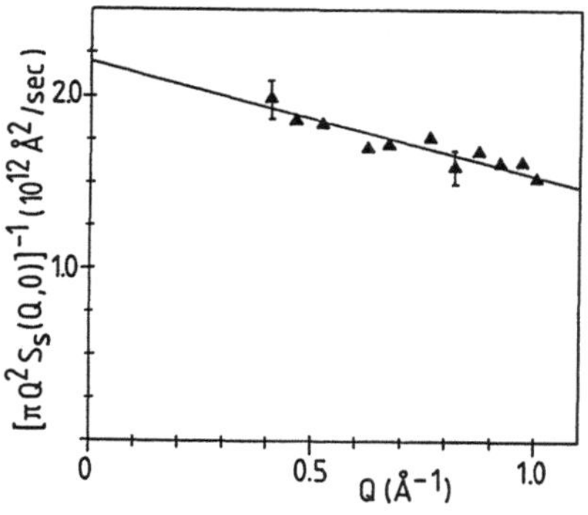

Fig.3
D"(Q,0) vs Q (▲ : experiment, straight line: MCT)

as $D'(Q,\omega = 0) = 0$. Thus the Q-dependent diffusion coefficient D"(Q,0) is an easily accessible quantity, which is plotted in Fig. 3.

In contrast to $\Sigma(Q)$ (Fig. 1), D"(Q,0) shows a very strict linear Q dependence throughout the whole interval up to $Q = 1.2$ Å^{-1}. This is in full agreement with mode coupling theory, which predicts /3/

$$D''(Q,0) = D(1 - bQ) \quad . \tag{4}$$

For liquid sodium at T = 800 K the non-linear term contributing to D"(Q,0) /3/ is nearly three orders of magnitude smaller than the linear one. Thus (4) can be viewed here in contrast to (2) as exact without higher-order corrections in the whole measured Q range.

The strict linearity in Fig. 3 allows the determination of D by simple linear extrapolation according to the Green-Kubo relation

$$D = \lim_{Q \to 0} D''(Q,0) \quad . \tag{5}$$

A similar linear extrapolation of $\Sigma(Q)$ to Q = 0 (Fig. 1) is possible only in a smaller interval $Q < 1$ Å^{-1}, because of the increasing importance of non-linear terms in (2) above this limit.

From a simple linear least squares fit, a diffusion constant $D = 2.17 \cdot 10^{-4}$ cm^2/sec is extrapolated in good agreement with the literature value /16/ ($D = 2.20 \cdot 10^{-4}$ cm^2/sec, T = 803 K). Moreover, from the slope $b = (2D/\omega_0)^{1/2}$ a value of $\omega_0 = 5.24 \cdot 10^{13}$ sec^{-1} is evaluated, corresponding to a shear viscosity $\eta = 2.33 \cdot 10^{-3}$ g/(cm sec). Thus two transport coefficients of the liquid are determined by a simple linear least squares fit in the hydrodynamic region of single particle motion.

III. The Frequency Distribution $z(\omega)$

Whereas the well-known Egelstaff formula for the evaluation of the frequency distribution /17/

$$z(\omega) = \lim_{Q \to 0} \frac{\omega^2}{Q^2} S_s(Q,\omega) \quad (6)$$

can be shown to be valid only for $x \gg 1$ with $x = \omega/DQ^2$, the hydrodynamic limit $\omega \to 0$, $Q \to 0$, $x \leq 1$ requires a low-frequency extension of (6), removing the singularity for $Q \to 0$.

Taking into account the low ω diffusion contribution to $z(\omega)$, an improved extrapolation formula can be derived, which is valid in the hydrodynamic limit as well /14/:

$$z(Q,\omega) = \left(\frac{\omega^2}{Q^2} + D^2Q^2\right) S_s(Q,\omega) - cQ \sqrt{\frac{1 + \sqrt{1 + x^2}}{1 + x^2}} \,,$$

$$z(\omega) = \lim_{Q \to 0} z(Q,\omega) \quad ; \quad c = \frac{2D}{\pi} \sqrt{\frac{D}{\omega_0}} \,, \quad (7)$$

where the term $D^2Q^2S_s(Q,\omega)$ has been derived by Carneiro in an empirical way earlier /18/. An evaluation according to (7) with D and ω_0 known from Fig. 3 is shown in Fig. 4, where the Q-dependent frequency distribution $z(Q,\omega)$, varying smoothly in the low Q interval, is extrapolated linearly to $z(\omega)$ at Q=0. The linear extrapolation is supported further by the agreement of $z(Q,\omega=0)$ (7) in the low Q limit with the exact linear relation for $\pi^{-1}D''(Q,0)$ (4), as $z(Q,0) = \pi^{-1} D''(Q,0)$.

In Fig. 5, $z(\omega)$ is plotted vs $\omega^{1/2}$, revealing clearly a low-frequency square root cusp in the frequency distribution of liquid sodium at high temperature, as mode coupling theory gives /1-4, 10-14/

$$z(\omega) = \frac{D}{\pi} \left(1 - \sqrt{\frac{\omega}{\omega_0}}\right) \,, \quad (\omega \to 0) \,. \quad (8)$$

From the linear fit (straight line in Fig. 5) the values $D = 2.18 \cdot 10^{-4}$ cm²/sec and $\omega_0 = 5.42 \cdot 10^{13}$ sec^{-1} are evaluated here.

As a direct consequence of the $t^{-3/2}$ long time tail in the velocity autocorrelation function $\Phi(t) = \langle \vec{v}_i(t) \cdot \vec{v}_i(0) \rangle$, the Fourier transform of which is $z(\omega)$, the square root behavior in Fig. 5 indicates the coupling of shear modes to the single particle diffusion mode in the liquid. These viscous transverse modes show a diffusive decay and lead to a $t^{-3/2}$ long time behavior of $\Phi(t)$ and a $\omega^{1/2}$ cusp in $z(\omega)$ at low ω.

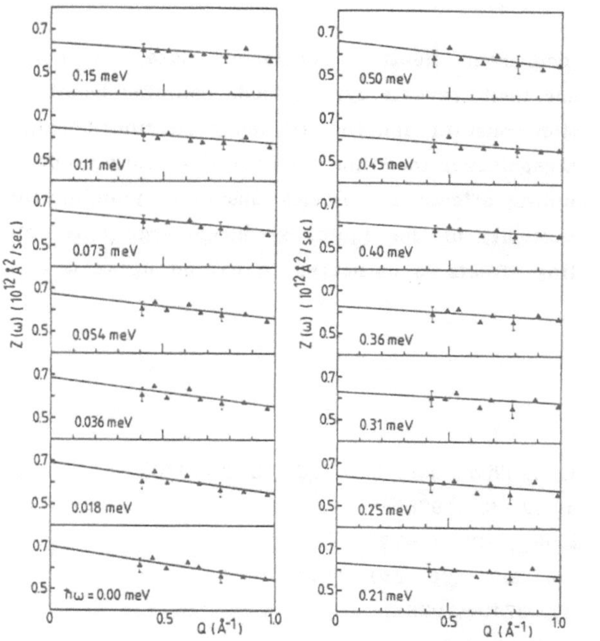

Fig.4 Linear extrapolation of z(Q,ω) to Q = 0 for a series of ħω values

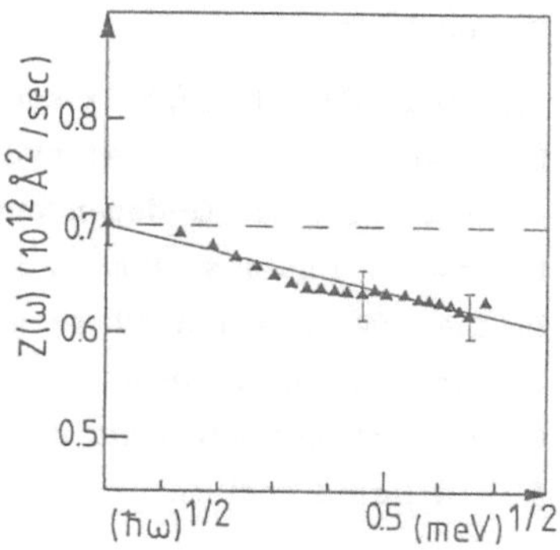

Fig.5
The frequency distribution z(ω) vs $\omega^{1/2}$ indicating the low-frequency cusp

IV. Comparison with Dense Hydrogen Gas

As mode coupling effects in single-particle dynamics have been observed in dense hydrogen gas too /10/, a comparison of liquid sodium (T = 800 K) with dense hydrogen gas (120 K, 730 bar) /10/ is interesting. Whereas the parameters a, b and ω_o (2) seem to indicate close similarity between both the systems, the parameters D and τ_E are crucial and quite different (τ_E: Enskog mean collision time).

These two parameters allow an estimation for the Q range of mode coupling effects. They are observed for $\tau_{obs} \geq 10\ \tau_E$, as molecular dynamics simulation work of Alder and Wainwright /7/ shows. With $\tau_{obs} \approx 1/DQ^2$, in the hydrodynamic region follows $1/DQ^2 \geq 10\ \tau_E$. Thus the Q range of mode coupling effects is restricted to

$$Q \leq Q^* = (10\ D\tau_E)^{-1/2}\ . \qquad (9)$$

For liquid sodium ($D = 2.2 \cdot 10^{-4}$ cm²/sec, $\tau_E = 0.043$ psec) $Q^* = 1.0$ Å^{-1}, whereas for dense H_2 gas /10/ ($D = 6.1$ cm²/sec, $\tau_E = 0.064$ psec) $Q^* = 0.50$ Å^{-1} is estimated. Thus the hydrodynamic mode coupling region for liquid sodium seems to be larger than for dense hydrogen gas and a hydrodynamic analysis, as shown in Figs. 1-4 using linear extrapolation, is restricted to smaller Q values in the case of dense hydrogen gas, because of its much larger diffusion constant.

V. Conclusions

An analysis of high resolution incoherent neutron scattering data, extending Egelstaff's formula to the hydrodynamic limit, reveals typical mode coupling behavior, especially for the Q-dependent diffusion constant and the frequency distribution z(ω). In case of liquid sodium, a linear extrapolation technique is shown to work properly in the low Q region where mode coupling effects arc clearly observed, yielding the coefficients for diffusion and shear viscosity of the liquid. By simple arguments, the Q range of hydrodynamic mode coupling effects is estimated to extend up to Q = 1 $Å^{-1}$ in liquid sodium.

References

1 Ernst, M., Hauge, E., Van Leeuwen, J.; Phys. Rev. Lett. 25, 1254 (1970)

2 Pomeau, Y., Résibois, P.; Phys. Rep. 19, 63 (1975)

3 de Schepper, I., Ernst, M.; Physica 98A, 189, (1979)

4 Bosse, J., Götze, W., Lücke, M.; Phys. Rev. A20, 1603 (1979)

5 Gaskell, T., Miller, S.; J. Phys. C 11, 3749 (1978)

6 Wahnström, G., Sjögren, L.; J. Phys. C 15, 401, (1982)

7 Alder, B.J., Wainwright, T.E.; Phys. Rev. A1, 18 (1970)

8 Alley, W.E., Alder, B.J.; Phys. Rev. A27, 3158 (1983)

9 Erpenbeck, J., Wood, W.; Phys. Rev. A32, 412 (1985)

10 Verkerk, P., Builtjes, J., de Schepper, I., Phys. Rev. A31, 1731 (1985)

11 Montfrooy, W., de Schepper, I., Bosse, J., Gläser, W., Morkel, Chr.; Phys. Rev. A33, 1405 (1986)

12 Morkel, Chr., Gronemeyer, Chr., Gläser, W., Bosse, J.; Phys. Rev. Lett. 58, 1873 (1987)

13 Gronemeyer, Chr.; Ph.D.Thesis, TU München (1986)

14 Morkel, Chr.; Gronemeyer, Chr.; Zeitschr. Phys. B - Cond. Matt. 72, 433 (1988)

15 Morkel, Chr., Gläser, W.; Phys. Rev. A33, 3386 (1986)

16 Faust, O.; "Sodium Engineering Handbook", Vol. I (Gordon and Breach, New York 1972)

17 Egelstaff, P.A.; "Introduction to the Liquid State" (Academic, London 1967)

18 Carneiro, K.; Phys. Rev. A14, 517 (1976)

Part II

Theory of Liquids

The Statistical Mechanics of the Ionic Equilibrium of Water. IV. The Force Law

S. Bratos, B. Guillot, and Y. Guissani

Laboratoire de Physique Théorique des Liquides*,
Université Pierre et Marie Curie, 4, Place Jussieu,
F-75252 Paris Cédex, France
*Equipe Associée au CNRS

A theory is presented to calculate the pH value of water. The calculations are executed by combining the molecular dynamics and test particle methods. The corresponding force law is studied in detail. Quantum mechanical theories and semi-empirical models of H-bonding are discussed and compared. A good agreement between theory and experiment is emphasized.

1. INTRODUCTION

An increasing interest has been manifested in the last decade in the computer simulation study of chemically reactive liquids. The motivation of these studies is that a majority of chemical reactions take place in the liquid phase. The liquid medium strongly influences the chemical behavior. Extended areas of chemistry will thus remain semi-empirical in character as long as statistical mechanics of chemically reactive liquids is insufficiently developed. The following classes of reactions have so far been investigated theoretically : conformational changes [1-3], ion pair formation [4], nucleophilic S_N2 substitution [5-7], aqueous Fe^{II} - Fe^{III} electron transfer [8] and the photolytic dissociation of I_2 [9]. This field was reviewed in Refs.[10,11].

The present authors recently published a computer simulation study of the ionic equilibrium of water [12-14]. The water molecules of pure water undergo a spontaneous fragmentation in two ions, H_3O^+ and HO^- ; the negative logarithm of the concentration of the H_3O^+ ions represents the pH value of water. Although this quantity is one of the most important quantities of chemistry, no attempt to determine it a priori was ever reported ; this is a consequence of the intrinsic difficulty of the problem. The theory of the present authors is a semi-classical theory in which translational and rotational degrees of freedom are classical, and vibrational degrees of freedom are quantum-mechanical. The equilibrium constant $K(p,T)$ governing the ionic equilibrium of water was calculated by combining the molecular dynamics and test particle methods. The theory predicted a pH value of a correct order of magnitude and showed that the polarization effects are surprisingly large. This shows the usefulness of statistical studies of elementary chemical processes in liquids : the crucial role of the polarization effects in the ionic equilibrium of water was never suspected !

The purpose of the present paper is to reexamine one of the central questions of this problem, the force law of the system. The presence of weak H bonds between water molecules, and of strong H-bonds around oxonium or hydroxyl ions makes this problem exceptionally difficult. The paper is organized as follows: Essential elements of the theory are briefly sketched in Section 2 which also describes its main results. The discussion of the force law governing the H-bonded systems is the subject of Section 3. Finally, the conclusions are collected in the last Section.

Springer Proceedings in Physics, Vol. 40
Editors: M. Davidović and A.K. Soper
Static and Dynamic Properties of Liquids

2. THEORY

2.1 Description of the model

The system submitted to investigation contains N water molecules, n oxonium ions and n hydroxyl ions. The following model is used in this study. (i) Molecular translations and rotations are governed by classical mechanics, and molecular vibrations obey quantum mechanics. (ii) Molecular rotators are all rigid. Their geometry is the mean geometry of these species in the liquid phase, and differs from that of the free species. (iii) The vibrational frequencies are mean frequencies in liquid water. Heavily afffected by hydrogen bonding, they differ noticeably from the corresponding gas phase frequencies. For justification of this model, see Ref.[12].

2.2 Principle of the calculation

The most direct way to study a chemical equilibrium is to realize a molecular dynamics, or Monte Carlo, simulation of a mixture of reacting molecules. Supposing that intermolecular potentials are known, the computer generates the equilibrium situation automatically. Unfortunately, this direct method is inapplicable in the present case. The fraction of dissociated water molecules is of the order of 10^{-9}. The simulation would thus require samples comprising 10^{10}-10^{11} molecules which is not envisageable.

The method employed in this calculation is thus indirect. It is based on the mass action law according to which the activities a_{H_2O}, $a_{H_3O^+}$ and a_{HO^-} of H_2O, H_3O^+ and HO^- are related to each other by the formula

$$\frac{a_{H_3O^+}\ a_{HO^-}}{a^2_{H_2O}} = K(p,T) \qquad (1)$$

where $K(p,T)$ represents the chemical equilibrium constant. Then, as the activity $a_{H_3O^+}$ is equal to the concentration n/N at high dilutions, the pH value can be calculated from Eqn(1) if $K(p,T)$ is known. From the other side, $K(p,T)$ can be expressed in terms of chemical potentials $\mu_i = \mu_i^{id} + \mu_i^{ex}$ of the reaction partners i, μ_i^{id} and μ_i^{ex} representing the ideal and excess part of μ_i, respectively. It is thus possible to write :

$$K(p,T) = K^{id}(p,T)\exp\left[-\beta(\mu^{ex}_{H_3O^+} + \mu^{ex}_{OH^-} - 2\mu^{ex}_{H_2O})\right] = K^{id}(p,T)\ K^{ex}(p,T) \qquad (2)$$

$$\beta\mu_i^{ex} = -\ln\langle\exp(-\beta U_i)\rangle \qquad (3)$$

In Eqs(2,3) $K^{id}(p,T)$ denotes the equilibrium constant of an ideal gas, the exponential factor $K^{ex}(p,T)$ describes the effect of intermolecular forces and $i = H_3O^+$, HO^-, H_2O. Eqn(3) is the well known WIDOM formula [15] in which U_i represents a many-body potential of interaction between the particle, the chemical potential of which is desired, and the remaining particles of the system. The average is over a canonical ensemble of the system in absence of the extra particle. Although still very difficult, this indirect method represents a practicable route to calculate the pH of water.

2.3 Practical realization of calculations

The first step of this calculation consists in determining $K^{id}(p,T)$, the dissociation equilibrium constant of an hypothetical system of water molecules in which molecular interactions are absent. This calculation is easy, but requires the knowledge of mean molecular geometries and of mean vibrational frequencies. These data are collected in Fig.1 and Table 1.

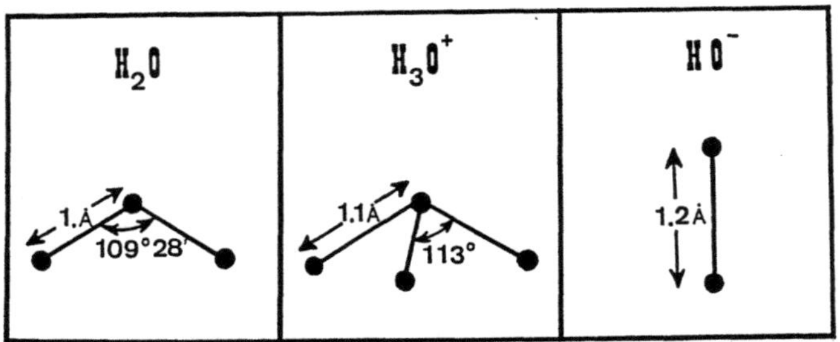

Fig.1 Model geometries of H_2O, H_3O^+ and HO^-

Table 1. Model frequencies of H_2O, H_3O^+ and HO^-

	$\nu_1(cm^{-1})$	$\nu_2(cm^{-1})$	$\nu_3(cm^{-1})$	$\nu_4(cm^{-1})$
H_2O	1650	3300	3500	
H_3O^+	2300	1200	2300	1750
HO^-	2300			

The next step concerns the study of the force law. The presence of weak H-bonds between water molecules, and of strong H-bonds around the ions makes this study particularly difficult. The potentials which are involved are the N-body potentials $U_{w/w}$, $U^{\pm}_{i/w}$ representing the interaction between the extra, or test, particle and a collection of N water molecules. The extra particle is either H_2O, H_3O^+ or HO^-. They were written as a sum of expressions describing the Lennard-Jones interactions, the coulombic interactions, the interactions due to the polarization of water molecules by the ions, and those due to the polarization of strong H-bonds by water molecules. Strong H-bonds in fact are highly polarizable objects ; their polarizability, called the ZUNDEL polarizability [16], exceeds that of water by one or two orders of magnitude. The resulting expressions are

$$U_{w/w} = \sum_i 4\epsilon \left[\left(\frac{\sigma}{R_{0i}}\right)^{12} - \left(\frac{\sigma}{R_{0i}}\right)^{6} \right] + \sum_{0'} \sum_{i} \frac{q_{0'} q_i}{R_{0'i}} \tag{4}$$

$$U^{\pm}_{i/w} = \sum_i 4\epsilon \left[\left(\frac{\sigma}{R_{0i}}\right)^{12} - \left(\frac{\sigma}{R_{0i}}\right)^{6} \right] + \sum_{0'} \sum_{i} \frac{q_{0'} q_i}{R_{0'i}} - \sum_i \frac{1}{2} \alpha \vec{F}_i^2 - \frac{S}{2} \vec{F}' \overleftrightarrow{\alpha}' \vec{F}' \tag{5}$$

$$S = \left[1 + e^{\gamma(u-u_0)} \right]^{-1} \tag{6}$$

The notation is as follows: The indices 0 and 0' refer to the extra particle and the index i to the water molecules ; ϵ and σ describe the Lennard-Jones interaction between the oxygen atom of the extra particle 0 and those of the remaining water molecules i ; $\vec{F}_i$ and $\vec{F}'$ are electric fields at the center of gravity of the i'th water molecule and of the ion, respectively ; and α is the mean polarizability of H_2O whereas $\overleftrightarrow{\alpha}'$ is the effective polarizability tensor of H_3O^+, or HO^-, in a perfectly aligned hydrate $H_3O^+(H_2O)_n$, or $HO^-(H_2O)_n$. Finally, S is the species specification operator for these hydrates ; it is equal to 1 for low energy configurations where they exist and to 0 for high energy configurations where they are absent. The quantity u entering into (6) comprises the Lennard-Jones, coulombic and water polarization energies of the ion and of its five nearest neighbours ; the quantity u_0 discriminates strongly polarizable H-bonds from the others. The parameters of $U_{w/w}$ were taken from BERENDSEN et al [17], whereas those associated with H_3O^+ and HO^- were fitted to reproduce experimental gas phase energies [18-20] and calculated quantum-mechanical geometries [21,22] of $H_3O^+(H_2O)_n$, $HO^-(H_2O)_n$, n = 1 to 4. For more details, see Ref (12). The potentials (4-6) have the important, and non-trivial, property of tri-coordination of the inner coordination shell.

The last step of calculation is the determination of the excess chemical potentials $\mu^{ex}_{H_2O}$, $\mu^{ex}_{H_3O^+}$, $\mu^{ex}_{HO^-}$. This problem was solved by combining the molecular dynamics and test particle methods [23]. The procedure consists in calculating the average in Eqn. (3) by repeatedly inserting the extra particle, H_3O^+, HO^- or H_2O, into the computer generated configurations of pure water. The details of this calculations are given elsewhere and are not reproduced here. It will only be mentioned that 216 water molecules were considered, and that a number of insertions as large as 10^{10} was required to perform the average in Eqn. (3). This calculation was thus close to the feasability limit with a Cray 1 computer.

2.4 Results and discussion

The present theory led to the following main results. (i) The calculated pH value of water at T = 293 K and $\rho = 0.999$ g/cm^3 is equal to 6.5. A closer agreement with experiment could have been obtained by choosing the vibrational frequencies in an ad hoc way ; however, such an agreement would be entirely artificial. The magic number pH = 7 is thus only approached, but not exactly reproduced. The temperature and the density dependence of the pH are also correctly described (Table 2). (ii) The isotope effect on the pH, consecutive to the H → D, or H → T, substitution, was calculated by replacing m_H by m_D, or m_T, and by reducing all vibrational frequencies of Table 1 by the factor $(m_D/m_H)^{1/2}$, or $(m_T/m_H)^{1/2}$. Theory and experiment agree fairly well. (iii) The polarization contribution to the pH is unexpectedly large (Table 3). If molecular interactions were absent, the pH of water would be 82.4 instead of 6.5 . Considering Lennard-Jones and coulombic forces reduces the pH from 82.4 to 56.7, a value still far from 6.5 . The

Table 2. Calculated and experimental pH values of liquid H_2O

T(K)	ρ(g/cm^3)	pH (th)	pH (exp)
293	0.998	6.5	6.97
473	0.865	3.6	5.45
593	0.667	3.8	5.87

Table 3. Contribution of different types of intermolecular forces to the pH value of water in room conditions

Intermolecular forces considered	pH
None	82.4
LJ, Coulomb	56.7
LJ, Coulomb, water polarization	22.1
LJ, Coulomb, water and Zundel polarizations	6.5

dominating effect is the polarization of water molecules by the ions ; taking it into account reduces the pH from 56.7 to 22.1 ! Finally, the ZUNDEL polarization of strong H-bonds around the ions brings the pH down from 22.1 to 6.5 ! These results emphasize, as do many other studies, the prominent role of modern methods of statistical mechanics in studying elementary chemical processes in liquids : the effect of polarization on the pH value of water was never suspected. Statistical mechanics of chemically reactive liquids certainly merits attention of the liquid state community !

3. THE FORCE LAW

3.1 Nature of hydrogen bonding

Once the basic elements of the theory have been sketched, the characteristics of the force law for liquid water will be discussed in some detail. Molecular dynamics is governed in this system by hydrogen bonding, a special type of molecular interaction between molecules containing a polar group like OH, NH or FH, and molecules including an electronegative atom like O, N or F. One generally distinguishes between weak, moderately strong and strong H-bonds. The energy, physico-chemical characteristics and spectral signatures in IR are very different for these three classes of H-bonds. The difficulty is that two of them are simultaneously present in the liquid water : weak H-bonds link two water molecules, whereas strong H-bonds act between H_3O^+ and H_2O, or HO^- and H_2O. This is why the force law is particularly complex in the present case.

It is generally considered that H-bonding represents a complicated superposition of the following five "elementary" interactions : Coulomb, exchange, polarization, charge transfer and dispersion interactions. This partitioning was first suggested by the early theoretical work in this field [24]. It was also confirmed by the recent investigations ; two methods were used for that purpose. (i) The perturbation theory was applied to molecular interactions. Starting from approximate solutions of the Schrödinger equation of the isolated subsystems, individual contributions to the energy of interaction were calculated by different modifications of the Rayleigh-Schrödinger perturbation theory [25].(ii) The energy of interaction, calculated by a full SCF calculation, was split into various contributions by different model assumptions [26,27]. Within the Hartree-Fock approximation,no dispersion energy can be obtained.

It results from these calculations that, in weak and moderately strong H-bonds, the first order coulombic and exchange repulsion contributions to the interaction

Table 4. Contribution of the polarization energy ΔE_{POL} to the total energy of molecular interaction ΔE.

System	ΔE(kcal/mol)	$\Delta E_{POL}/\Delta E$
$Li^+ - H_2O$	- 33.7	0.38
$Na^+ - H_2O$	- 27.4	0.19

SCF theories [29]

System	ΔE(kcal/mol)	$\Delta E_{POL}/\Delta E$
$H_3O^+ - H_2O$	- 34.7	0.47

Semi-empirical models [12]

energy dominate over those of the three remaining second-order terms. However, as they are of opposite sign and they cancel each other to a considerable extent, the polarization, charge transfer and dispersion forces still play a non-negligible role. In particular, they are responsible for the fact that the equilibrium OH...O distances in a H-bonded complex are smaller than the sum of the corresponding Van der Waals radii. To our knowledge, no study of the partitioning of the energy is available for the strong H-bonds in $H_3O^+ - H_2O$ and $HO^- - H_2O$. Nevertheless, a similar study for the systems like $Na^+ - H_2O$, $Li^+ - H_2O$, $Mg^{2+} - H_2O$ etc brought out a spectacular increase of the polarization energy (Table 4). Something similar is happening in hydrated oxonium and hydroxyl ions. It was also found that these hydrated species have a highly flexible structure, a circumstance influencing the outcome of the present calculation. For reviews covering this field, see e.g. Refs [28,29].

3.2 Water-water interactions

A large number of empirical expressions have been employed to describe weak H-bonding between water molecules. The most successful potentials are the potentials proposed by STILLINGER and RAHMAN [30] (potential ST2), MATSUOKA, CLEMENTI and YOSHIMINE [31] (potential MCY), BERENDSEN, POSTMA et al. [17] (potential SPC), JORGENSEN [32] (potentials TIPS), LEMBERG and STILLINGER [33] (central force potential CF) and the potential due to BARNEY, FINNEY and NICHOLAS [34] (electropole polarizable potential, EP). Some of them are pair additive [17,30-32], and the others are not [33,34]. They differ in the number of interaction sites per molecule : one for the EP, three for the SPC, four for the MCY or the TIPS and, finally, five for the ST2 potential. The parameters which enter into their definition were determined either semi-empirically, by comparing the computer simulated and experimental thermodynamic quantities, or a priori, by calculating the potential energy of the water dimer in various geometrical configurations by the help of the configuration interaction method. All these models, except the CF model, are mainly adapted to the study of rigid water molecules.

The potential employed in this work is the BERENDSEN SPC potential. This particular choice is based on two arguments. The first is that this potential is particularly economical, which is of uttermost importance in a calculation which is close to the feasability limit of the computer. Its second advantage is that it allows a simple generalisation to the case of strong H-bonds. As easily seen in Eqn. (4), the three terms of the BERENDSEN potential describe the exchange, dispersion and Coulomb interactions, respectively. This conforms to what is known theoretically about the partitioning of the energy of weak H-bonds. The polarization and charge-transfer interactions are not considered directly ; nevertheless, the polarization effects are included at least indirectly. In fact, the point charges q_O, q_H correspond to an effective dipole moment of the water molecule equal to 2.27 D, rather than to its gas phase dipole moment equal to 1.85 D. Then, recalling that the use of effective dipole moment permits a rough simulation of the polarizability effects justifies the above statement.

3.3 Ion-water interactions

Only a limited number of semi-empirical potentials have been proposed to describe the $H_3O^+(H_2O)_n$ and $HO^-(H_2O)_n$ interactions. They are due to STILLINGER and DAVID [35], BUFFEY and BYERS-BROWN [36], KOCHANSKI [37] and to FORNILI, MIGLIORE et al. [38,39]. The first of them is just the CF potential cited earlier in Section 3.2. The others are either TIPS type potentials comprising the Coulomb, exchange and dispersion energies [37], or are MCY-like including a term to express the polarization of water molecules by the ions [36], or else they represent complicated combinations of inverse powers of intermolecular distance [38,39]. They refer either to $H_3O^+(H_2O)$ [36-38], or to $HO^-(H_2O)$ interactions [39], or to both of them [35]. The quality of these potentials is inferior to those describing the interactions between water molecules. In particular, the non-additivity of successive hydration energies was not explored carefully enough.

The potentials employed in this work are similar to the BERENDSEN potential, but contain two new terms. (i) The first is $1/2 \sum_i \alpha \vec{F}_i^2$ and describes the polarization of water molecules by ions. It is present in the BUFFEY and BYERS-BROWN expression for the oxonium-water interactions. (ii) The second term is $1/2\ S\vec{F}'\vec{\alpha}'\vec{F}'$ and describes the polarization of strong H-bonds around the ions by water molecules. Its counterpart in the SCF theory is the quantity $\Delta E_{REL} = E_I - E_{II}$, where E_I is the SCF energy of the complex in a fully optimized geometry, and E_{II} that of a geometry optimized under the constraint that the partners conserve their free molecule configurations. It should also be noted, from a technical point of view, that introducing into the Hamiltonian an extra polarizability represents an approximate technique for studying the non-rigidity effects in the rigid molecule computer simulations. The potentials (5,6) are not pair additive.

4. CONCLUSIONS

It results from discussions of this paper that the present semi-classical model is the simplest model envisageable to study the pH of water. To go further, a fully quantum mechanical theory would be required, particularly if a complete description of the effects of molecular non-rigidity is desired. Still, it can safely be stated that modern methods of statistical mechanics are now powerful enough to permit the calculation of the majority of basic quantities of the liquid state chemistry.

5. REFERENCES

1. J.P. Ryckaerts, A. Bellemans: Chem. Phys. Lett. 30 (1975), 123
2. D. Chandler: J. Chem. Phys. 68 (1978), 2959
3. B. Bigot, W.L. Jorgensen: J. Chem. Phys. 75 (1981), 1944
4. A. Karim, J.A. Mc Cammon: Chem. Phys. Lett. 132 (1986), 219
5. J. Chandrasekhar, S.F. Smith, W.L. Jorgensen: JACS 106 (1984), 3049
6. J. Chandrasekhar, S.F. Smith, W.L. Jorgensen: JACS 107 (1985), 154
7. J.P. Bergsma, B.J. Bradley, K.R. Wilson, J.T. Hynes: J. Chem. Phys. 86 (1987), 1356
8. D. Chandler, R.A. Kuharski: to appear in Faraday Discuss. Chem. Soc 85 (1988)
9. Ph. Bado, P.H. Berens, K.R. Wilson: Proc. Soc. Photo Opt. Instrum. Eng 322 (1982), 230
10. R. Kapral: Adv. Chem. Phys. 50 (1981), 71
11. Réactivité Chimique en Phase Liquide: Rapport de Prospective, S. Bratos Editor, CNRS, Paris 1986
12. Y. Guissani, B. Guillot, S. Bratos: J. Chem. Phys. 88 (1988), 5850
13. S. Bratos, Y. Guissani, B. Guillot: in Chemical Reactivity of Liquids Fundamental Aspects, M. Moreau and P. Turq Editors, Plenum, New York 1988, p.241
14. B. Guillot, Y. Guissani, S. Bratos: to appear in Synergetics, Order and Chaos, M. G. Velarde Editor, World Scientific Press, Singapore 1988
15. B. Widom: J. Chem. Phys. 39 (1963), 2808
16. G. Zundel: in The Hydrogen Bond, P. Schuster, G. Zundel and G. Sandorfy Editors, North Holland, Amsterdam 1976, Vol. 2, p. 683
17. H.J.C. Berendsen, J.P.M. Postma, W.F. Van Gunsteren, J. Hermans: in Intermolecular Forces, B. Pullman Editor, Reidel, Dordrecht 1981; p. 331
18. P. Kebarle, S.K. Searles, A. Zolla, J. Scarborough, M. Arshadi: JACS 89 (1967), 6393
19. M. De Paz, J.J. Leventhal, L. Friedman: J. Chem. Phys. 51 (1969), 3748
20. M. De Paz, A.G. Guidoni, L. Friedman: J. Chem. Phys. 52 (1970), 687
21. M.D. Newton, S. Ehrenson: JACS 93 (1971), 4971
22. M.D. Newton, J. Chem. Phys. 67 (1977), 5535
23. J.G. Powles: Mol. Phys. 41 (1980), 715
24. S. Bratos: in Advances in Quantum Chemistry, P.O. Löwdin Editor, Academic Press, New York 1967, Vol.3, p.209

25. J.N. Murrell, G. Shaw: J. Chem. Phys. 46 (1967), 1768
26. M. Dreyfus, A. Pullman: Theoretica Chimica Acta 19 (1970a), 20
27. K. Morokuma: J. Chem. Phys. 55 (1971), 1236
28. P. Schuster: in The Hydrogen Bond, P. Schuster, G. Zundel and C. Sandorfy Editors, North Holland Amsterdam 1976, Vol. 1, p.25
29. A. Karpfen, P. Schuster: in The Chemical Physics of Solvation, R.R. Dogonadze, E. Kalman, A.A. Kornyshev and Ulstrup Editors, Elsevier, Amsterdam 1985, Vol.A, p.265
30. F.H. Stillinger, A. Rahman: J. Chem. Phys. 60 (1974), 1545
31. O. Matsuoka, E. Clementi, M. Yoshimine: J. Chem. Phys. 64 (1976), 1351
32. W.L. Jorgensen: J. Chem. Phys. 77 (1982), 4156
33. H.L. Lemberg, F.H. Stillinger: J. Chem. Phys. 62 (1975), 1677
34. P. Barnes, J.L. Finney, J.D. Nicholas, J.E. Quinn: Nature 282 (1978), 452
35. F.H. Stillinger, O.W. David: J. Chem. Phys. 69 (1978), 1473
36. I.P. Buffey, W. Byers-Brown: Chem. Phys. Lett. 109 (1984), 59
37. E. Kochanski: Nouv. J. Chim. 8 (1984), 605
38. S. Fornili, M. Migliore, M.A. Palazzo: Chem. Phys. Lett. 125 (1986), 419
39. G. Andaloro, M.A. Palazzo, M. Migliore, S. Fornili: Chem. Phys. Lett 149 (1988), 201

Part III

Liquids at Surfaces

Liquid Surfaces and the Effect of Premelting

P. von Blanckenhagen and W. Schommers

Kernforschungszentrum Karlsruhe, Institut für Nukleare Festkörperphysik, P.O. Box 3640, D-7500 Karlsruhe, Fed. Rep. of Germany

Due to the reduced strength of the interatomic interaction, the reduced symmetry and the increased anharmonicity, the structure and dynamics of surfaces are quite different from the corresponding bulk properties. Distribution and correlation functions, frequency spectra and mean-square displacements are used for the structural and dynamical description of liquid surfaces. For example, single-particle distribution functions (density profiles) for the characterization of liquid-vapor interfaces are discussed in connection with experimental results.

The effect of premelting is briefly reviewed for recent results obtained by molecular-dynamics, model calculations, and experiments. More detailed results are given for the Kr(100) and the Pb(110) surface.

I. Introduction

During the last two decades increasing attention has been paid to surface phenomena. Enormous progress was made in the microscopic analysis of solid surfaces. In contrast to this the study of liquid surfaces and of the solid-liquid phase transition at surfaces is still at the beginning. However, this rapidly growing field is now benefitting from the development of new experimental and theoretical methods in surface science; for example, computer simulations [Monte-Carlo (MC) and molecular-dynamic (MD) calculations] and the application of the synchrotron radiation in grazing incidence X-ray experiments.

At liquid surfaces the density distribution across the liquid-vapor interface and the surface roughness have, for example, already been studied for several liquids. Increasing interest is also devoted to liquid crystal surfaces [1], to liquid-solid interfaces in connection with epitaxial crystal growth [2] and to wetting phenomena [3]. The latter three subjects will not be considered here. In the development of the theory of melting, the following question arose: Does there exist a liquid nucleation centre at the surface of solids below the melting temperature? Such a *premelting* effect could explain the absence of superheating under normal conditions 4 . Indeed premelting effects were observed by MD calculations for noble gases in the case of four layers on a rigid lattice [5] and also within full MD calculations [6,7]. Only recently, a liquid-like structure *and* diffusion were observed experimentally at the Pb(110) surface.

This article is organized as follows: In the next section correlation and distribution functions are briefly introduced followed by a discussion of typical results for liquid surfaces. Then, essential results on premelting are given, and the surface properties related to the premelting effect are discussed for the Kr(100)- and the Pb(110) surface. Finally an outlook on future developments is given.

II. Distribution and Correlation Functions

The structure of liquid surfaces is described in terms of *distribution* and *correlation* functions [8]. The n-particle distribution function $\rho^{(n)}(\vec{r}_1,\vec{r}_2,..,\vec{r}_n)$ has the following meaning:

$$\rho^{(n)}(\vec{r}_1,\vec{r}_2,...,\vec{r}_n)d\vec{r}_1 d\vec{r}_2 \; ... \; d\vec{r}_n$$

Springer Proceedings in Physics, Vol. 40
Editors: M. Davidović and A.K. Soper

Static and Dynamic Properties of Liquids

is the probability that one atom of the system will be in $d\vec{r}_1$ at $\vec{r}_1$, another in $d\vec{r}_2$ at $\vec{r}_2$, ... The n-particle correlation function $g_n(\vec{r}_1,\vec{r}_2,\ldots,\vec{r}_n)$ is connected to $\rho^{(n)}$ by [8]

$$\rho^{(n)}(\vec{r}_1,\vec{r}_2,\ldots,\vec{r}_n)=[\rho^{(1)}(\vec{r}_1)\rho^{(1)}(\vec{r}_2)\ldots\rho^{(1)}(\vec{r}_n)]g_n(\vec{r}_1,\vec{r}_2,\ldots,\vec{r}_n), \qquad (1)$$

where $\rho^{(1)}$ is the single-particle distribution function.

Let us consider a semi-infinite system with a surface, which separates the liquid from the vapor phase, defining the x,y-plane. In the surface region the liquid is no longer isotropic and homogeneous as in the bulk liquid and, therefore, distribution functions and integral relations (see, for example, Ref. 8), cannot be described by scalar functions of the interatomic separation r, but these quantities and relations depend upon the vector field $\vec{r} = \vec{r}_{12} = \vec{r}_1-\vec{r}_2$ and on the distance z as measured along the axis perpendicular to the surface.

In the transition zone between the bulk liquid and the bulk vapor the single-particle distribution function $\rho^{(1)}(\vec{r})$ is no longer a simple constant, given by the $N/V = \rho_L$, but is dependent on the coordinate z (not on x and y):

$$\rho^{(1)}(\vec{r}) = \rho^{(1)}(z) \, . \qquad (2)$$

Furthermore, the two-particle distribution function must be expressed as

$$\rho^{(2)}(\vec{r}_1,\vec{r}_2) = \rho^{(2)}(z_1,z_2, |\vec{r}_1-\vec{r}_2|) \qquad (3)$$

and is not only simply dependent on $r = |\vec{r}_1-\vec{r}_2|$ as in the case of the bulk liquid.

The simplest non-trivial equation relating distribution functions and the pair potential is given by [8]

$$k_BT \frac{\partial}{\partial\vec{r}_1} \rho^{(1)}(z_1) + \int \frac{\partial}{\partial\vec{r}_1} \nu(r_{12})\rho^{(2)}(z_1,z_2,r_{12})d\vec{r}_2 = 0 \, . \qquad (4)$$

In the derivation of (4) it was assumed that the pair potential $\nu(r)$ remains unchanged as we pass through the surface zone between the liquid and vapor phases. This is certainly a good approximation for noble gases, but probably not an appropriate procedure in a liquid metal where the pair potential is dependent on the density.

Equation (4) yields a starting point for the theoretical determination of $\rho^{(1)}(z)$ from a given pair potential $\nu(r)$ provided that $\rho^{(2)}(z_1,z_2,r_{12})$ can be treated reliably. As examples, in Ref. [8] statistical mechanical models are discussed.

III. Results for Liquid Surfaces

The density distribution normal to the surface of simple dielectric and metallic liquids has been studied by *model calculations* and *computer simulations* (MC and MD) [8-15]. In Lennard-Jones liquids the density distribution function (Fig. 1a) falls monotonically from the value characteristic for the bulk liquid to that of the vapor. The width of the distribution increases with temperature and is of the order of two molecular diameters near the triple point [10]. The previously observed density oscillations for these systems are possibly an artefact of the simulation procedure [9-12,14]. Experimental information for noble gases would be desirable, since there are indications that the Lennard-Jones potential is not adequate for the description of noble gases (see below).

In the case of simple metals, stable density oscillations (see Fig. 1b) extending several atomic layers into the bulk liquid have been observed by MC simulations on the basis of electron-ion pseudopotentials and considering a metal-nonmetal transition in the transversal of the liquid-vapor interface [12]. The spacing of this structure parallel to the surface is about one atomic diameter. The layering of the liquid metal surface may be a relatively general phenomenon.

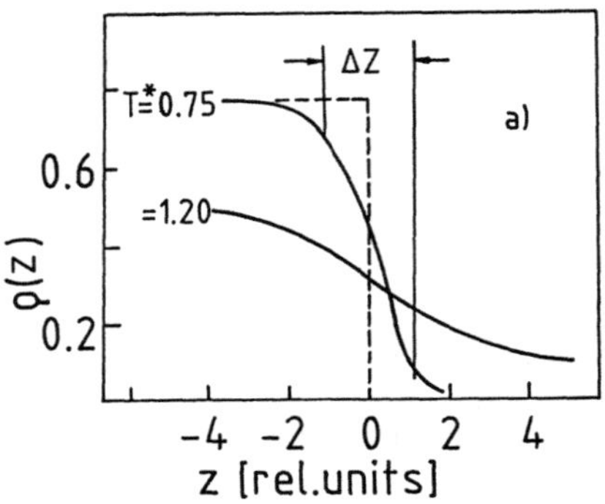

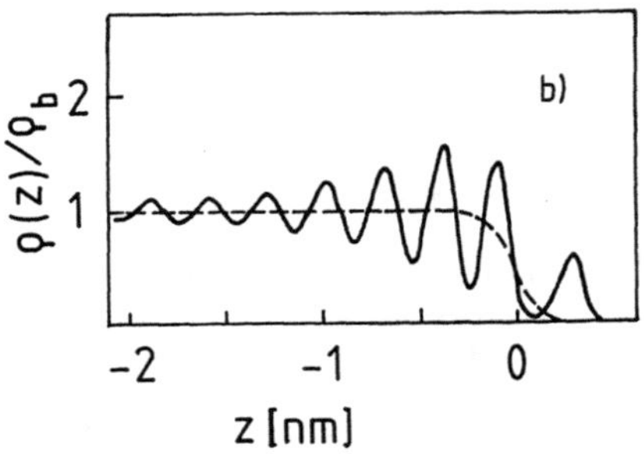

Fig. 1 Density distribution functions at the liquid-vapor interface: a) For a Lennard-Jones system at two different reduced temperatures T, Δz indicates the 10-90% width of the distribution and the dashed line shows a steplike interface. The monotonic density distribution has found to be a function with a tanh shape [15]. b) For a metal obtained by Monte-Carlo simulations and application of the pseudoatom theory for Hg (solid line) [12a]. Dashed line: average reference density profile. The outermost peak contains atoms and ions and is ascribed to a metal non-metal transition.*

It develops as a consequence of the confining effects of the conduction electrons [12,13]. Model calculations have shown that density oscillations perpendicular to the surface influence the transverse structure factor $S_{||}(Q)$ (Fourier transform of the pair correlation function parallel to the surface) [15]. In connection with $S_{||}(Q)$ it should be emphasized: Theories for the long range transversal correlations at a liquid surface (including the capillary-wave theory) reveal a characteristic Q^{-2} divergence for $S_{||}(Q)$ as $Q \rightarrow 0$ [16] (for other aspects of the transverse correlations and the capillary wave theory see [10,17,23]).

Experimental results for the properties of liquid surfaces discussed above are obtained by ellipsometry and X-ray reflectivity measurements. However, under special circumstances also neutrons, electrons and He-atoms could be used as probes for the structure and dynamics at the surface of liquids [18]. The width of the density distribution of the liquid vapor interface of liquid Ar has been measured by ellipsometry to be 7.9 Å at 90 K and 15.2 Å at 120 K, and this is in accordance with theoretical predictions; the discrepancies were less than 30% [19]. The angular dependence of the X-ray reflectivity near the critical angle for total reflection has been measured at the surface of different liquids (e.g. Na [20], Cs [21], Hg [22-24], Ga [23] and H_2O [25]). From these data, information concerning the density distribution functions could be deduced. All results for liquid metals confirm the oscillatory behavior of the density distribution as predicted by the MC simulations (e.g. see Fig. 1b). Steplike and monotonous distribution functions could be excluded although the conclusion is rather indirect (see [22] for the case of Hg). More direct results may be obtained by the X-ray fluorescence method [26,27] and probably also by diffraction experiments. The transversal structure factor $S_{||}(Q)$ for the surface of liquid Hg has recently been measured by grazing incidence X-ray diffraction and compared with results for $S_{||}(Q)$ from model calculations using different density distributions and with the structure factor for the bulk liquid [16]. No conclusion could be drawn about the longitudinal density distribution, however, the $Q \rightarrow 0$ divergence of $S_{||}(Q)$ is in agreement with theoretical predictions. From the X-ray reflectivity of the pure water surface, the root-mean-square roughness of the water surface was obtained to be 3.2 Å in good agreement with the theory for thermally activated capillary waves. The electron cloud of the water molecules would cause a root-mean-square roughness of about 1.9 Å only [25].

In conclusion, we note that the results of model calculations and simulations for the structure of the surface of liquids are by no means conclusively proven by experiments. Furthermore, the dynamics at the surface as discussed in [8] has not yet been proven experimentally.

IV. Premelting

There are many experimental indications and models for the effect of premelting [4,28,29] but conclusive microscopic data became available only recently. On the basis of these experimental and theoretical data, premelting is obviously a real effect and elucidates the melting process. For the present, let us consider any local reduction of the bulk melting temperature T_m as premelting, regardless of the reason for the effect. The main characteristics of premelting are a reduction of the atomic order (breaking of bonds) and an increase in the mobility of the atoms. The disorder may cause, for example, an anomalous increase of electrical resistivity (Fig. 2) [30] and of heat capacity [31,32] by approaching T_m. The increase in the mobility is, for example, manifested in the strongly enhanced diffusion coefficient at grain boundaries and at free surfaces near T_m [33]. The bulk melting temperature may be locally reduced in the vicinity of impurities [34], at grain boundaries (grain boundary melting) [35-37], in small clusters [38], in thin films [39,40] and at free surfaces (surface premelting [4,8,28,29]). By any type of premelting, superheating may be suppressed. Recently it was demonstrated that Ag single crystals (T_m = 1234 K) without free surfaces (∿0.15 mm diameter spheres, coated with Au (T_m = 1336 K)) could be superheated for periods of the order of 1 min by up to 24 K above the melting point [41]. Such superheating experiments and the observed decrease of the melting temperature for small clusters with decreasing diameter (e.g., about 30% for Pb particles of 20-30 Å diameter) demonstrate unambiguously the rule of surfaces in the melting process [42]. As a consequence of the empirical Lindemann criterion [4,43], according to which melting is a vibrational instability released when the root-mean-square amplitude of the lattice vibrations reaches a critical fraction of the interatomic distance, crystal surfaces should generally melt prior to the bulk because the vibrational amplitudes at surfaces are in general larger than in the bulk (see, e.g., Fig. 4a).

In the following, let us concentrate our discussion on surface premelting. Referring to the characterization of premelting given above, the transverse structure factor $S_{\|}(Q)$ (or the pair correlation function) and the surface diffusion coefficient D_S (or the mean square displacement $\langle r^2(t)\rangle$ of the atoms) would be the most important quantities for the description of surface premelting: they can be measured with different probes by scattering experiments. $S_{\|}(Q)$ for a few liquid

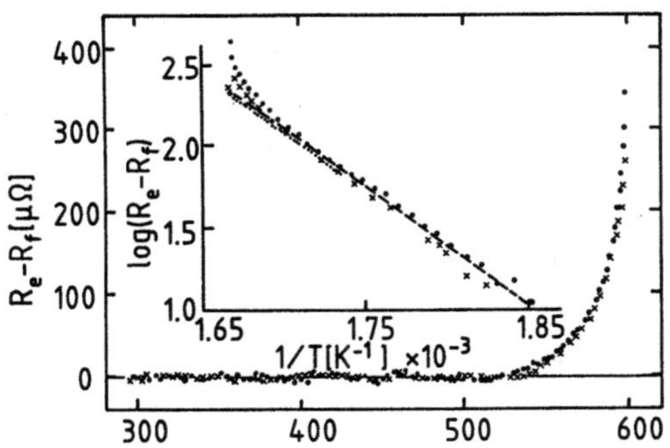

Fig. 2 Indications for premelting effects in polycrystalline lead (T_m = 600.7 K) from electrical resistance measurements [30]:
The difference between the measured resistance $R_e(T)$ and the fitted theoretical values $R_f(T)$ is plotted as a function of temperature (crosses: increasing T, full circles: decreasing T). The inset shows log(R_e-R_f) as a function of 1/T, the deviation from the linear law appearing at high temperatures are presumably due to premelting effects.

Table 1

Experimental results for order-disorder transitions and premelting.
D_S: surface diffusion coefficient; D_B: diffusion coefficient of the bulk liquid;
T_o: temperature where the order-disorder transition starts;
od: order-disorder transition; pm: premelting; ro: roughening;
sro: short range order; sd: surface diffusion

Sample	Method	Phenomenon	T_o/T_m	Reference
Pb polycr.	$\rho(T)$	pm	0.9	30
Pb films	X-ray diffr.	od	0.9	40
Pb films	LEED	od	0.83	44
Ge/Pb/Ge films	TEM	od	0.93	56
Pb(Hg)	EXAFS	od	0.7	34
Pb(110)	RHEED	od	0.75	45
Pb(110)	RBS	od	0.75	45-48
Pb(110)	X-ray diffr.	sro,od	0.5	49
Pb(110)	LEED	od	0.9	50
Pb(110)	He-sc.	sd($D_S \gtrsim D_B$)	0.75	51
Pb(110)	PES	od	0.75	52a
Al(110)	LEED	od, sro	0.73	54
Al(100)	PES	od	0.73	52b
Cu single cryst.	light em.	pm	0.998	55
Cu(110)	X-ray	ro	0.74	58
Cu(100)	He-diffr.	ro	0.56	59
Ge(111)	LEED	od	0.88	53
Ar(111)/Graphite	cp	pm, ro	0.96	32
CH_4(100)/MgO	n-sc.	sd($D_S \sim D_B$)	0.67	39
Ice I	κ,α	pm	-	57

Table 2

Theoretical results (MD, MMC) for order-disorder transitions and premelting.
LJ: Lennard-Jones system; GT: general theories; pm: premelting; ro: roughening;
od: order-disorder transition; gm: grain boundary melting;
lr: lattice relaxation; pt: phase transitions

Sample	Method	Result	Reference
LJ(100),(110),(111)	MD	pm	5,7,36,69,75,77
LJ	MD	gm	35-37
LJ(110)	MD	ro	60
LJ, metals, GT	MD, MMC	pm	66
Kr(100),(111)	MD	pm	6,8,61-64
Cu(100),(110),(111)	MMC	pm, lr	67
Au(111)	MD	pm, lr	68
Al(110)	MD	pm	73
GT	MMC	pt	65,71
GT	MMC,MD	pm	74,78,79

layers on top of the ordered crystal will be quite different from the structure factor of a bulk liquid because the corrugation of the surface of the parent substrate crystal is impressed on the liquid structure, leading to a much more structurized and anisotropic $S_{\parallel}(Q)$. As a precondition for understanding the appearance of the surface premelting effect, the lattice relaxation and its temperature dependence should be known (see below). Also the local fluctuation of the density distribution perpendicular to the surface (roughening) could influence the order-disorder transition. Whereas order-disorder transitions have been detected at few surfaces (Table 1), only for Pb(110) and CH_4(100) has the diffusion coefficient been measured as yet, yielding evidence for a liquid-like diffusion at the disordered surface. MMC- and MD-studies and microscopic model calculations (MMC) revealed surface premelting at Lennard-Jones (LJ)-, at realistic noble gas - as well as at metal-crystal surfaces (see Table 2). The temperature range of premelting and the role of roughening for the effect of premelting is still somewhat controversal in those experimental and theoretical studies [60,76]. In the following two sections we present two sets of results in more detail in order to visualize the surface premelting effect on a microscopic scale.

The Kr(100) surface

The MD results [6,8,61-64] shown in Fig. 3 and 4 were obtained with a realistic pair potential for the Kr atoms instead of the commonly used LJ-potential. The importance of using a realistic potential in such studies follows from the fact that surface properties are very sensitive to small variations in the potential [62]. For example, the LJ-potential yields an unrealistic lattice expansion at low temperatures [61]. This experience is also important for premelting studies, since the lattice relaxation has a strong influence on the premelting effect (see below). The pair correlation function for the outermost layer (Fig. 3a) indicates a disordered structure, but not disordered to such an extent as in bulk liquids. The disordering is accompanied by large lattice relaxations (Fig. 3b) and mean square amplitudes (Fig. 4a). The lattice relaxation is strongly temperature dependent. The experiments as well as the MD calculations yield a coefficient of thermal expansion which is three times larger at the surface than in the bulk [61].

The frequency spectrum $f_{\parallel}(\omega)$ parallel to the surface at T = 102 K is strongly softened in comparison to that at T = 7 K (Fig. 4b). In particular, the finite value at $\omega = 0$ is due to the diffusive motion in the outermost layer at T = 102 K (for comparison, the melting temperature T_m = 116 K).

The time dependence of the mean square displacements $\langle r^2(t)\rangle$ shows clearly a diffusive behavior. On the other hand, if the mean z coordinates of the atoms are kept fixed, no diffusive motion of the atoms is observed (Fig. 4c). The diffusive particle motion is also hindered when the actual lattice relaxation is replaced by the relaxation at low temperatures (7 K) or if the layers beneath the outermost layer are transformed to a rigid Kr lattice [64b]. From these results we may conclude that the effect of premelting sensitively depends on the lattice relaxation and its temperature dependence as well as on the vibrational amplitudes perpendicular to the surface. It should be mentioned that at the surface the interatomic interaction is much more anharmonic than in the bulk, as shown by MD calculations [8,62]. This fact is responsible for the enhanced thermal expansion and the large vibrational amplitudes normal to the surface. It seems to be that both effects are prerequisites for the surface premelting.

In summary, the effect of surface premelting is conclusively proved by these MD results. The main features of this effect are a short-range-ordered structure and a high mobility of the particles in the outermost layer. In the next section it is shown that there are also experimental results showing these features.

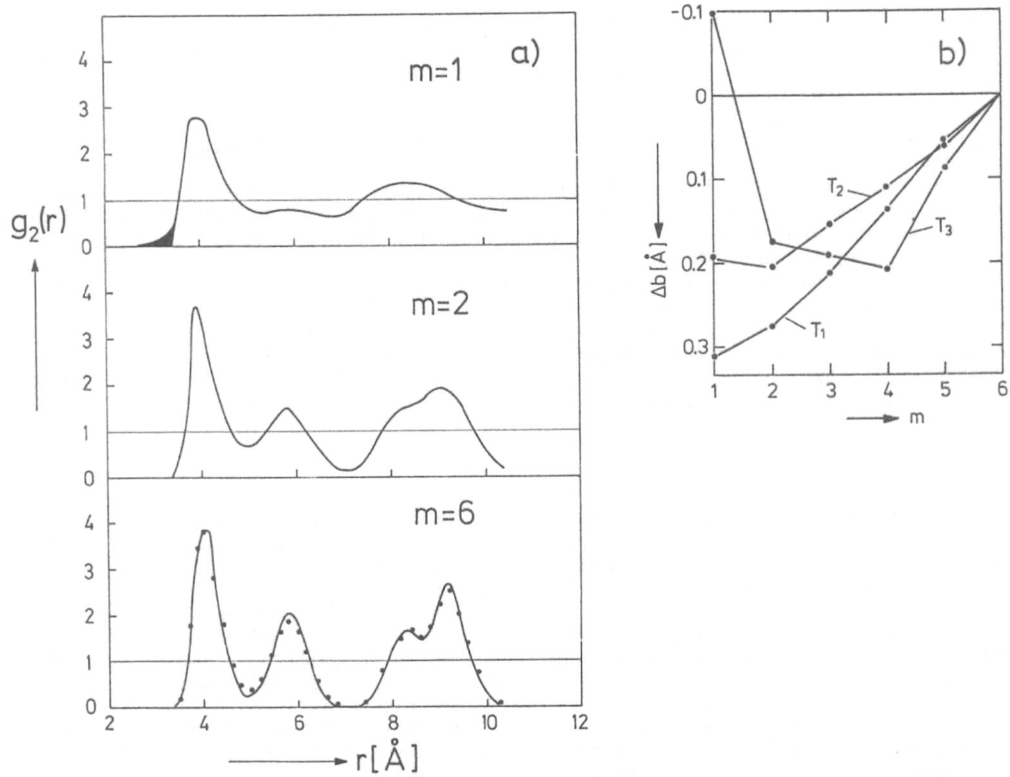

Fig. 3 Structure at the Kr(100) surface: Results of MD calculations for a 550 particle slab having two free surfaces and 11 layers [8]. m labels the layers starting with m = 1 for the outermost layer.
a) Pair correlation function $g_2(r)$ parallel to the surface for T = 102 K, (——) surface calculation, (···) bulk calculation. A comparison of $g_2(r)$ for m = 1 and m = 6 (the innermost layer) illustrates clearly a disordering of the outermost layer [62].
b) Deviations Δb of the mean positions of the layers from the mean positions that these layers would have in the bulk of the crystal, indicating a lattice contraction at low temperatures and a strong thermal expansion for the outer layers (T_1 = 7 K, T_2 = 70 K, T_3 = 102 K, melting temperature T_m = 116 K)[6].

The Pb(110) surface

For this surface most of the relevant properties discussed above for the Kr(100) surface have already been studied experimentally (see Fig. 5). The structural disordering is reflected in the decrease of the intensity of the LEED spots and in the increase of the diffuse X-ray intensity (corresponding to the first peak of $S_{\parallel}(Q)$) with increasing temperature. The ion backscattering data indicate also the increasing disorder with increasing temperature (Fig. 5) and they show a crystal-face dependence [47] of the disordering and an enhanced thermal expansion near the melting point (Fig. 5, inset). The ion backscattering data also yield the thickness of the disordered layer as a function of the temperature [45].

The continuous order-disorder transition observed by different methods at the (110) surface of the lead crystal may be described in the framework of the Landau theory: Lipowsky and Speth [65] have shown that a semi-infinite system undergoing

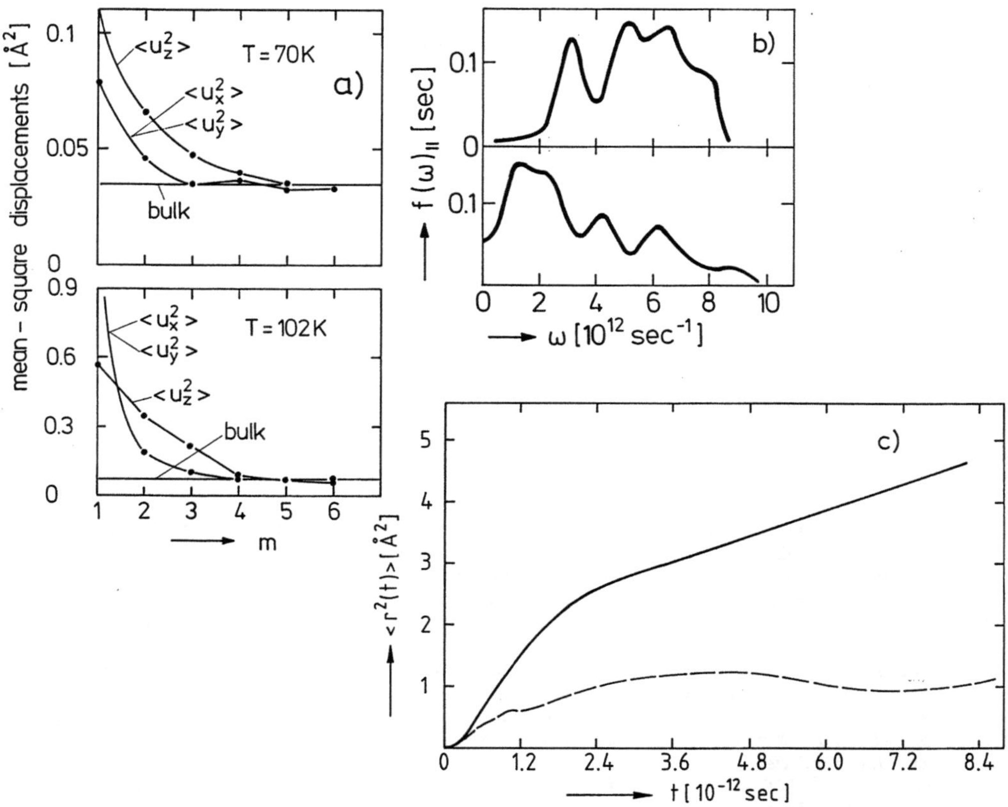

Fig. 4 Lattice dynamics at the Kr(100) surface. Results from MD calculations (see also caption of Fig. 3):
a) Mean-square displacements for two temperatures. The strong anisotropy and temperature dependence of the displacements are remarkable [62].
b) Frequency distribution for the lattice vibrations parallel to the surface (for m=1) at 7 K (top) and 102 K (bottom). By increasing the temperature the distribution is shifted to lower frequencies and f(0) becomes nonzero indicating surface diffusion [62].
c) Mean-Square displacement parallel to the surface for m=1 at 102 K as a function of time. For particles with unrestricted coordinates (——) $\langle r^2(t)\rangle$ is proportional to t at large t. From the slope of the linear part of $\langle r^2(t)\rangle$ the surface diffusion coefficient D_S has been derived to be $0.85\cdot10^{-5}$ cm²/sec. This value is in agreement with the value obtained from f(0) (see (b)) and is of the order of the diffusion coefficient in bulk liquids. For a system in which the mean particle positions normal to the surface are fixed (---) no surface diffusion appears [64a].

a first-order transition in the bulk may exhibit critical behavior at its surface (surface quantities may behave continuously even though bulk quantities are discontinuous). This theory predicts for the thickness of the disordered layer $\ell=\ell_0\cdot\ln[T_0/(T_m-T)]$, with constants ℓ_0 and T_0, as T_m is approached from below. The ion backscattering data are consistent with such behavior [45]. The disordered layer starts to grow at 0.75 T_m.

The order-disorder transition at the Pb(110) surface is also reflected in the temperature dependence of the photoemission peaks due to the p bands of Pb [52a]. The authors conclude that no evidence of molten or fully disordered layers has

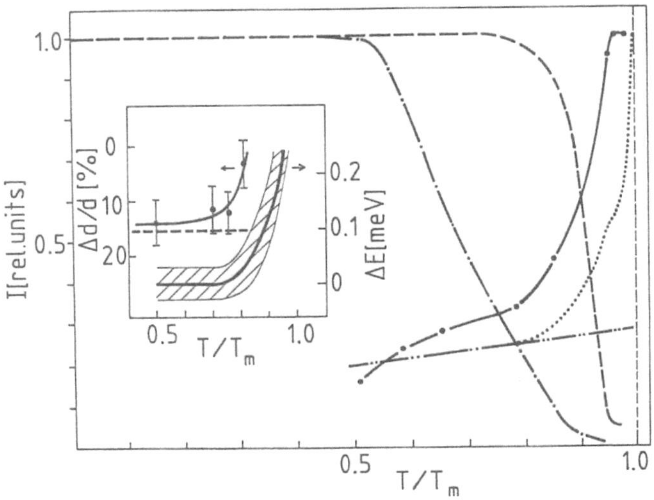

Fig. 5 Indications for surface premelting from surface scattering experiments. By approaching the melting temperature T_m the following effects have been observed:
- a decrease of LEED intensities after corrections for background and Debye-Waller factor (—·—·(01) - spot from the Al(110) surface [54], ---(10) - spot from the Pb(110) surface [50]) indicating the decrease of the long range order far below the melting point;
- an increase of the surface peak (SP) intensity in ion backscattering experiments with secondary energy analysis on the Pb(110) surface (... SP intensity, —··—·· expected SP intensity for an ordered surface with realistic mean vibrational amplitudes obtained by Monte-Carlo simulations) reflecting the appearance of atomic disorder at the surface [45];
- an increasing diffuse X-ray scattering intensity at the Pb(110) surface indicating a strong and anisotropic short range order in the disordered surface layers [49] (●-●-● integrated intensity of the first diffuse peak at $Q = 2\ \mathring{A}^{-1}$);
- a broadening of the quasielastic energy distribution (ΔE) of He atoms scattered from the Pb(110) surface giving the evidence for a large diffusion coefficient at the disordered surface near T_m comparable to the diffusion coefficient for the bulk liquid [51] (inset: ——, the hatched area represents the error bars) and
- an enhanced thermal expansion near T_m derived from results of ion backscattering experiments (inset: ●-●-● relative lattice relaxation Δd of the outermost layer with respect to its bulk-truncated position as a function of temperature, the dashed line represents the lattice expansion expected for the bulk expansion coefficient [48]).

been observed at the Pb(110) surface. However, as the X-ray diffraction experiments show [49], the disorder at the Pb(110) surface is not fully developed but exhibits a relatively strong short range order. The photoemission signal shows an unusual temperature dependence near the melting point (much stronger than the temperature dependence governed by the Debye-Waller factor) obviously due to the order-disorder transition as directly detected by the diffraction experiments.

From quasielastic scattering of low energy He-atoms, it is inferred that the disordered layer is really molten. The resolution-corrected width ΔE of the energy distributions of the scattered He-atoms (Fig. 5, inset) yields the surface diffusion coefficient D_s. It can be derived by the relation $\Delta E = 2\hbar D_s Q^2$, where $\hbar Q$ is the momentum transfer parallel to the surface. Already 50 K below T_m the measured D_s is equal to the bulk diffusion coefficient of liquid Pb [51].

The ion backscattering experiments have shown a crystal-face dependence of the surface disordering: The (110) face of the Pb crystal starts to disorder already about 150 K below T_m, but the (111) face does not disorder significantly at all [47]. It is expected that a maximum vibrational instability is associated with the least densely packed planes, i.e., the (110) planes of fcc crystals. Such a trend has also been observed by optical emission measurements on spherical Cu single crystals [55]. However, the strong face dependence of the disorder observed by ion scattering experiments contradicts the Lindemann criterion. Recently, X-ray diffraction experiments [49] have shown weak diffuse scattering due to the disorder which appears at the Pb(111) surface. In general, experimental results for the Pb(110) surface are a confirmation of the surface premelting hypothesis. The key experiment was the measurement of the surface diffusion coefficient.

The order-disorder transition observed for Al(110) by LEED experiments (see Fig. 5) is in good agreement with results of MD calculations for the Al(110) surface [73]. But the diffusion coefficient has still to be determined, experimentally as well as theoretically.

In summarizing the results on the surface premelting effect we may note: Surface premelting was observed in computer simulations for noble gases and also - with all expected features - by scattering experiments at the Pb(110) surface. Indications for surface premelting (order-disorder transitions) were also found, among others, for the following surfaces: Al(110), Ar(111), Cu(110) and Ge(111). The full analysis of the structure and dynamics at such crystal surfaces has still to be done; in particular, the role of the surface roughening in connection with premelting has to be elucidated. Regarding the melting theory, we may conclude that surface premelting is one of the possible reasons for the absence of superheating under normal conditions.

5. Outlook

The liquid-vapor interface is of fundamental importance for chemistry and biology. There are only very few experimental techniques successfully applied for the analysis of the surface of liquids. It would be very desirable to have methods for a more direct measurement of the density profile (for metallic and nonmetallic liquids). The dynamics at the surface is still an open field. For the theory, the main problem for the near future may be the development of realistic interatomic potentials for metal surfaces which should correctly include the temperature dependence. In this connection it should be emphasized that the determination of the pair potential from structure data in the bulk liquid can be done most reliably by MD [88,81]. Such a method should be applied to liquid surfaces.

Realistic interatomic potentials are also needed for the study of the effect of premelting at the surface of metal crystals. A simultaneous investigation of the temperature dependence of surface properties such as lattice structure, thermal expansion, self-diffusion, lattice vibrations and electronic structure seems to be a challenge for the future. Such studies would be useful not only for the elucidation of the effect of surface premelting but also for the solution of problems in connection with interface formations. The general premelting problem including grain-boundary melting etc. deserves more attention in order to understand the melting from a more general viewpoint.

References

[1] J. Als-Nielsen, *Solid and Liquid Surfaces Studied by Synchrotron X-ray Diffraction*, in Structure and Dynamics of Surfaces II, ed. by W. Schommers and P. von Blanckenhagen, Topics Curr. Phys., Vol. 43 (Springer, Berlin, Heidelberg 1987), p. 181

[2] W.O. Luedtke, U. Landman, M.W. Ribarsky, R.N. Barnett, and C.L. Cleveland, Phys. Rev. B 37, 4647 (1988)

[3] P.G. de Gennes, Rev. Mod. Phys. 57, 827 (1985)

[4] A.R. Ubbelohde, The Molten State of Matter (Wiley, New York 1978)

[5] J.Q. Broughton and L.V. Woodcock, J. Phys. C 11, 2743 (1978)

[6] W. Schommers and P. von Blanckenhagen, Vacuum 33, 733 (1983)

[7] J.Q. Broughton, G.H. Gilmer, J. Chem. Phys. 79, 5119 (1983)

[8] W. Schommers, *Statistical Mechanics of the Liquid Surface and the Effect of Premelting*, in Structure and Dynamics of Surfaces II, ed. by W. Schommers and P.v. Blanckenhagen, Topics Curr. Phys. Vol. 43 (Springer, Berlin, Heidelberg 1987) p. 223
[9] F.F. Abraham, CRC Crit. Res. Solid State Mat. Sci. 10, 169 (1981)
[10] R. Evans, Adv. Phys. 28, 143 (1979)
[11] C.A. Croxton, Statistical Mechanics of the Liquid Surface (Wiley, New York 1980)
[12] M.P. D'Evelyn and S.A. Rice, J. Chem. Phys. 78, 5081 (1983)
M.P. D'Evelyn and S.A. Rice, J. Chem. Phys. 78, 5225 (1983)
[13] J.P. Badiali, M.L. Rosinberg, D. Levesque, and J.J. Weis, J. Phys. C 16, 21 (1983)
[14] G.A. Chapela, G. Saville, S.M. Thompson, J.S.Rowlinson, J. Chem. Soc. Faraday Trans. II, 73, 1133 (1977)
[15] S. Toxvaerd, J. Chem. Phys. 55, 3116 (1971)
[16] N.T. Thomas, S.W. Barton, F. Novak, and S.A. Rice, J. Chem. Phys. 86, 1036 (1987)
[17] D. Bedeaux, J.D. Weeks, J. Chem. Phys. 82, 972 (1985)
[18] P. von Blanckenhagen and W. Schommers, *Surface Phenomena and their Analysis by Scattering Experiments*, in Structure and Dynamics of Surfaces II, ed. by W. Schommers and P.v. Blanckenhagen, Topics in Curr. Phys., Vol. 43 (Springer, Berlin, Heidelberg 1987)
[19] D. Beaglehole, Phys. Rev. Lett. 27, 2016 (1979)
[20] J. Gryko, S.A. Rice, J. Non-Cryst. Solids 61&62, 703 (1984)
[21] D. Sluis, M.P. D'Evelyn, and S.A. Rice, J. Chem. Phys. 78, 1611 (1983)
D. Sluis and S.A. Rice, J. Chem. Phys. 79, 5659 (1983)
[22] L. Bosio, M. Oumezine, J. Chem. Phys. 80, 959 (1984)
[23] L. Bosio, R. Cortes, A. Defrain, and M. Oumezine, J. Non-Cryst. Solids 16/62, 697 (1984)
[24] B.C. Lu and S.A. Rice, J. Chem. Phys. 68, 5558 (1978)
[25] A. Braslan, M. Deutsch, P.S. Pershan, and A.H. Weiss, J. Als-Nielsen, and J. Bohr, Phys. Rev. Lett. 54, 114 (1985)
[26] J.M. Bloch, M. Sansone, F. Roudelez, D.E. Pfeifer, P. Pincus, M.W. Kim, P.M. Eisenberger, Phys. Rev. Lett. 54, 1039 (1983)
[27] B.N. Thomas and S.A. Rice, J. Chem. Phys. 86, 3655 (1987)
[28] D. Nenow, *Surface Premelting*, in Prog. Crystal Growth and Charact. 9, 185 (1984)
[29] W. Heyes, *Premelting*, Contemp. Phys. 27, 519 (1986)
[30] M. Pokorny and G. Grimvall, J. Phys. F 14, 931 (1984)
[31] G. Fritsch, H. Diletti, E. Lüscher, Phil. Mag. A 50, 545 (1984)
[32] Da-Ming Zhu and J.G. Dash, Phys. Rev. Lett. 57, 2959 (1986)
[33] P. von Blanckenhagen, *Surface Diffusion and Layer Growth*, in Structure and Dynamics of Surfaces II, Topics in Curr. Phys., Vol. 43, ed. by W. Schommers and P.v. Blanckenhagen (Springer, Berlin, Heidelberg 1987) p. 73
[34] E.A. Stern and Ke Zhang, Phys. Rev. Lett. 60, 1872 (1988)
[35] T. Hguyen, P.S. Ho, T. Kwok, C. Nitta, S. Yip, Phys. Rev. Lett. 57, 1919 (1986)
[36] J.Q. Broughton and G.H. Gilmer, Phys. Rev. Lett. 56, 2692 (1986)
[37] O.G. Mouritsen, M.J. Zuckermann, Phys. Rev. Lett. 58, 389 (1987)
[38] K. Hoshino, K. Somorgai, Phil. Mag. A 40, 137 (1979)
C. Solliard, Solid State Commun. 51, 947 (1984)
H.L. Davis, J. Jellinek, R.S. Berry, J. Chem. Phys. 86, 6456 (1987)
[39] M. Bienfait, Europhys. Lett. 4, 79 (1987)
[40] R.H. Willens, A. Kornbilt, L.R. Testardi, and S. Nakahara, Phys. Rev. B 25, 290 (1982)
P.H. Brennan, P.H. Fuoss, and P. Eisenberger, in The Structure of Surfaces, ed. by M.A. van Hove and S.Y. Tong (Springer-Verlag, Berlin 1985) p. 421
[41] J. Dräges, H. Gleiter, J.H. Perepezko, Mat. Res. Soc. Symp. Proc. 57 (1986)
[42] R.W. Cahn, Nature 323, 668 (1986)
[43] F.A. Lindemann, Z. Phys. 11, 609 (1910)
[44] J. Heurion, G.E. Rhead, Surf. Sci. 29, 20 (1972)
[45] J.W.M. Frenken, P.M.J. Marée, and J.F. van der Veen, Phys. Rev. B 34, 7506 (1986)

[46] J.W.M. Frenken and J.F. van der Veen, Phys. Rev. Lett. 54, 134 (1985)
[47] B. Pluis, A.W. Duier van der Gon, J.W.M. Frenken, J.F. van der Veen, Phys. Rev. Lett. 59, 2687 (1987)
[48] J.W.M. Frenken, F. Hunssen, and J.F. van der Veen, Phys. Rev. Lett. 58, 401 (1987)
[49] P.H. Fuoss, L.J. Norten, and S. Brennan, Phys. Rev. Lett. 60, 2046 (1988)
[50] K.C. Prince, U. Breuer, and H.P. Bonzel, Phys. Rev. Lett. 60, 1146 (1988)
[51] J.W.M. Frenken, J.P. Toennies, and Ch. Wöll, Phys. Rev. Lett. 60, 1727 (1988)
[52] G. Jezequel, P. Thiry, G. Rossi, K. Hricovini, and Y. Petroff, Surf. Sci. 189/190, 605 (1987)
P. Thiry, G. Jezequel, and Y. Petroff, J. Vac. Sci. Technol. A 5, 892 (1987)
[53] E.G. McRae, R.A. Malic, Phys. Rev. Lett. 58, 1437 (1987)
[54] P.v. Blanckenhagen, W. Schommers, and V. Voegele, J. Vac. Sci. Technol. A 5, 649 (1987)
P. Martin, P. von Blanckenhagen, and W. Schommers, to be published
[55] K.D. Stock, Surf. Sci. 91, 655 (1980)
[56] G. Devaud and R.H. Willens, Phys. Rev. Lett. 57, 2683 (1986)
[57] Ph. Pruzan, D.H. Liebenberg, R.L. Mills, J. Phys. Chem. Solids 47, 949 (1986)
[58] S.G. Mochrin, Phys. Rev. Lett. 59, 304 (1987)
[59] G. Armand, D. Gorse, J. Lapujoulade, and J.R. Manson, Europhys. Lett. 3, 1113 (1987)
[60] V. Rosato, G. Ciccotti, V. Pontikis, Phys. Rev. B 33, 1860 (1986)
V. Pontikis and P. Sindzingre, Physica Scripta T19, 375 (1987)
[61] U. Romahn, W. Schommers, and P. von Blanckenhagen, to be published
[62] W. Schommers, Phys. Rev. B 32, 6845 (1985)
[63] W. Schommers, P. von Blanckenhagen, Surf. Sci. 162, 144 (1985)
[64] W. Schommers and P. von Blanckenhagen, J. Vac. Sci. Technol. A 5, 644 (1987)
W. Schommers, P. von Blanckenhagen, and U. Romahn, submitted to Modern Physics Letters
[65] R. Lipowsky and W. Speth, Phys. Rev. B 28, 3983 (1983)
[66] R.M. Cotterill, E.J. Jensen, and D.W.D. Kristensen, *Melting: Theories and Recent Computer Simulations*, in Anharmonic Lattices, Structural Transitions and Melting, ed. by. T. Riste (Noordhoff, Leiden 1974) p. 405
[67] C.S. Jayanthi, E. Tosatti, L. Pietromero, Phys. Rev. B 31, 3456 (1985)
C.S. Jayanthi, E. Tosatti, and A. Fasolino, Phys. Rev. B 31, 470 (1985)
[68] P. Carnevali, F. Ercolessi, E. Tosatti, Phys. Rev. B 36, 6701 (1987)
[69] A. Trayanov, E. Tosatti, Phys. Rev. Lett. 59, 2207 (1987)
[70] L. Pietronero, E. Tosatti, Solid State Commun. 32, 255 (1979)
[71] R. Lipowski, J. Appl. Phys. 55, 2485 (1984)
[72] C.S. Jayanthi, E. Tosatti, A. Fasoline, L. Pietronero, S. Sci. 152/153, 155 (1985)
[73] P. Stoltze, J.K. Norskov, U. Landmann, Phys. Rev. Lett. 61, 440 (1988)
[74] A.C. Levi and E. Tosatti, Surf. Sci. 178, 425 (1986)
[75] M.R. Mruzik and G.M. Pound, J. Phys. F. 11, 1403 (1981)
[76] H. van Beijeren and I. Nolden, *The Roughening Transition*, in Structure and Dynamics of Surfaces II, ed. by P. von Blanckenhagen and W. Schommers, Topics in Curr. Phys., Vol. 43 (Springer, Berlin, Heidelberg 1987) p. 259
[77] S. Valkealahti, R.M. Nieminen, Physica Scripta 36, 646 (1987)
[78] Changde Gong, Dingyu Xing, A. Holz, Jinzhong Pang, Phys. Rev. B 36, 5517 (1987)
[79] E. Tosatti, *Theory of Surface Melting and Non-Melting,* in Structure of Surfaces II, ed. by J.F. van der Veen and M.A. Van Hove, Springer Series in Surface Science, Vol. 11 (Springer, Berlin, 1988) p. 535
[80] W. Schommers, Phys. Rev. A 20, 3599 (1983)
[81] W. Schommers, Phys. Rev. Lett. 58, 427 (1987)

Monte Carlo Simulation of a Lennard-Jones Fluid near a Hard and a Soft Wall

G. Balabanić[1], *B. Borštnik*[2], *R. Milčić*[3], *A. Rubčić*[4], *and F. Sokolić*[5]

[1]Civil Engineering Institute, Rijeka, Yugoslavia
[2]"Boris Kidrič" Institute of Chemistry, Zagreb, Yugoslavia
[3]Civil Engineering Institute, Zagreb, Yugoslavia
[4]Faculty of Science, University of Zagreb, Zagreb, Yugoslavia
[5]"Rudjer Bosković" Institute, B.O. Box 1016, Zagreb, Yugoslavia

We have calculated density profiles and pair correlation functions for a Lennard-Jones fluid near a hard and a soft wall by Monte Carlo simulation. Our results are compared with those obtained previously by different methods. It is shown that at higher densities the fluid layer adjacent to the wall shows two-dimensional character which is reflected in the increase of the second neighbours' distance to the reference atom. When comparing the pair correlation function for the atoms within the first layer parallel to the wall with the bulk pair correlation function, the following differences were observed: in the case of the soft wall the maxima of pair correlation function for the first layer are enhanced, while in the case of the hard wall the maxima are lower than that for the bulk pair correlation function.

1. Introduction

There is considerable interest in the structure of fluids in the inhomogeneous region. The questions which are not yet fully settled refer to the details of the density profile and the pair correlation function (PCF) within the interface. This is true for liquid-vapour and liquid-crystal interfaces as well as for the model interfaces in the case of the simple fluid in contact with a hard or a soft wall.

Two models of liquid-solid interface exist, concerning the structure of the wall. In one of them the wall consists of particles, while in the other the wall is structureless. Although the first one is more realistic, the other is a good approximation for a wide range of phenomena, i.e. interaction of colloids in solutions and liquids in pores. The structureless wall was often used in solving the integral equations or the simulations of the inhomogeneous liquids.

The fluid near a plane wall shows two distinct behaviours depending on the nature of the particle-wall interaction and the

Springer Proceedings in Physics, Vol. 40
Editors: M. Davidović and A.K. Soper

Static and Dynamic Properties of Liquids

thermodynamic conditions. For a fluid near a repulsive wall at low density and/or low temperature there is a continuous rise of density towards the bulk value as a function of the distance from the wall. In the high density and/or high temperature region, the density profile has an oscillatory character, indicating the existence of ordered layers near a wall. An interpretation was given in the sense that a crystal-like structure of a fluid is induced by the presence of the wall. Confirmation of this crystal-like ordering may be obtained by the study of the PCF.

Several papers have been devoted to the study of the Lennard-Jones (LJ) fluid in contact with a soft and a hard wall. Nevertheless, the complete physical picture of these systems has not yet been reached.

First Monte Carlo (MC) simulation results on a LJ fluid-hard wall (LJ/HW) system were reported by ABRAHAM [1] and the most recent ones by POWLES et al. [2]. SNOOK and VAN MEGEN [3] performed the MC calculations for different kinds of particle-particle and particle-wall potentials. In these references the results for density profile at different thermodynamic points are reported. Much less attention has been paid to the calculation of the PCF, and only one simulation result exists, that in ref. [3] for LJ fluid near a soft wall (LJ/StW).

In the same time, integral equation theories were developed to treat the inhomogeneous fluids. BORŠTNIK et al. [4,5] have applied the BGYB hierarchy of equations to a LJ/HW system, calculating also the PCF in [5], but without pointing out any specific features of these functions. MEISTER and KROLL [6] reported the results of the density functional theory showing the drastic change in the form of the density profile for a slight change in the density in the vicinity of the triple point. The density functional method is also used in refs. [2] and [7].

PLISCHKE and HENDERSON [8] have recently reported the results of Percus-Yevick (PY) theory applied to the LJ/HW system. Two approaches were presented: singlet PY and self-consistent PY. In both cases the density profiles were determined, while PCF was calculated only in the self-consistent approximation. As they pointed out [8], the self-consistent theory is more precise. This was also recently confirmed by BALABANIĆ et al. [9] by MC simulation.

In this work we want to shed more light on the behaviour of LJ fluid near a hard and a soft wall. We present the numerical evidence about the correlation between the details of the density profile and the details of the PCF. We explore how the qualitative features of the density profile are realised in terms of two particle distributions. The central question refers to the dilemma whether the characteristic behaviour of the density profiles are accompanied with the corresponding characteristics of the PCF. We present evidence that the above-mentioned relations are heavily influenced by the form of the particle-wall potential.

2. Results and Discussion

In our calculations two parallel walls were situated at z=0 and z=3L, where L determines the basic MC cell: L x L x 3L. Periodic boundary conditions are employed in the x and y directions. Cut-off radius r_C=L/2 was used. A standard MC technique for calculation of the density profile and the PCF, as described by SNOOK and HENDERSON [10], was adopted. Our system contained 384 particles, which is twice that used in [9]. The equilibration run contained 5 to 10 million configurations, and the productive run up to 30 million.

Our system was sufficiently large to make negligible the influence of the walls on the central part of the cell. This was confirmed by the comparison of the PCF in the central layer of the MC cell with the bulk PCF calculated by us and those given by VERLET [11].

Already existing MC results [1-3,11] are in quantitative agreement with those reported here. Some differences will be discussed in more detail later.

The results are also checked by analysis of the contact density. Pressure deduced from it is in good agreement with that obtained for the bulk at the same thermodynamic conditions [12].

Only minor changes are observed in comparison to the results obtained on a system half the size [9].

We conclude that our system is large enough to observe the behaviour of a fluid near a single wall.

The present MC results for density profiles at three points at the reduced temperature T=1.35, and the comparison with those in ref. [8] are given in Figs.1,2 and 3. The densities in Figs.1 and 3 are chosen such as to represent the typical examples of continuous and oscillatory density profiles, respectively. In Fig.2 the density profile from the transitional region is given, where the contact density is approximately equal to bulk density. There is a good agreement with the self-consistent PY theory, as was expected, while the singlet PY approximation gives considerable deviations from our MC results, especially for medium densities. In Fig.3, our

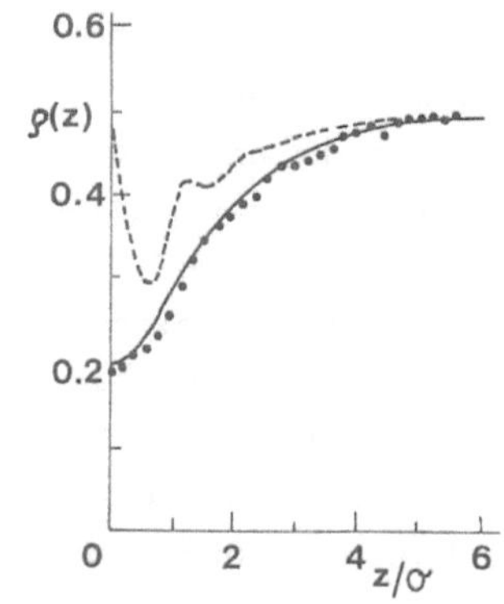

Fig.1 The reduced density ρ(z) for ρ=0.5 and reduced temperature T=1.35. The solid curve: self-consistent PY theory, the dashed curve: singlet PY theory. The dots are the present MC results.

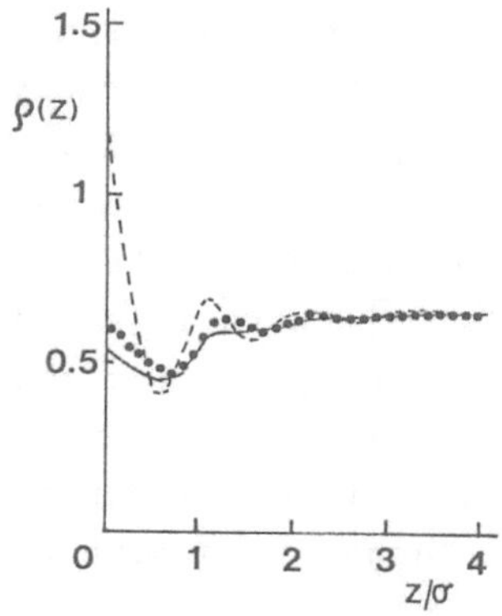

Fig.2 Same as Fig.1, but for ρ=0.65.

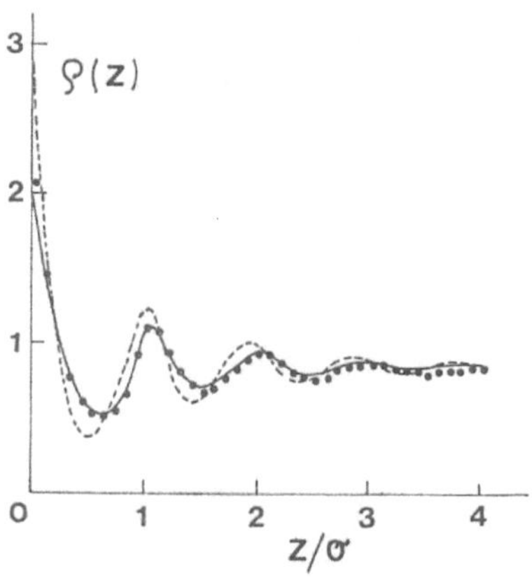

Fig.3 Same as Fig.1 but for ρ=0.85. The dots are present MC results for ρ=0.82.

bulk density is ρ=0.82, less than that in ref. [8], where ρ=0.85. The reason for this is the "a posteriori" determination of ρ in a MC run and it sometimes happens that the starting average density is not properly chosen. Nevertheless, the contact density in Fig.3 is larger than that in ref. [8] as can be understood from Fig.1 of ref. [8].

The central issue of the work is the pair correlation functions. The analysis of the particle distribution was aimed towards elucidation of the inhomogeneities of the two particle correlations. For this purpose the bulk PCF was calculated and also the

PCF in the subsequent layers next to the wall. The results are shown in Figs. 4,5 and 6. The thermodynamic conditions are the same as in Figs. 1,2 and 3, respectively. MC results in Fig.4 are in nice agreement with those of Fig.7 in ref. [8]. However, there are some differences between our results in Fig.6 and those in Fig.8 of ref. [8]. While in ref. [8] there is a shift of second and subsequent maxima (and minima) to larger r for z=0, for z=0.75 there is a slight shift in a reverse direction. After that, for still larger z, the bulk PCF is reproduced. This behaviour is the opposite to our results where the shift to larger r is most pronounced for z=0, but then continu-

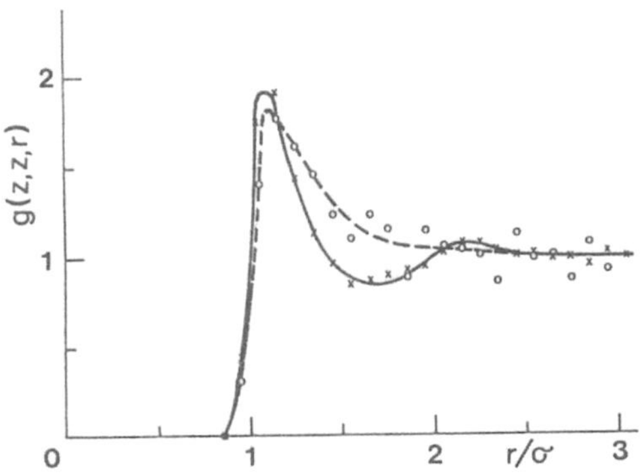

Fig.4 Pair correlation function g(z,z,r) for ρ=0.5 and T=1.35. Circles are for panel at z=0 and crosses for z=10. Full line is the bulk pair correlation function. Dashed line represents an eye-guide passing through circles.

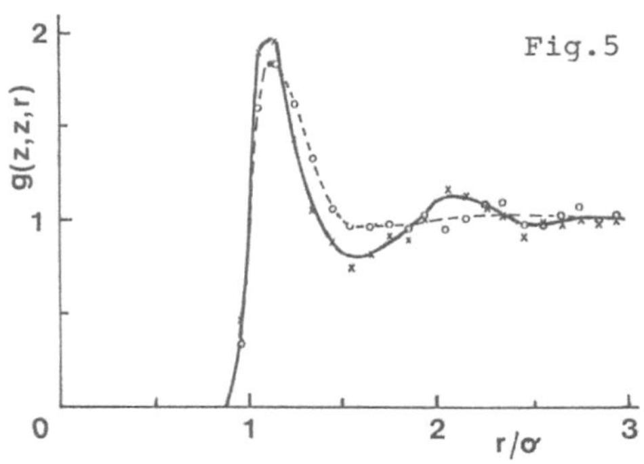

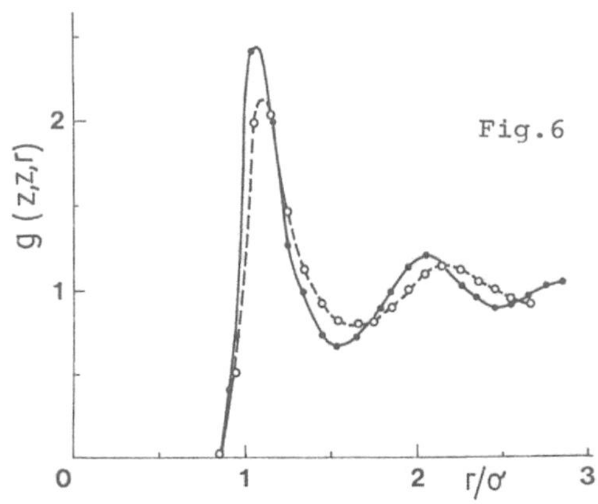

Fig.5 Pair correlation function g(z,z,r) for ρ=0.65 and T=1.35. Circles are for panel at z=0 and crosses for z=10. Solid and dashed lines are the eye-guides for crosses and circles, respectively.

Fig.6 Pair correlation function g(z,z,r) for ρ=0.82 and T=1.35. Circles are for panel at z=0 and dots for bulk. Solid and dashed lines are the eye-guides for dots and circles, respectively.

ously moves back to the bulk values for large z, i.e. for the panels far from the wall.

In Figs. 4,5 and 6 it may be seen that the values of the maxima and minima of the PCF are reduced. This may be an indication of a decrease of the short range order on approaching the wall, for all densities. It is seen from Fig.4 that the PCF in the contact layer is of the gas-like form, indicating its gaseous nature.

The increase of the oscillation length or stretching of the PCF in the panels near the wall was demonstrated by SNOOK and HENDERSON [10] and also recently by HENDERSON and VAN SWOL [13] for the hard sphere - hard wall (HS/HW) system. In this system, contrary to the LJ/HW system, the values of maxima and minima increase relative to those of the bulk on approaching the wall, which may be an indication of an increase of order in the panel close to the wall.

The similar behaviour concerning the shift and the opposite behaviour concerning the height of maxima and minima in two different systems motivated us to study the LJ fluid near the wall with Steele´s 10-4-3 potential. This system was already investigated using MC simulation by SNOOK and VAN MEGEN [3]. Our result for the density profile at the same thermodynamic point is given in Fig.7. It shows greater oscillations than the density profile in the nearby thermodynamic point for the LJ/HW system. In ref. [3] (Fig.10) an increase of the maxima and minima of the PCF of the contact layer was noticed, which was confirmed by our calculations (see Fig.8).

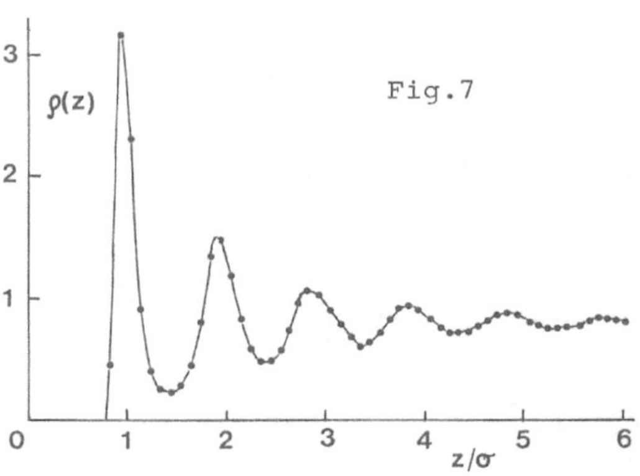

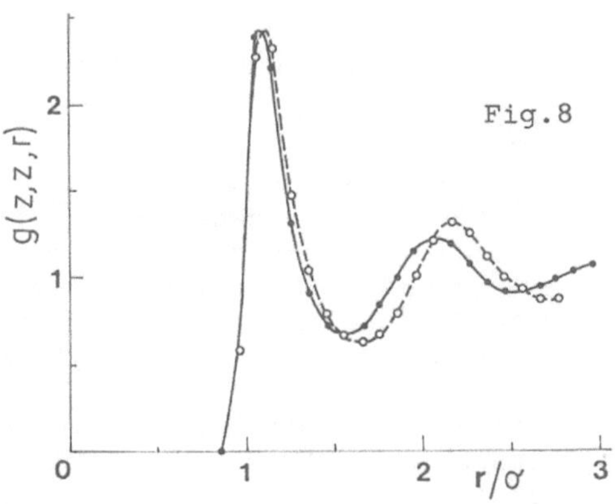

Fig.7 The reduced density $\rho(z)$ for LJ/StW system, at $\rho=0.82$ and T=1.17.

Fig.8 Pair correlation function g(z,z,r) for LJ/StW system for $\rho=0.82$ and T=1.17. Circles are for panel at z=0 and dots for bulk. Solid and dashed lines are the eye-guides for dots and circles, respectively.

This is opposite to what is found in the LJ/HW system. But, interestingly enough, they have not observed any shift in the position of maxima and minima of the PCF, as we did.

The layer adjacent to the wall may be considered as a quasi two-dimensional (2D) system. This was pointed out for the HS/HW system already by SNOOK and HENDERSON [10]. The PCF for the first liquid layer behaves like 2D hard-disc bulk and is different from the 3D hard-sphere bulk. The maxima and minima beyond the first maximum of the 2D fluid are shifted toward the higher r compared to the 3D bulk fluid. For the estimation of this shift we may use a very simple geometrical model. Let us consider the 2D and 3D hexagonal close-packed lattice. The second neighbour distance in 2D is $\sqrt{3}\sigma$ and in 3D is $2\sqrt{2/3}\,\sigma$. The difference is 0.1σ. This is in good agreement with all observed shifts for the second maximum of the PCF at high densities, regardless of the form of the particle-particle and particle--wall interaction potential. It is expected that all subsequent maxima and minima are proportionally shifted to the right. Such a stretching of the PCF can be seen in the case of the PCF for the HS/HW system obtained by MD simulation by HENDERSON and VAN SWOL [13].

In conclusion, we may say that a plane wall causes an increase in the oscillation length of the PCF for all systems, as shown here by simulations for the LJ/HW and LJ/StW systems, and in [10] and [13] for the HS/HW system. The change in magnitudes of the maxima and minima of the contact layer PCF with respect to the bulk PCF suggests an increase of the short range order near the wall in the HS/HW system. Further investigation of this behaviour is in progress and will be published elsewhere.

References

1. F.F.Abraham: J.Chem.Phys. 68, 3713 (1978).
2. J.G.Powles, G.Rickayzen, M.L.Williams: Mol.Phys. 64, 33 (1988).
3. I.Snook, W.van Megen: J.Chem.Phys. 70, 3099 (1979).
4. B.Borštnik, D.Janežič, A.Ažman: Mol.Phys. 42, 721 (1981).
5. B.Borštnik, D.Janežič: Mol.Phys. 50, 1199 (1983).
6. T.F.Meister, D.M.Kroll: Phys.Rev. A31, 4055 (1985).
7. J.P.R.B.Walton, N.Quirke: Chem.Phys.Lett. 129, 382 (1986).
8. M.Plischke, D.Henderson: J.Chem.Phys. 84, 2846 (1986).
9. G.Balabanić, F.Sokolić, R.Milčić, A.Rubčić: Farm.Vestnik 38, 187 (1987).
10. I.K.Snook, D.Henderson: J.Chem.Phys. 68, 2134 (1978).
11. L.Verlet: Phys.Rev. 165, 201 (1968).
12. A.Barker, D.Henderson: Rev.Mod.Phys. 48, 578 (1976).
13. J.R.Henderson, F.van Swol: Mol.Phys. 51, 991 (1984).

RISM Study of the Structure of Water at the Hydrophobic Wall

D. Janežič and B. Borštnik

Boris Kidrič Institute of Chemistry, P.O. Box 30,
YU-61115 Ljubljana, Yugoslavia

The RISM integral equation method was applied to the study of the structure of liquid water at a soft hydrophobic wall. The calculation was designed as a study of a two-component mixture with water as a solvent and spherical Lennard-Jones particles as a solute. The potential between water molecules was represented as a sum of the interactions between three interaction sites on each water molecule with dispersion and Coulomb terms. Water-solute interaction was of the Lennard-Jones form. The correlation functions for the pairs of sites residing on water molecules were treated within the formalism of renormalization of long range forces. The numerical procedure brought forward the total and direct correlation function for five pairs of sites: OO, OH, HH, OW and HW where O, H and W stand for oxygen, hydrogen and solute, respectively. The calculations were performed for several steadily increasing solute radii. Finally, a limiting procedure was performed and h_{OW} and h_{HW} were determined for infinite radius of the solute. This means that limiting functions (h_{OW} + 1) and (h_{HW} + 1) represented the density profile of O and H atoms of water molecules at a plane wall. We used this information in order to generate the orientational distribution function of water molecules at the wall. The results show that water molecules are orientationally ordered with hydrogen atoms pointing preferentially towards the interior of the bulk. Within a few molecular diameters the orientational order is lost.

1. Introduction

The studies of the interfaces where liquid water medium resides at one side of the interface are attracting more and more intense attention. The motivation lies in the wide range of fields where solvation, and in particular hydration, phenomena are important, such as colloid science and many branches of life sciences. The influence of water as a solvent can be roughly classified in terms of hydrophobic and hydrophilic interactions. The hydrophilic interactions are based on the hydrogen bonded network and electrostatic interactions which are present at the interface. This phenomenon is always encountered in the solvation of biological macromolecules, which are usually charged and have hydrogen donor and acceptor groups which facilitate the hydrogen bonding. Besides the hydrophilic interactions, a phenomenon exists which is usually called hydrophobic bonding. This phenomenon has its roots in the fact that the cavity surface decreases after two cavities fuse, which brings together two hydrophobic particles.

For a better understanding of the above-mentioned phenomena, one can undertake various approaches, such as experimental studies of solubility, surface and interface tensions and theoretical approaches in the form of computer simulation and integral equation studies. In this work we report on the RISM study of the interface between liquid water at the soft hydrophobic wall. It is shown that a reasonable amount of information about the interfacial structure can be obtained in this way.

2. Description of the Calculations

The fundamental equations of the reference interaction site model are the Ornstein-Zernike equations which connect the direct correlation function and the total correlation function of the reference interaction sites of the molecules [1]

Springer Proceedings in Physics, Vol. 40
Editors: M. Davidović and A.K. Soper

Static and Dynamic Properties of Liquids

$$\varrho^2 \boldsymbol{h} = \boldsymbol{\omega} * \boldsymbol{c} * \boldsymbol{\omega} + \varrho^2 \boldsymbol{\omega} * \boldsymbol{c} * \boldsymbol{h} \quad . \tag{1}$$

They should be supplemented with the closure equations, which could be of the hypernetted chain or Percus-Yevick form. We will use the HNC-like closure:

$$c_{ij}(r) = \exp\left[- \beta u_{ij}(r) + t_{ij}(r)\right] - t_{ij}(r) - 1 \ , \tag{2}$$

$$\boldsymbol{t} = \boldsymbol{h} - \boldsymbol{c} \quad .$$

Equations (1) and (2) form a closed set of equations on which basis one can calculate all the distribution functions. This formalism was successfully applied to the study of liquid water structure by renormalization of long range forces of the electrostatic type [2,3]. Due to this variant, a set of nonlinear integral equations is obtained for direct and total correlation functions. The bare Coulomb potential does not appear in these equations. The resulting correlation functions reproduce fairly well the experimentally determined correlation functions.

Equations (1) and (2) can also be applied to multicomponent systems [4]. In such a case one obtains a more complex system of interconnected equations. In the limiting case, when a component is present at infinite dilution, the equations simplify to a certain extent since all the terms which are multiplied by the density of the solute vanish as the density of the solute tends to infinity. In the remaining equations the number of terms is reduced. In the case of a symmetric triatomic molecule such as water (HOH) which is mixed with a monoatomic (W) there are six distinct site-site pairs: OO, HH, OH, OW, HW, and WW. In the limit of low density of the solute, WW pairs are neglected. The equations for the remaining five pairs become uncoupled and the equations obtain the following form: Three equations of the form (1) are coupled and refer to the correlation functions for the O and H sites. They can be solved independently of the remaining two equations. We solved them following the renormalization approach of Pettitt and Rossky. The last two equations which follow from (1) and (2) for a two-component system have the following form:

$$\hat{h}_{\mathrm{HW}}(k) = \hat{c}_{\mathrm{OW}}(k)\left[s + \varrho\hat{h}_{\mathrm{OH}}(k)\right] + \hat{c}_{\mathrm{HW}}(k)\left[1 + S + 2\varrho\hat{h}_{\mathrm{HH}}(k)\right] \quad , \tag{3}$$

$$\hat{h}_{\mathrm{OW}}(k) = \hat{c}_{\mathrm{OW}}(k)\left[1 + \varrho\hat{h}_{\mathrm{OO}}(k)\right] + 2\hat{c}_{\mathrm{HW}}(k)\left[s + \varrho\hat{h}_{\mathrm{OH}}(k)\right] \quad , \tag{4}$$

$$s = \left[\sin(kR_{\mathrm{OH}})\right]/kR_{\mathrm{OH}} \quad ,$$
$$S = \left[\sin(kR_{\mathrm{HH}})\right]/kR_{\mathrm{HH}} \quad ,$$

where R_{HH} and R_{OH} are the intramolecular site-site distances in the HOH molecule. These two equations can be solved after the OO, HH and OH site-site correlation functions have been determined. In our case the solute-solvent interactions are of Lennard-Jones type and therefore there is no need to apply renormalization formalism in (3) and (4). The solution of (3) and (4) provides us with the correlation function of oxygen and hydrogen atoms relative to the solute particles. In order to get the distribution functions of water molecules at the plane wall, one can use a standard procedure [5] due to which a series of calculations has to be performed at increasing values of solute radius and finally, the limiting functions towards which the distribution functions converge are determined. In this procedure we kept the width of the Lennard-Jones well, which determines the solute-solvent interaction, constant as the radius of the solute increased in size.

3. Results and Discussion

The calculations were performed for the TIPS 3-site water model. The model site-site interactions are parametrized as

$$U_{ij}(r_{ij}) = A_{ij}/r_{ij}^{12} - C_{ij}/r_{ij}^{6} + q_i q_j/r_{ij}$$

with A_{OO} = 580000 kcal $Å^{12}$/mol, C_{OO} = 525 kcal $Å^6$/mol, A_{OH} = 225 kcal $Å^{12}$/mol and $q_O = -2q_H = -0.8e$. This set of parameters is the one proposed by JORGENSEN [6] except for the addition of A_{OH} which was suggested by PETTITT and ROSSKY [2] as a means to overcome the infinite negative interaction energy between an O and an H as their separation approaches zero.

The bulk density of water was taken to be 1 g/cm^3 and the temperature T = 300 K. Our correlation functions which determine the bulk water site-site correlation agreed satisfactorily with the ones obtained by other authors. Solute-water distribution functions were calculated for several values of solute diameters. The results are shown in Figs. 1 and 2. Besides the calculated correlation functions Figs. 1 and 2 also depict the functions which are obtained by the extrapolation towards zero curvature of the solute. The extrapolated solute-solvent correlation functions have the meaning of oxygen and hydrogen density profiles at the plane wall. We can see that the profiles are oscillatory with rather well-expressed maxima and minima, especially in the immediate vicinity of the wall. We can compare the heights of the peaks with the density profiles generated by the Monte Carlo simulation of LEE et al. [7]. It turns out that our profiles are much more structured than the profiles provided by the simulation. Some caution is necessary when doing the comparison since the two calculations have been done by different types and different parametrization of water-water and wall-water potentials. The density profiles which are depicted in Figs. 1 and 2 belong to the wall-water potential with the depth of the potential well equal to $6k_B$K. The calculations were also performed for a ten times deeper well and the density profiles turn out to be more structured. In the Monte Carlo simulation of LEE et al., the wall-water potential was deeper but still the oscillations of the profile were weaker. We can thus conclude that our approach overemphasizes the structuring effect of water layer at the hydrophobic wall.

An indication in favour of the relevance of our approach is the fact that much better agreement with the Monte Carlo results is obtained for the non-polar model of water. For such a model LEE et al. obtain a much more structured density profile and also our results give the oscillations with similar amplitudes which are higher than in the case of polar water.

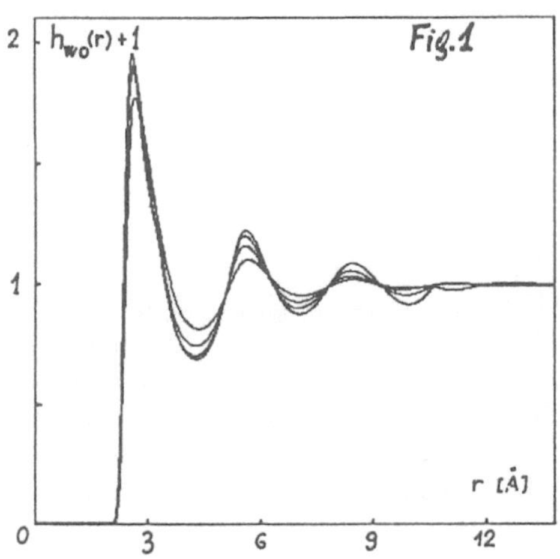

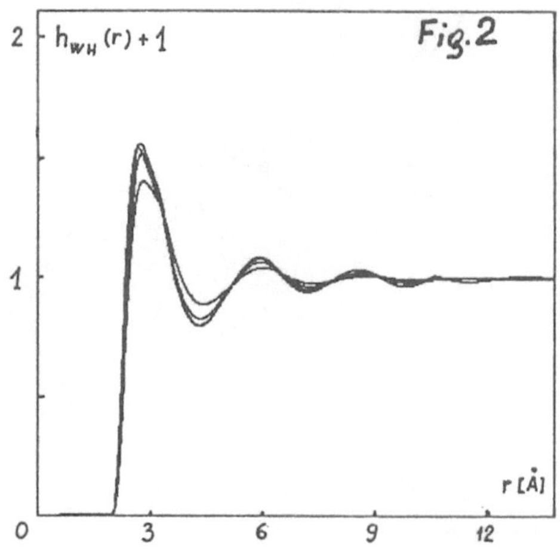

Fig. 1. Solute-oxygen pair correlation functions for the diameter of the solute particle being 3, 6 and 9 Å. The curve with the greatest oscillations is the result of the extrapolation procedure and corresponds to the solute with infinite diameter. The depth of the potential describing solute-oxygen and solute-hydrogen interaction was $6k_B$K

Fig. 2. Solute-hydrogen pair correlation function. See also the caption of Fig. 1

The shortcoming of the RISM type calculations is the inability to give direct information about the orientational distribution functions of polyatomic molecules at the liquid surface. The function which would contain full information about the distribution of nonlinear triatomic molecules at a plane wall would be a function of four variables: three Euler angles and the distance from the wall. In the case of water molecules, which possess a symmetry axis, one could be satisfied with the information about the distribution with respect to the cosine of the angle between the molecule symmetry axis and the normal to the plane wall.

On the basis of the density profiles of O and H atoms, it is possible to retrieve partial information about the angular distributions. A straightforward approach would be to express the angular distribution function $f(z, \cos\theta)$ by means of the product of density profiles as follows:

$$F(z, \cos\vartheta) \propto \varrho_{\mathrm{OW}}(z) \int_0^\pi \varrho_{\mathrm{HW}}(z')\varrho_{\mathrm{HW}}(z'')d\varphi \tag{5}$$

where z' and z'' are the coordinates of hydrogen atoms. Equation (5) simply expresses the supposition that the probability that a molecule will be located at a certain position and will have a certain orientation is proportional to the product of probabilities to find an O atom at z and another two hydrogens at z' and z''.

The resulting angular distribution functions are depicted in Fig. 3. The first maximum next to the wall at $\cos\theta \approx 1$ corresponds to the water molecules which are pointing away from the wall. The next maximum towards the interior of bulk water is lower and corresponds to the molecules which have approximately opposite orientation with regard to the molecules which reside at the first maximum. Later on, the inhomogeneity with regard to $\cos\theta$ dies out. The results are partly uncertain due to the fact that (5) is only a rough approximation. Qualitatively, other authors [7,8] also obtain similar results for water or other simple polar fluids.

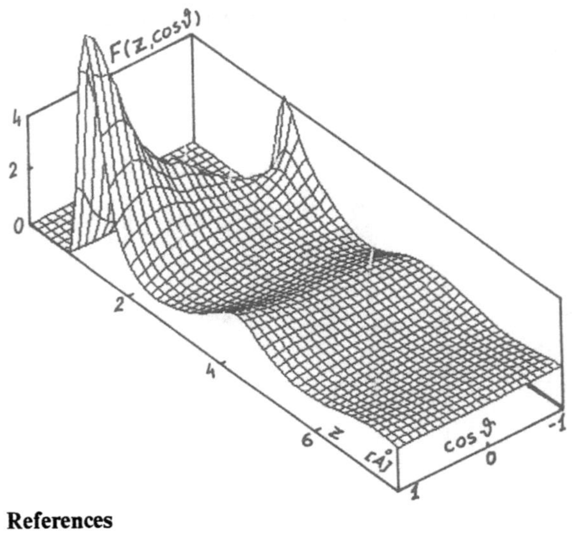

Fig. 3. Angular distribution function for water molecules at the wall. z measures the distance of oxygen atoms from the wall and θ is the angle between the molecule symmetry axis and normal to the wall. See also the caption of Fig. 1

References

1. D. Chandler, H.C. Andersen: J. Chem. Phys. **57**, 1930 (1972).
2. B.M. Pettitt, P.J. Rossky: J. Chem. Phys. **77**, 1451 (1982).
3. P.J. Rossky: Pure & Appl. Chem. **57**, 1043 (1985).
4. B.M. Pettitt, M. Karplus: Chem. Phys. Lett. **121**, 194 (1985).
5. D.J. Mitchell, B.W. Ninhan, B.A. Pailthorpe: J. Chem. Soc. Faraday II, **74**, 1089 (1978).
6. W.L. Jorgensen: J. Am. Chem. Soc. **103**, 335 (1981).
7. C.Y. Lee, McCammon, P.J. Rossky: J. Chem. Phys. **80**, 4448 (1984).
8. J. Eggebrecht, K.E. Gubbins, S.M. Thompson: J. Chem. Phys. **86**, 2286 (1987).

Molecular Dynamics of Wetting and Drying at Solid-Fluid Interfaces

M.J.P. Nijmeijer [1], *C. Bruin* [1], *A.F. Bakker* [1], *and J.M.J. van Leeuwen* [2]

[1]Laboratorium voor Technische Natuurkunde, Technische Universiteit Delft, NL-2600 GA Delft, The Netherlands

[2]Instituut Lorentz, Rijksuniversiteit Leiden, Nieuwsteeg 18, NL-2311 SB Leiden, The Netherlands

The adsorption of a fluid at a structured solid substrate is studied with the aid of the Delft Molecular Dynamics Processor (DMDP). 2904 particles form a solid substrate and some 8500 particles are in a fluid phase. All particles interact through 12-6 Lennard-Jones (L-J) potentials with a cut-off at 2.5 σ. The solid particles are heavier and interact more strongly than the fluid particles. The mutual fluid-substrate interaction is varied while the temperature is held constant. Density profiles, coverages and surface tensions are measured as functions of the fluid-wall affinity. Two first-order transitions separate the dry, the partially dry (or partially wet) and the wet regimes. In separate simulations contact angles are determined and the results are compared with the surface tension data. In the wet region a search for prewetting has been started, but the results obtained so far do not show unambiguous evidence for prewetting.

1 INTRODUCTION

The wetting phase transition has been predicted theoretically by Cahn [1] and by Ebner and Saam [2] and it has been observed experimentally by Moldover and Cahn [3]. If the transition is of first order, which is usually the case, a prewetting transition should occur. For prewetting there is no experimental evidence as yet. Also, the wetting phenomenon is sensitive to the details of the microscopic interactions. So computer simulations could give an important contribution, although the systems have to be large and the relaxation times are considerable. The availability of the DMDP has enabled us to overcome these difficulties.

2 THE PARALLEL WETTING EXPERIMENT

The system [4] is enclosed in a cubic box of length L with periodic boundary conditions. Perpendicular to the z-direction we built a solid wall (2904 particles) of 3 layers as thick as three 100-planes of a fcc lattice. The space between the wall and its periodic image is filled with some 8500 fluid particles. The interaction between the particles i and j is given by the L-J potential as a function of the distance $r_{ij} = |\mathbf{r}_i - \mathbf{r}_j|$

$$\phi_{AB}(r_{ij}) = 4\,\varepsilon_{AB}\,\{\,(\sigma_{AB}/r_{ij})^{12} - (\sigma_{AB}/r_{ij})^{6}\} \tag{1}$$

with a cut-off at 2.5 σ and A and B referring to substrate (s) or fluid (f) particles. The substrate particles are more massive ($m_s = 3\,m_f$), interact more strongly ($\varepsilon_{ss} = 50\,\varepsilon_{ff}$) and are smaller ($\sigma_{ss} = 0.847\,\sigma_{ff}$, $\sigma_{sf} = 0.912\,\sigma_{ff}$) than the fluid particles. The affinity of the fluid to the wall $\varepsilon_r = \varepsilon_{sf}/\varepsilon_{ff}$ is the main parameter in our simulations. Varying ε_r allows us to keep the temperature $T^* = k_BT/\varepsilon_{ff}$ fixed at the value 0.9 (between the critical and the triple-point temperatures).

The equations of motion for all the particles (the wall included) are solved numerically with a time step $\Delta t^* = \Delta t\sqrt{\varepsilon_{ff}}/(\sigma_{ff}\sqrt{m_f}) = 0.01$. Runs consisted of 10^4 timesteps for equilibration and 0.25 - $1 * 10^5$ timesteps for measurements.

Transverse and normal pressure are measured by evaluating

$$p_T(z) = n(z)\,k_BT - \Big\langle \sum_{i\neq j} \Big(x_{ij}\frac{\partial\phi}{\partial x_{ij}} + y_{ij}\frac{\partial\phi}{\partial y_{ij}}\Big)\,\delta(z - z_i)\Big\rangle / 4A\,, \tag{2a}$$

$$p_N(z) = n(z)\,k_BT - \Big\langle \sum_{i\neq j} z_{ij}\frac{\partial\phi}{\partial z_{ij}}\,\delta(z - z_i)\Big\rangle / 2A\,, \tag{2b}$$

where n(z) is the density at z and $A = L^2$ is the area of the substrate. In the middle of a bulk phase the pressure becomes isotropic $p_T(z) = p_N(z)$. In the fluid phases this condition is fairly well achieved. This allows us to find not only the total surface tension

$$s_{tot} = \int_{-L/2}^{L/2} dz\, [\, p_N(z) - p_T(z)\,] = \partial F / \partial A \tag{3}$$

but also that of individual interfaces, always parallel to the solid wall. Inside the wall the pressure is not isotropic (though the stress is minimized by putting $L = 29.1\ \sigma_{ff}$). The contributions of the middle layer are equally attributed to the surface tensions of the two solid-fluid interfaces. For an interface involving a solid, the surface free energy γ may differ from the surface tension s according to

$$s = \gamma + A\ \partial\gamma/\partial A. \tag{4}$$

We observed three regimes, varying ε_r:
i) For low ε_r the liquid is pushed away from the substrate (completely dry phase), yielding a symmetric density profile (see Fig. 1). One has for the total surface free energy $\gamma_{tot} = 2\ \gamma_{sv} + 2\ \gamma_{lv}$ from the two solid-vapor and liquid-vapor interfaces.
ii) For intermediate ε_r the profile is asymmetric (partially dry/wet phase). One has $\gamma_{tot} = \gamma_{sv} + \gamma_{sl} + \gamma_{lv}$ from the solid-liquid, liquid-vapor and solid-vapor interfaces.
iii) For high ε_r the vapor is pushed away from the substrate (completely wet phase). The density profile is symmetric and one has $\gamma_{tot} = 2\ \gamma_{sl} + 2\ \gamma_{lv}$ from two solid-liquid and liquid-vapor interfaces.

From the behavior of the profiles and the number of particles Γ_1 in the first absorbed layer near the wall (see Fig. 2) we located the drying transition $\varepsilon_d \approx 0.2$ and the wetting transition $\varepsilon_w = 0.8$. From the surface tensions (see Fig. 3) we find $\varepsilon_d = 0.53$ and $\varepsilon_w = 0.8$. Near the wetting transition a considerable hysteresis is observed. In the partially dry/wet region one has for the contact angle $\cos\theta = (\gamma_{sv} - \gamma_{sl})/\gamma_{lv}$. Inserting the surface tension per interface instead of the surface free energies we can compute $\cos\theta$ (see Fig. 4, open circles), giving again $\varepsilon_d = 0.53$ and $\varepsilon_w = 0.8$. This substitution can be justified since we checked that $\delta(\gamma_{sv} - \gamma_{sl})/\delta A$ is negligible, although $\delta\gamma_{sv}/\delta A$ and $\delta\gamma_{sl}/\delta A$ themselves are definitely not.

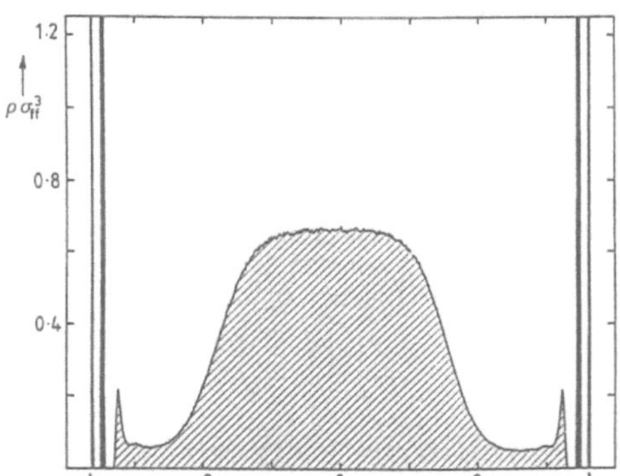

Fig. 1 Density profile of a symmetric (completely dry) configuration at $\varepsilon_r = 0.3$. Dark peaks correspond to the substrate layers. The arrows denote the periodic boundary

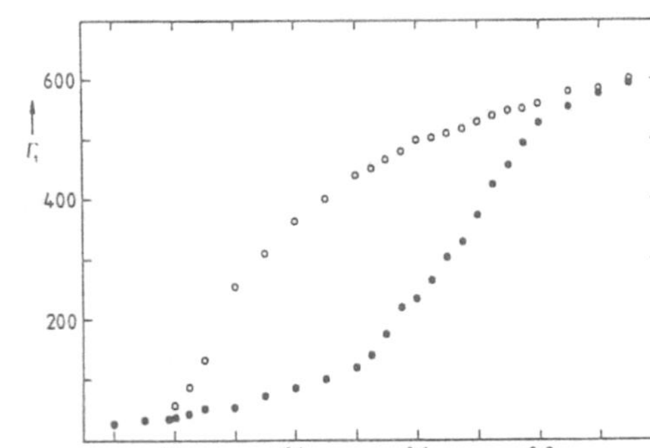

Fig. 2 The coverage Γ_1 in the first absorbed atomic layer at the solid-liquid interface (open circles) and at the solid-vapor interface (filled circles)

3 PERPENDICULAR WETTING EXPERIMENT

In view of the discrepancy between ε_d from the surface tension data as related to contact angles and ε_d from the behavior of the density profiles, we have set up an experiment in which we measure the contact angle directly. The same substrate is used but the fluid is prepared as a liquid slab perpendicular to the substrate walls [5]. One can derive that the liquid-vapor interface is part of a circle (see Fig. 5) so one has $\cos\theta = (L - 2\delta)/2r$, with r the radius of the circle and δ the distance between the first fluid layer

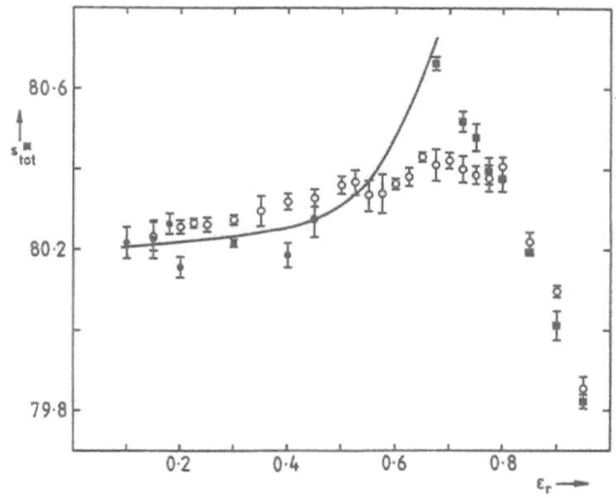

Fig. 3 The reduced total surface tension s^*_{tot} (in units of $\varepsilon_{ff}/\sigma_{ff}^2$) versus ε_r. Open circles refer to asymmetric profiles, filled squares to completely wet substrates and filled circles to completely dry substrates. The solid line is constructed from contact angles

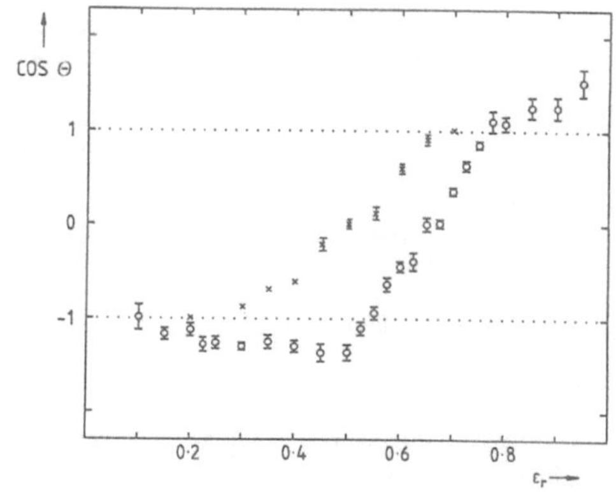

Fig. 4 Cos θ versus ε_r. Circles from asymmetric profiles in the parallel experiment. Crosses from direct measurement in the perpendicular experiment

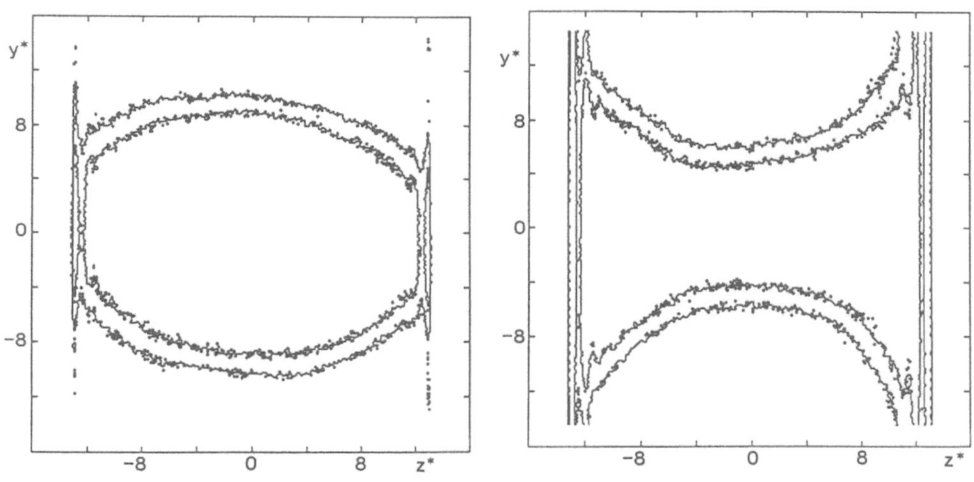

Fig. 5 Lines of constant density $\rho_{ff}\sigma^3 = 0.4, 0.2$ between the liquid phase (around $y^* = 0$) and the vapor phase, for $\varepsilon_r = 0.35$ (left) and $\varepsilon_r = 0.65$ (right)

and the center of the wall. The results for cos θ, varying ε_r at fixed temperature $T^* = 0.9$, are shown in Fig. 4 (crosses) and are compatible with the transition data from density profiles and coverages. Thus locating the drying transition at $\varepsilon_d = 0.2$ we have here hysteresis as well.

4 PREWETTING EXPERIMENT

In a first-order wetting transition one has a jump from a finite coverage below the wetting point to an infinite (or macroscopic) coverage above the wetting point. This infinite jump occurs at the vapor coexistence density for the fluid. For a density slightly below coexistence, thermodynamic continuity of the surface free energy requires that the coverage must exhibit a finite discontinuity near the wetting point. Moving away from the coexistence density the discontinuity decreases and vanishes in a surface critical point. The difficulty with the prewetting phenomenon is that it happens close to coexistence and for a relatively small range of ε_r values near ε_w.

We have tried to observe the prewetting transition by gradually extracting particles out of the vapor phase and measuring the total surface tension (for both $\varepsilon_r = 0.85$ and $\varepsilon_r = 0.8$). As long as the liquid phase is present the total surface tension is constant. It starts to change when the liquid slabs near the wall have disappeared. The total surface tension goes through a maximum with decreasing vapor density ending at the value for the wall in vacuum. When the system goes through the prewetting transition the slope of the surface tension as a function of the vapor density should change. No such behavior has been observed.

However, for some combinations of the vapor density n_v (near the coexistence density) and ε_r, a systematic large deviation occurs between the two coverages on both sides of the wall. This could be interpreted as prewetting. We are checking this phenomenon for a system with a two times larger substrate area, since deviations also occur for $\varepsilon_r = 0.7$, which is well below ε_w.

REFERENCES

1. J. W. Cahn: J. Chem. Phys. 66, 3667 (1977)
2. C. Ebner, W. F. Saam: Phys. Rev. Lett. 38, 1486 (1977)
3. M. R. Moldover, J. W. Cahn: Science 207, 1073 (1980)
4. J. H. Sikkenk, J. O. Indekeu, J. M. J. van Leeuwen, E. O. Vossnack: Phys. Rev. Lett. 59, 98 (1987)
5. G. Saville: J. Chem. Soc. Faraday Trans. 12, 1122 (1977)

Brillouin Scattering near a Solid-Liquid Interface

E.F. Gramsbergen and G.H. Wegdam

Laboratory for Physical Chemistry, University of Amsterdam, Nieuwe Achtergracht 127, NL-1018 WS Amsterdam, The Netherlands

Using Brillouin scattering from an evanescent wave, an interfacial fluid mode has been observed at a solid-liquid interface. Its dispersion relation depends on the component of the wave vector parallel to the interface, and is, within the accuracy of the experiment, linear with a speed equal to the bulk sound speed. We also observed a damped Rayleigh wave.

1. Introduction

Interfaces induce long-range correlations in liquids which decay exponentially from the interface into the bulk (1,2). Fluctuating hydrodynamics predicts such correlations at a liquid-solid interface, as a result of thermally excited interfacial fluid modes (1). The dispersion relation of the fluid modes near a liquid-solid interface depends on the component of the wave vector parallel to the interface, $k_{//}$. It is linear in the small-$k_{//}$ limit where coupling with viscous and heat modes is negligible. In our regime of densities and sound velocities, the velocity c_i of the fluid interface mode is equal to the bulk velocity c. We measured the dynamic structure factor $S(\vec{k},\omega)$ for different values of $\vec{k}$, which enables us to verify whether the dispersion relation is linear in $k_{//}$ and independent of $k_{\perp}$ as predicted.

2. Experimental

An argon-ion laser beam is focussed onto the center of the base of a smoothly polished quartz semicylinder which is immersed in a large bath filled with liquid (see Fig. 1, inset). The angle of incidence on the quartz base, ϕ, is slightly below the critical angle of total reflection to generate an evanescent wave in the liquid with a typical penetration depth of 1 μm. The evanescent field, with a real wavevector k_i parallel to the interface, acts as a source for scattering. The scattered radiation with wavevector $\vec{k}_f$ is analyzed by a triple-pass Fabry-Perot interferometer and detected by a photomultiplier in order to obtain the dynamic structure factor $S(\vec{k}_f-\vec{k}_i,\ \omega)$. Interface roughness contributes to an intense Rayleigh peak at ω=0. The contrast of the triple-pass Fabry-Perot is enough to suppress the wings of this peak in the Brillouin region. Because Brillouin scattering of the incident and totally reflected beams in the quartz itself also contributes to the measured spectra, the scattering angles Θ must be carefully chosen to keep these peaks away from the frequency region of interest.

3. Results and discussion

A typical spectrum obtained at a propanol-quartz interface is shown in Fig. 1. The low frequency side is dominated by the intense Rayleigh peak at ω=0 and the two Brillouin peaks from scattering inside the quartz. The remaining two peaks are the bulk Brillouin peak of the

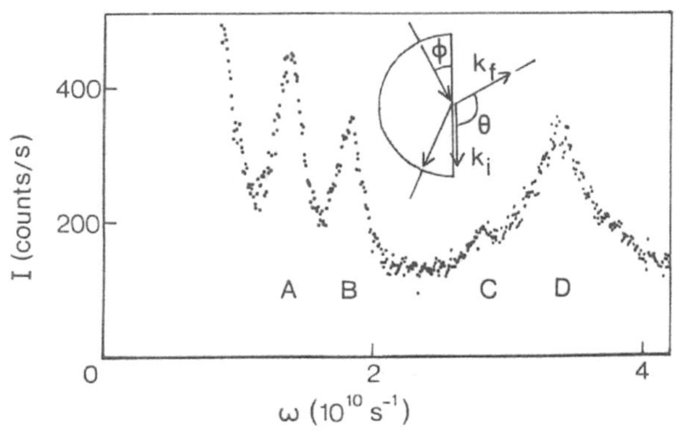

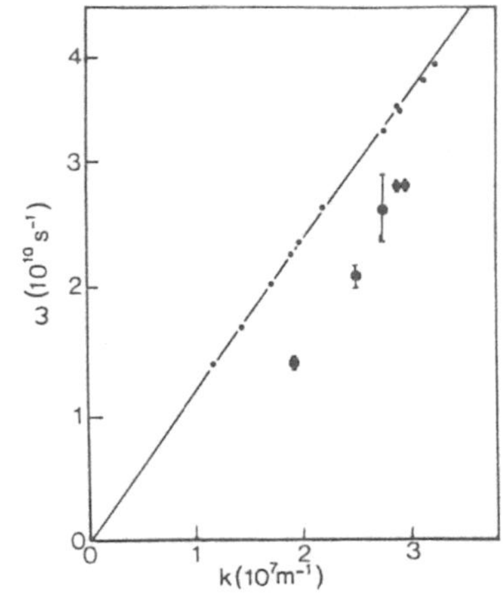

Fig. 1
Typical spectrum from a quartz-propanol interface, Θ=120°. Inset: the scattering geometry, C: scattering from the interface mode, D: the bulk propanol Brillouin peak, A and B: direct and totally reflected quartz Brillouin peaks. The high intensity at zero frequency ($5*10^6$ counts/s) is due to interface imperfections.

Fig. 2
Dispersion relation ω vs. k of propanol. Small dots: bulk, large dots: interface propanol-quartz. The straight line has a slope equal to the bulk sound velocity c=1210 m/s.

liquid and, on its low frequency wing, a weak peak from the fluid interface mode. When the sample was turned over 0.5° to create a homogeneous plane wave in the liquid, the interface peak disappeared and the bulk Brillouin peak gained one to two orders of magnitude in intensity. Fig. 2 demonstrates the usual linear dispersion(ω=ck) of the bulk sound mode for propanol. The interface points appear below this line. For the dispersion relation of these points only the $k_{//}$ component of the wavevector is relevant, with the linear relation $\omega=ck_{//}$ (Fig. 3). Other

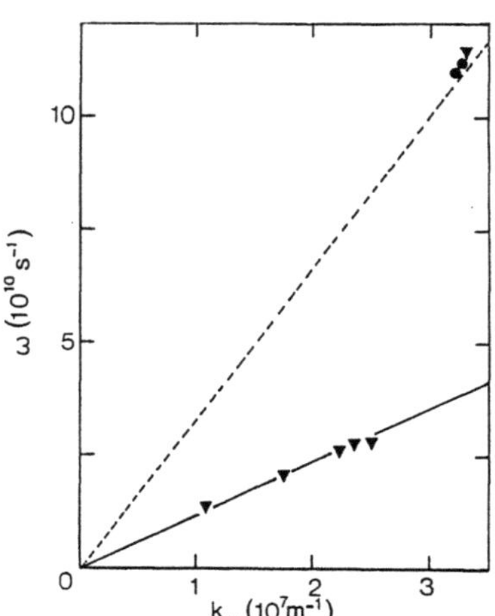

Fig. 3
Dispersion relation ω vs. $k_{//}$ of ethanol/quartz (triangles) and propanol/quartz (circles) interface modes. Solid line: fluid interface mode, broken line: Rayleigh wave.

points belong to Rayleigh waves on the quartz surface: $\omega=c_R k_{//}$ with $c_R=3340$ m/s. The peaks produced by these Rayleigh waves are slightly broadened compared to the bulk quartz peaks, as predicted by mode-coupling theory (1,3). This is the first sign of coupling between the Rayleigh wave and the fluid interface modes while the sound velocity is still unaffected.

References

(1) D. Gutkowicz-Krusin and I. Procaccia, Phys. Rev. Lett. 48, 417 (1982); Phys. Rev. A 27, 2585 (1983); D. Gutkowicz-Krusin, Phys. Rev. A 28, 2802 (1983)

(2) J.G. Dil, Rep. Prog. Phys. 45, 285 (1982)

(3) L.M. Brekhovskikh, "Waves in Layered Media" (Academic Press, New York, 1980)

Part IV

Supercooled Liquids and the Glass Transition

Liquid-Glass Transitions

A. Sjölander

Institute of Theoretical Physics, Chalmers University of Technology, S-41296 Göteborg, Sweden

1 Introduction

It is well known that a liquid can be supercooled below its ordinary melting point. Since crystalization is a fairly slow process on a microscopic time scale, it is possible to bypass this by rapid cooling and the liquid then goes over to a disordered solid, a glass, at some critical temperature T_g. Actually, for many fluids a cooling rate of only a few degrees per minute is sufficent, whereas a cooling rate of the order 10^{11} K/s or more is required for simple Lennard-Jones fluids. The latter can only be achieved in computer simulations. Once the glass is formed it may stay in this phase for a very long time, although it is not in a thermal equlibrium state. In some sense one can view the glass as an extremely viscous liquid, where the macroscopic flow is too slow to be observed over any reasonable experimental time. The glass transition temperature is conventionally defined where the shear viscosity exceeds a value 10^{13}P, whereas for ordinary simple liquids a characteristic value is 10^{-2}P.

Considering the glass as just a highly viscous fluid, it is surprising that certain experimentally observed quantities seem to change discontinuously when passing through the glass transition point. We may ask ourselves whether a liquid can exist in two different phases, analogous to what happens for liquid helium. The entropy and volume change continuously, whereas the corresponding second thermodynamic derivatives - heat capacity C_p, thermal expansion coefficient α_p, and isothermal compressibility κ_T - change essentially discontinuously. In this respect it looks like an ordinary second-order phase transition in the Ehrenfest sense. However, the transition is not sharp and it depends on the cooling rate. A higher rate implies a larger transition region. Furthermore, characteristic hysteresis effects are observed for the heat capacity. Measurements of the Prigogine-Defay ratio [1]

$$\Pi = \frac{\Delta C_p \Delta \kappa_T}{TV(\Delta \alpha_p)^2}, \tag{1}$$

where the changes in C_p, κ_T and α_p at the transition point enter, always give a value larger than one. A transition from one equlibrium state to another would give the value unity. This seems to imply that the transition is basically of a dynamical origin, where certain relaxations are simply too slow to be observed during typical measuring time. Measurements of the shear viscosity show that the time for structural relaxations increases dramatically as we approach the glass transition point. It has been common practice to fit the experimental data to the Vogel-Fulcher law

$$\eta = A \exp\left[\frac{B}{T - T_0}\right], \tag{2}$$

although no real justification for this particular form has been given so far. Actually, one often finds a better agreement with the ordinary Arrhenius law, i.e. $T_0 = 0$, when η becomes very large. For viscosities of to 10^2P or so one finds a good fit to a power law [2]

$$\eta = A(T - T_0)^{-\gamma}. \tag{3}$$

Springer Proceedings in Physics, Vol. 40 Editors: M. Davidović and A.K. Soper

Static and Dynamic Properties of Liquids

This only shows that we lack a deeper understanding of what is going on in the glass transition region. Even more dramatic and puzzling results are obtained when one considers the dynamics in more detail. This will be discussed in the next section.

2 Experimental Results on the Dynamics

A variety of experiments, such as neutron scattering, light scattering, dielectric loss experiments, etc., are all related to fluctuations in the microscopic density or partial densities of the system. The experimental data are often expressed in terms of the dynamical structure factor $S(q,\omega)$ or the corresponding intermediate scattering function $F(q,t)$. For liquids the latter tends to zero for large times. If it stays finite for $t \to \infty$, it implies that the atoms do not diffuse a macroscopic distance at any time, but that they are localized and form a solid. For a disordered system this would lead to a strictly elastic peak in $S(q,\omega)$ for all q.

The experiments reveal some remarkable universial behavior for a large variety of glass-forming liquids. The dynamical structure factor shows an elastic peak on the glass side, and on the liquid side this broadens to a quasi-elastic peak due to structural relaxation, so-called α-relaxation. For the intermediate scattering function it leads for large times to a non-exponential decay of the characteristic form (Kohlrausch-Williams-Watts law)

$$\exp[-(t/\tau)^{\beta}] \qquad 0 < \beta < 1, \tag{4}$$

with τ increasing rapidly as we approach the glass transition point from above. Besides the above peak in $S(q,\omega)$, one finds in neutron scattering experiments a broad inelastic spectrum similar to the phonon spectrum in polycrystalline samples. In $F(q,t)$ it manifests itself through a rapid decrease within a time of the order the inverse Debye frequency. Particularly in dielectric loss experiments, where one measures essentially the quantity $\omega S(q,\omega)$, one observes close to the glass transition point a sharp very low frequency α-peak and another, called the β-relaxation peak, which lies far below the phonon spectrum. This latter peak is found on both sides of the glass transition point and it changes smoothly over the transition region. Figure 1 shows typical experimental data for a system slightly above the transition point. The appearance of two different slow relaxation processes seems to be a universial behaviour for large classes of glass-forming systems. We may have to distinguish between glasses like SiO_2 with strong network bonds and fragile glasses [3]. The former show essentially an Arrhenius temperature dependence of the shear viscosity in the whole supercooled region, whereas the other class shows a strong deviation from this. We consider here the latter group of systems. The observations on these imply that the intermediate scattering function decreases within a microscopic time due to phonon-like vibrations, then relaxes on a much slower time scale without showing any such significant structural relaxation within the experimental time, leading to an essentially elastic peak in $S(q,\omega)$.

New and remarkable results from recent neutron scattering experiments [4-7] show that all the essential features mentioned above appear also quite far above T_g, where the viscosity is only of the order $10^2 - 10^3$P. The α-relaxation region then stretches up to frequencies of the order 10^9p/s and the β-region is around 10^{11}p/s. It appears as if the mechanism for freezing begins already in this viscosity range and that only the time scale changes to longer times as we lower the temperature and hence increase the viscosity. Deep inside the glass, the measured mean square displacements of the atoms are essentially the same as in a crystal and follow a linear temperature dependence. A deviation from this is observed already below T_g and one obtains a finite value also above T_g up to a temperature T_0, where the structural relaxations begin to be observed in the neutron scattering experiments [5,6]. This indicates again that in this viscosity range the liquid behaves as a frozen system over a period of time of the order 10^{-10}s or so, which is large on a typical microscopic time scale.

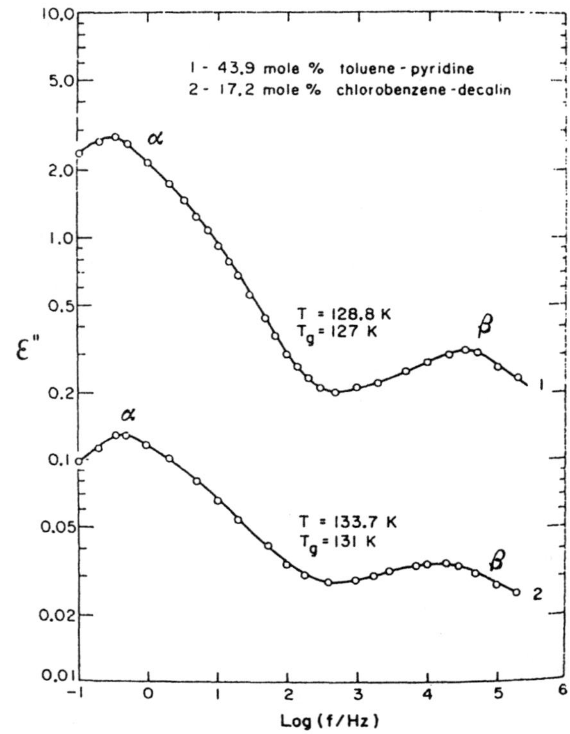

Fig. 1: Results for $\epsilon'' \propto \omega S(q,\omega)$ from dielectric loss experiments (taken from [22]).

The fact that some quantities show a significant dependence on the cooling rate implies that long memory effects are expected near the glass transition point. This should invalidate ordinary hydrodynamics. The non-exponential decay in both the α and β relaxation regions tells us that any conventional visco-elastic theory would fail.

3 Generalized Hydrodynamics

Even when interpreting neutron scattering data just for ordinary liquids, it became necessary to generalize the hydrodynamic equations in order to make them valid also for rapid and short wavelength fluctuations. A convenient way of doing this was introduced by Mori [8], where he modified earlier ideas of Zwanzig [9]. For the Fourier-Laplace transform of the intermediate scattering function

$$F(q,w) = i\int_0^\infty dt\, e^{i\omega t}\langle n(q,t)n(-q,0)\rangle, \tag{5}$$

one obtains the following exact expression:

$$F(q,\omega) = -S(q)\frac{\omega + q^2 m(q,\omega)}{\omega^2 - (q^2/m\beta S(q)) + \omega q^2 m(q,\omega)}, \qquad S(q,\omega) = \frac{1}{\pi}\mathrm{Im}F(q,\omega), \tag{6}$$

where $\beta = 1/k_B T$ is the inverse temperature in energy units and $S(q)$ is the equlibrium liquid static structure factor. The quantity $m(q,\omega)$ is a generalized friction, which for $q,\omega \to 0$ goes over to the familiar hydrodynamic form

$$m(q,\omega) \to i\frac{\frac{4}{3}\eta + \zeta}{\rho} - \frac{T}{\rho C_V}\left(\frac{\partial p}{\partial T}\right)^2 \frac{1}{\omega + iq^2\lambda_T/C_V} \tag{7}$$

with conventional notation. The last term arises from the coupling between density and temperature fluctuations and it leads to a shift of the hydrodynamic sound velocity to the adiabatic one. Since this part does not play any crucial role for our discussion, it will be left out in what follows. In doing this we lose the quasi-elastic heat diffusion peak in $S(q,\omega)$ with the width proportional to $q^2\lambda_T/C_V$. The main interest lies in the first term in (7) and its generalization to all q and ω. Using the Mori formalism, one can obtain a formally exact expression for $m(q,\omega)$, valid for any q and ω, but one has to take approximations in order to obtain any quantitative predictions. For a recent review of this we refer to [10]. It should be stressed that (6) has no direct relation to hydrodynamics and it contains a description of the atomic motion on a microscopic scale. This is why it can be used as a basis for analyzing neutron scattering data. In particular, it contains the so-called "cage effect," which is an important aspect of ordinary liquid dynamics and should be even more important in the supercooled region. It implies that each atom is caught for some time in the field of the surrounding atoms and that this restricts its possibility of diffusing away.

4 Mode Coupling Approach for Liquid-Glass Transitions

Relaxation towards an equilibrium liquid state will always occur, if $m(q,t) \to 0$ for $t \to \infty$. The appearance of a non-equlibrium frozen state implies therefore that $m(q,t)$ stays finite over the experimental time, which for practical purposes, we can set to $t = \infty$. If so, a liquid-to-glass transition is characterized by $m(q,t=\infty)$ becoming finite, and this implies that a low-frequency singularity appears in $m(q,\omega)$ of the form

$$m(q,\omega) \to \frac{m(q,t=\infty)}{\omega}. \tag{8}$$

We assume that crystalization occurs at a much slower rate than the approach towards the supercooled liquid state and that the latter is characterized by the static structure factor $S(q)$.

It is shown by Bengtzelius and Sjögren [11] that $m(q,t=\infty) \neq 0$ yields a Prigogine-Defay ratio larger than one, and approximate calculations gave reasonable agreement with observations. Inserting (8) in (6) leads to a change in the sound velocity on freezing. They were able to show that the heat capacity, the thermal expansion coefficent, and the compressibility change discontinuously [11]. In order to make any quantitative predictions and to be able to show that $m(q,\omega)$ has the above low-frequency, we have to provide an explicit expression for $m(q,t)$.

The basic underlying assumption will be that the slow decay of $m(q,t)$ in the strongly supercooled region is due to slow relaxations of the local structure, appearing through the microscopic density $n(q,t)$. We shall assume that beyond microscopic times of the order of inverse Debye frequency, $m(q,t)$ can be written as a functional of $F(q,t)$, where within the frozen state $F(q,t) \to S(q)f(q) \neq 0$ for $t \to \infty$. This, together with (6), leads to an equation for $f(q)$ of the form

$$\frac{f(q)}{1-f(q)} = m\beta S(q)m[f(q')], \tag{9}$$

where the solution depends on the explicit form of $m[f]$.

A particularly simple, but still reasonable physical form was analyzed in detail in the original work of Leutheusser [12] and of Bengtzelius, Götze and Sjölander [13], and it gave a $f(q) \neq 0$, when the factor $m\beta S(q)$ exceeds a certain critical value. The following more realistic expression was also given in [13] and was based on previous work on liquids [14]:

$$m[F(q,t)] = m_0(q,t) + \frac{n}{2m\beta q^2}\int \frac{d\mathbf{q}'}{(2\pi)^3}\left[V(\mathbf{q},\mathbf{q}')\right]^2 F(q,t)F(|\mathbf{q}-\mathbf{q}'|,t) \tag{10}$$

with

$$V(\mathbf{q},\mathbf{q}') = (\hat{\mathbf{q}}\cdot\mathbf{q}')c(q') + [\hat{\mathbf{q}}\cdot(\mathbf{q}-\mathbf{q}')]\,c(|\mathbf{q}-\mathbf{q}'|), \qquad \hat{\mathbf{q}} = \mathbf{q}/q, \tag{11}$$

where

$$c(q) = [S(q)-1]/nS(q) \tag{12}$$

is the direct correlation function. The term $m_0(q,t)$ is supposed to contain a regular contribution to the viscosity and it can for our purpose be taken as a constant $i\nu_0$. The above model was analysed numerically in great detail by Bengtzelius [15] and a generalization thereof was analysed analytically by Götze [16]. Other versions of the same kind of theory have since then been considered by various authors, particularly models where the q-dependence has been ignored for simplicity [17-21]. It seems that the q-dependence is not a crucial point, when considering the basic dynamical aspects of the liquid-glass transition within this model. However, it should be stressed that the microscopic fluctuations are the basic cause of the transition.

The words "mode coupling theory" have sometimes been misinterpreted to be restricted to such situations where the "modes" represent well-defined resonances. In the present context it simply means non-linear coupling of a set of correlation functions, irrespective of whether they contain any resonances or not. The physics involved determines which kind of "modes" one should include and it determines also the appropriate form of the generalized viscosity. The expression above with only a pair of density correlation functions appearing seems to be the simplest one that can lead to a liquid-glass transition.

When considering the self-motion of an atom, a similiar set of equations can be derived for the Van Hove self-correlation function $S^s(q,\omega)$, where the corresponding generalized viscosity $m^s(q,t)$ now has one of the F's replaced by the self-part $F^s(q,t)$. The whole integral represents then the situation where a tagged particle, represented by F^s, moves through the medium and interacts with the surrounding fluid, which is represented by F. For the coherent $S(q,\omega)$ no particle is singled out and for this reason the two F's appear in (10). A simple approximation of $m(q,t)$ would be to assume an exponential decrease in time. This leads to the conventional visco-elastic theory, which we already earlier concluded is not able to explain the experimental observations near a liquid-glass transition. Any reasonable theory has to predict a decrease of the self-diffusion constant D_s as we supercool the liquid. For D_s very small it leads to a slow asymptotic decay of $F(q,t)$ and of $m(q,t)$, and hence to a large enhancement of the low-frequency part of $m(q,\omega)$. For this reason one would expect to obtain a sharp peak in $S(q,\omega)$ around $\omega = 0$ due to structural relaxation, arising from the slow particle diffusion. However, one could not beforehand expect to obtain a strong non-exponential decay of $F(q,t)$ and an anomalous temperature dependence of the hydrodynamic viscosity. Nor would one anticipate another slow relaxation process, analogous to the observed β-relaxation. This is actually what the mode coupling model above predicts.

5 Theoretical Applications

The liquid static structure factor $S(q)$ enters as an input in the theory and it can be assumed to vary smoothly around the glass transition point. It reflects the local static ordering of the equilibrium liquid and it depends on the mean density and the temperature. We notice that the interaction potential enters only through $S(q)$ and the constant ν_0. Explicit calculations have been carried out for a hard sphere system and Lennard-Jones systems [15], where $S(q)$ was calculated separately, and it gave a liquid-glass transition at a certain critical value of the density, n_c, or the temperature, T_c. Above this density or below this temperature the dynamical structure factor has the form

$$S(q,w) = S(q)f(q)\delta(\omega) + S_{\rm inel}(q,\omega) \tag{13}$$

where the q-dependence of the form factor $S(q)f(q)$ reflects the local frozen structure. This can be measured in neutron scattering experiments. For $S^s(q,\omega)$ one obtains an analogous expression, where $f^s(q)$ gives information on the mean square displacement of the individual particles.

The mean square displacement $\langle u^2\rangle_T$ is of the same order of magnitude as in a crystal, but close to the transition point there is an anomalous temperature dependence of the form

$$\langle u^2\rangle_c - \langle u^2\rangle_T \propto (T_c - T)^{1/2} \tag{14}$$

with $\langle u^2\rangle_c$ being the value at T_c. The same kind of temperature anomaly is found for $f(q)$ and $f^s(q)$. This implies a finite value when approaching the transition point from below, but the slope becomes infinite. To this one should add a regular linear temperature dependence similar to that in crystals.

The macroscopic diffusion constant is strictly zero below T_c and it increases according to a power law above T_c. For the viscosity one finds close to the transition point

$$\eta \propto (T - T)^{-\gamma}, \tag{15}$$

i.e., a power law temperature dependence. It should be emphasized that the numerical value of γ and other quantities vary from system to system and cannot be predicted, except for the simplest kind of systems.

The analysis of Götze revealed some remarkable general features, which follow from the mode coupling equation above. Close to the transition point, the relaxation of $F(q,t)$ splits into three well-separated regions, characterized by three frequencies ω_D, ω_ϵ, and ω_ϵ'. Within a microscopic time domain of the order of the inverse Debye frequency $1/\omega_D$ there is a rapid decrease of $F(q,t)$ to a value near $S(q)f(q)$. This leads in $S(q,\omega)$ to a broad phonon-like spectrum which does not depend sensitively on temperature or density. The shape of this part of the spectrum depends on the details of the model and it cannot be calculated accurately without knowledge of $m_0(q,t)$ in (10).

After the initial drop, there is a slow decay on the time scale $1/\omega_\epsilon$, where $\omega_\epsilon \to 0$ as $T \to T_c$. Macroscopic diffusion plays no role here and the system behaves as if it is frozen. The dynamics is essentially the same on both sides of T_c. Within this so-called "first scaling region" we can write

$$F(q,t) = S(q)f(q) + \sqrt{|\epsilon|}h(q)f(t\omega_\epsilon) \qquad t\omega_\epsilon < 1, \tag{16}$$

where ϵ denotes the distance from the transition point, say $\epsilon = (n - n_c)/n_c$ or $\epsilon = (T - T_c)/T_c$. For the $S(q,\omega)$ it leads to a quasi-elastic peak, and in $\omega S(q,\omega)$ to a strong enhancement on the low-frequency side of the spectrum. Actually, from the calculations reported in [21] one finds a resonance peak below the main phonon part similiar to the observed β-peak (cf. Figs. 1 and 2). The form of the spectrum is independent of q and ϵ and only the amplitude depends on these quantities. The quantity $h(q)$ depends on the details of the model, but $f(\tau)$ satisfies a universal non-linear equation. The microscopic details enter only through a single parameter λ. Once this is fixed all the features of $f(\tau)$ follow uniquely from this equation, including the two scaling frequencies ω_ϵ and ω_ϵ'. For both $\tau \ll 1$ and $\tau \gg 1$, $f(\tau)$ is a power law τ^α, with different exponents, which are both related in a simple way to the parameter λ.

Beyond the "first scaling region" follows another relaxation process, which occurs on a time scale $1/\omega_\epsilon'$ with $\omega_\epsilon' \to 0$ for $T \to T_c$, but such that $\omega_\epsilon'/\omega_\epsilon \to 0$. Particle diffusion, and hence structural relaxation, plays a crucial role here. Below T_c, the function $F(q,t)$ tends to the finite asymptomatic value $S(q)f(q)$ and this leads to a strictly elastic peak in $S(q,\omega)$ with the intensity $S(q)f(q)$. Above T_c, $F(q,t)$ decreases slowly to zero and this leads to a broadening of the above elastic peak, but the area under the peak changes smoothly. An important point to notice is that the width of the peak does not go to zero for $q \to 0$, and it differs in this respect from an ordinary self-diffusion or heat diffusion peak. This is in agreement with the

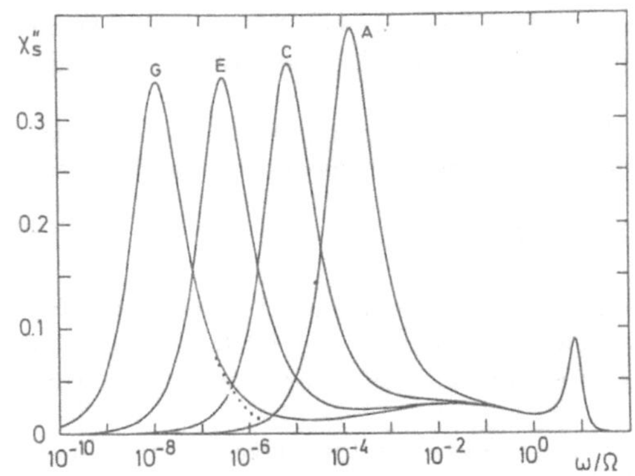

Fig. 2: $\chi_s'' \propto \omega S^s(q,\omega)$ in the supercooled liquid region from a mode coupling model. The curves $A \rightarrow G$ are for different distances from the glass transition point. The low frequency peak corresponds to the α-peak in Fig. 1 and the broad peak at higher frequencies to the β-peak. The narrow high-frequency peak (not seen in Fig. 1) is the phonon-like spectrum, which on a logarithmic frequency scale appears contracted. Notice that here a linear scale is used along the ordinate, whereas in Fig. 1 a logarithmic scale is used (taken from [21]).

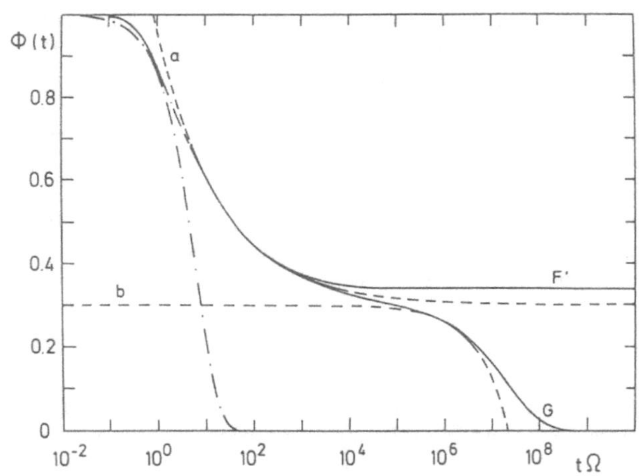

Fig. 3: Time dependence of the intermediate scattering function, normalized to unity for t=0 and here denoted by $\Phi(t)$. The microscopic time region extends to $t\Omega \approx 1$, and it is followed by the β-relaxation region extending to $t\Omega \approx 10^3$. Beyond this we have the α-relaxation region, where a large difference appears above and below the transition point (curves F' and G are for the same distance from the transition point but on opposite sides) (taken from [19c]).

experiments. The whole time dependence of $F(q,t)$ is illustrated in Figure 3. For $\omega S(q,\omega)$ we obtain a narrow low-frequency peak located below the β-peak that is analogous to the α-peak in the experiments (see Figs. 1 and 2). Numerical solutions [15b, 17] have shown that the time dependence follows very closely the Kohlrausch-William-Watts law:

$$F(q,t) = A(q) \exp\left[-(t/\tau_q)^{\beta_q}\right], \qquad t\omega'_\epsilon \geq 1, \tag{17}$$

where τ_q is proportional to $1/\omega'_\epsilon$ and goes to infinity for $T \to T_c$ from above. Calculation on a Lennard-Jones system gave a β_q around 0.7, depending somewhat on q. The numerical value is model dependent and should therefore vary from system to system. The solution of our mode coupling equation is still not fully understood in this structural relaxation region. This so-called "second scaling region" joins the first one for $1/\omega_\epsilon < t < 1/\omega'_\epsilon$, and the "first scaling region" joins the microscopic one for $1/\omega_D < t < 1/\omega_\epsilon$. It has been pointed out by Götze and Sjögren [19c] that the crucial point is not that one finds in $\omega S(q,\omega)$ two separate resonance peaks, the α- and β-peaks, but rather that there is a minimum in the spectrum separating the two regions. They make explicit predictions concerning the position of this minimum and the form of the spectrum on both sides of the minimum.

6 Comparison With Experimental Observations

It is evident that the essential features of the spectrum, predicted from the present mode coupling theory, agree with what is found experimentally. Neutron scattering has shown that these features do not disappear on passing T_g, but they also remain in the supercooled region where the viscosity is of the order $10^2 - 10^3$ poise. This is actually the reason why neutron scattering is such a powerful research tool in this context.

Taborek et al. [2] fitted viscosity data to a power law, $(T - T_c)^{-\gamma}$, for a series of glass-forming liquids. They obtained a good fit in the low viscosity region up to η of the order of $10^2 - 10^3$P, but not beyond this. However, they had to choose T_c significantly above the conventional glass transition point T_g. This proved that the present form of the mode coupling theory fails to explain the low-frequency dynamics in the large viscosity region and then obviously also inside the glass. The above authors interpreted this to mean that thermally activated relaxation starts to dominate around T_c, and that kind of process is not included in the theory. This explanation seems very plausible and it would at least qualitatively explain the discrepancies between theory and experiments at lower temperatures.

Krieger and Bosse [20b] analyzed the neutron scattering data of Mezei et al. [4] on $Ca_{0.4}K_{0.6}(NO_3)_{1.4}$, using a simple version of a two-component mode coupling theory. The parameter λ, which has been introduced before, was determined from fitting measured viscosity data to a power law. It gave $\gamma = 4.2$. The intermediate scattering function was then calculated (full curve in Fig. 4) and its form was compared with the experimental data (dots).

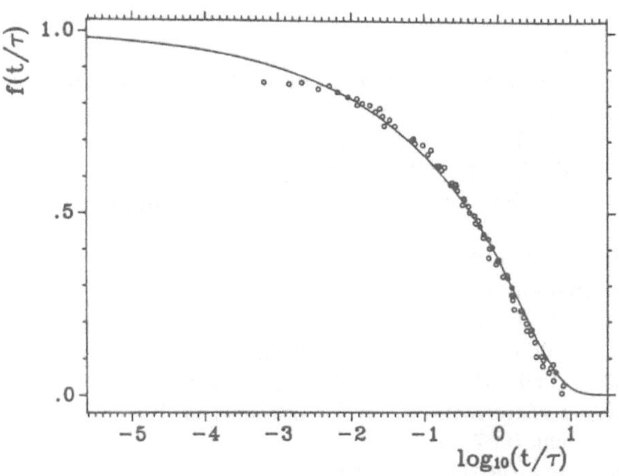

Fig. 4: Intermediate scattering function for $Ca_{0.4}K_{0.6}(NO_3)_{1.4}$, obtained from a mode coupling model, compared with experimental data of Mezei et al. (taken from [20b])

The agreement is quite remarkable. A more detailed comparison between theory and neutron scattering results is presented in a more recent paper [4c]. Measurements of the temperature dependence of $S(q)f(q)$ have been reported by several groups [4-7] and these results seem to show a singular behavior of the form $(T_c - T)^{1/2}$ with $T_c > T_g$, but no anomaly at T_g. Reference [7] is of particular interest in this connection. It should be noticed that as long as the width of the α-peak is less than the experimental resolution, one measures the area under this peak, *i.e.*, $S(q)f(q)$, and according to the theory this has the above-mentioned singularity below Tc and stays essentially constant above.

Götze and Sjögren [19a,c] have suggested a modification of the theory by adding a coupling to the current density. Through this, a new parameter δ enters, which has the effect of restoring ergodicity and hence erasing all unphysical mathematical infinities. Some defects of the earlier model are then removed. One can then also understand why some liquids like SiO_2, do not show any singular behaviour in the viscosity, for instance. However, there still remain some basic questions to be resolved.

7 Concluding Remarks

The aim of this paper has been to review the present status of mode coupling approach for supercooled liquids and for liquid-glass transitions and to relate it to what is observed experimentally. Even though the theory does not carry us all the way to the conventional glass transition region and below, it has predicted some unexpected behaviour of the supercooled liquid. Recent neutron scattering experiments, in particular, seem to confirm the essential predictions and that most of the dynamics, normally associated with the region very close to the glass transition point, are also observed quite far above.

Since the real glass-forming systems are far from simple, a quantative comparison between theory and experiments has not been possible. In this respect one may hope that computer simulations on simpler systems can be brought to such a stste that detailed tests of the theory can be carried out. Molecular dynamics calculations by Ullo and Yip [23], and particularly by the French group [24-26], on simpler systems are of great interest and call for continuation.

Concerning the theory, there remains the challenging task of incorporating thermally activated processes into the theory. Whether this can be possible within a mode coupling approach is an open question.

References

1. For a recent survey of the field see: S. Brawer: In *Relaxation in Viscous Liquids and Glasses*, Amer. Ceram. Soc. Inc. (Columbus, Ohio 1985)
2. P. Taborek, R.N. Kleinman, D.J. Bishop: Phys. Rev. B **34**, 1835 (1986).
3. C.A. Angell, W. Sichina: Ann. NY Acad. Sci **275**, 53 (1976).
4. F. Mezei, W. Knaak, B. Farago: (a) Phys. Rev. Lett. **58**, 571 (1987); (b) Physica Scr. T **19** 363 (1987); (c) preprint (1988).
5. F. Fujara, W. Petry: Europhys. Lett. **4**, 921 (1987).
6. B. Frick, D. Richter, W. Petry, U. Buchenau: Z. Phys. B **70**, 73 (1988).
7. D. Richter, B. Frick, B. Farago:Phys. Rev. Lett. **61**, 2465 (1988)
8. H. Mori: Progr. Theor. Phys. **33**, 423 (1965); *ibid.* **34**, 399 (1965).
9. R. Zwanzig: J. Chem. Phys. **33**, 1338 (1960); Phys. Rev. **124**, 983 (1961).
10. A. Sjölander: In *Amorphous and Liquid Materials*, ed. by E. Lüscher, G. Frisch and G. Jacucci (Martinus Nijhoff Publ., Dordrecht 1987) p. 239.
11. U. Bengtzelius, L. Sjögren: J. Chem. Phys. **84**, 1744 (1986).
12. E. Leutheusser: Phys. Rev. A **29**, 2765 (1984).
13. U. Bengtzelius, W. Götze, A. Sjölander: J. Phys. C **17**, 5915 (1984).

14. L. Sjögren: Phys. Rev. A **22**, 2883 (1980).
15. U. Bengtzelius: (a) Phys. Rev. A **33**, 3433 (1986); (b) A **34**, 5059 (1986).
16. W. Götze: Z. Phys. B **60**, 195 (1985); in *Amorphous and Liquid Materials,* ed. by E. Lüscher, G. Frisch and G. Jacucci (Martinus Nijhoff Publ., Dordrecht 1987) p. 34.
17. H. DeRaedt, W. Götze: J. Phys. C **19**, 2607 (1986).
18. J. Bosse, U. Krieger: J. Phys. C **19**, L609 (1986).
19. W. Götze, L. Sjögren: (a) Z. Phys. B **65**, 415 (1987); (b) J. Phys. C **20**, 879 (1987); (c) *ibid.* C **21**, 3407 (1988).
20. U. Krieger, J. Bosse: in *Amorphous and Liquid Materials,* ed. by E. Lüscher, G. Frisch and G. Jacucci (Martinus Nijhoff Publ., Dordrecht 1987) p.139; (b) Phys. Rev. Lett. **59**, 1601 (1987).
21. G. Buchalla, U. Dersch, W. Götze, L. Sjögren: J. Phys. C **21**, 4239 (1988).
22. G.P. Johari: Ann. NY Acad. Sci. **279**, 117 (1976).
23. J.J. Ullo, S. Yip, Phys. Rev. Lett. **54**, 1509 (1985).
24. B. Bernu, J.P. Hansen, Y. Hiwatari, G. Pastore: Phys. Rev. A **36**, 4891 (1987).
25. H. Miyagawa, Y. Hiwatari, B. Bernu, J.P. Hansen: J. Chem. Phys. **88**, 3879 (1988).
26. G. Pastore, B. Bernu, J.P. Hansen, Y. Hiwatari: Phys. Rev. A **38**, 454 (1988).

Scaling Behaviour of the Density Fluctuations near the Glass Transition: A Signature of Critical Instability

F. Mezei

Hahn-Meitner-Institut Berlin, Postfach 390128, D-1000 Berlin 39, Germany

1. Introduction

Recent theoretical considerations /1/ suggest that the liquid-glass transition might be related to a genuine instability, quite similarly to ordinary phase transitions. This instability has to be of dynamic character, since there definitely is no discontinuity in the thermodynamic parameters. A dynamic phase transition only displays one common feature with thermodynamic phase transitions: the non-ergodic nature of at least one of the phases involved. In ordinary phase transitions non-ergodicity is related to a symmetry breaking, e.g. the breaking of translational symmetry in a crystal implies that only one subset of equivalent states differing by an arbitrary translation is realized by the system. The broken symmetry also allows us to identify the order parameter(s), e.g. the Bragg peak structure-factors. In contrast, there is no difference in symmetry between the glass and the liquid phases. Therefore, no conventional order parameter distinguishes them from each other. In this situation, as it has been proposed for spin glasses /2/, the order parameter can be identified with the degree of non-ergodicity itself.

The non-ergodicity of the glass phase corresponds to its solidity, i.e. atoms remember their relative positions for a long time. This is mathematically expressed by the non-vanishing $t\to\infty$ limit of the density-density correlation function, and for convenience we shall consider the wavenumber dependent Fourier representation of this correlation function $\Phi(q,t)$. The liquid (or gas) state corresponds to $\Phi(q,t\to\infty)=0$ for $q\neq0$ and the solid state can be represented by $\Phi(q,t\to\infty)=f(q)\neq0$, at least for some non-vanishing q values. In a crystalline material $f(q)$ is non-vanishing at a few well-defined q values only, while in a glass $f(q)\neq0$ for any value of q. By various authors $f(q)$ is either called non-ergodicity parameter or elastic Debye-Waller factor. Its definition is the exact analogon of the Edwards-Anderson order parameter in spin glasses, and it plays a similar role. (Note that for convenience we take $\Phi(q,t)$ to be normalized as $\Phi(q,t=0)=1$. The full density-density correlation function is then given by $S(q)\Phi(q,t)$ where $S(q)$ is the static structure factor.) In a quantum theory $\Phi(q,t)$ is in general not real and can be replaced by the real Kubo relaxation function. Alternatively, we can consider the classical approximation of $\Phi(q,t)$, which is real and leads to symmetrized scattering functions, such as $S(q,t) = S(q)\int \Phi(q,t)\cos(\omega t)dt$, which violate detailed balance. This latter approach will be adopted here following the recipe of P. Schofield /3/.

Springer Proceedings in Physics, Vol. 40 Static and Dynamic Properties of Liquids
Editors: M. Davidović and A.K. Soper

With all thermodynamic parameters changing in a continuous fashion across the glass transition, we have to look for a signature of instability in the form of a singularity in the variation of various dynamic properties, similar in nature to the power laws and scaling behaviour characterizing the ordinary second order phase transitions. As we shall see in more detail below, recent theories have identified several features of this type in the temperature and time dependence of $\Phi(q,t)$, together with a number of far from trivial scaling relations between the various exponents involved. Because of the dynamic nature of the transition, the correlation function $\Phi(q,t)$ is the most directly relevant quantity to look at. This singles out neutron scattering as the best adapted method for such investigations. (Note that the coherent neutron scattering function for liquids containing several atomic species is not exactly the density-density correlation function. Instead, the contribution of each nucleus is weighted by its neutron scattering amplitude. This is not expected to change the scaling behaviour in an essential manner, and in what follows we shall tentatively identify $\Phi(q,t)$ with the correlation function relevant for coherent neutron scattering.)

In the next section I shall describe the results of Neutron Spin Echo (NSE) and Time-of-Flight (TOF) experiments in the ionic system $Ca_{0.4}K_{0.6}(NO_3)_{1.4}$. This system is an archetypical, fragile (non-network) glassformer, which has previously been studied with macroscopic ($q\sim0$) methods in great detail /4/. The presentation of the results will focus on signatures of scaling. This will be followed by a quantitative comparison to theory.

2. Experimental Results

Figure 1 shows the temperature dependence of $\Phi(q,t)$ measured around a fix q value (the q resolution is 20% FWHM) near the peak of the structure factor at $q_0 = 1.86$ Å^{-1} /5/. (The q dependence of $\Phi(q,t)$ was found to be slight and smoothly oscillatory /6/.) The data points are directly measured NSE results and the continuous lines represent the Fourier transformed symmetrized spectra in the time range corresponding to the energies explored in the TOF experiment. Note that at energies much smaller than kT, in particular in the range of the NSE scans, the spectra are symmetric anyway. Thus the Schofield recipe does not introduce any error in matching the TOF and NSE data. On the other hand, the normalization of $\Phi(q,t)$ to 1 at t=0 is strongly influenced by the symmetrization, and therefore Fig.1 does not reflect the real intensity relations between low and high energy parts of the spectra. The procedure of merging the NSE and TOF spectra in order to obtain an extremely wide dynamic range (nearly 5 orders of magnitude in Fig.1) is rather delicate and more details have been given elsewhere /5/. The rather flat sections of $\Phi(q,t)$ correspond to low spectral densities and can better be visualized in the energy domain or in the directly measured TOF spectra (Fig.2).

The spectra in Fig. 1 clearly show a tendency towards non-ergodicity: On cooling $\Phi(q,t)$ relaxes to zero at dramatically longer and longer times. At 110° C and at 130° C it is also clear that there are two relaxation steps in $\Phi(q,t)$, a faster ("first") one for times shorter than 10^{-12} sec, and a slower ("second") one which accounts for the major part (some 80%) of the amplitude.

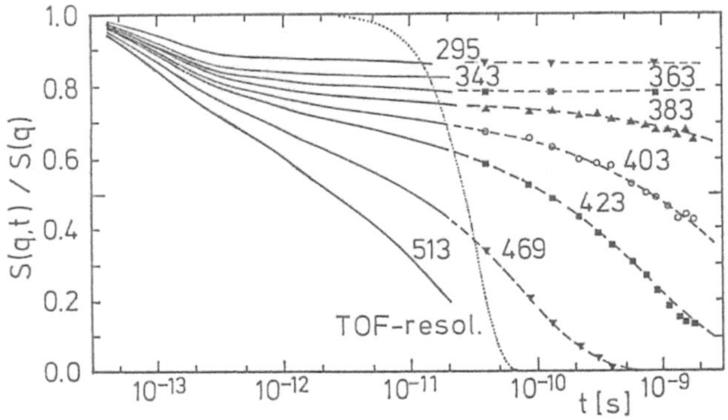

Fig. 1. Time dependence of the (neutron scattering) density-density correlations in $Ca_{0.4}K_{0.6}(NO_3)_{1.4}$ around the wave vector $q \simeq 1.9$ Å^{-1} at various temperatures indicated in K units /5/. The continuous lines are Fourier transformed symmetrized TOF spectra corrected for the resolution (dotted line) and the data points are directly measured NSE results. Above 370 K the dashed lines show the scaled Kohlrausch function representation of the results, below this temperature they are just guides to the eye.

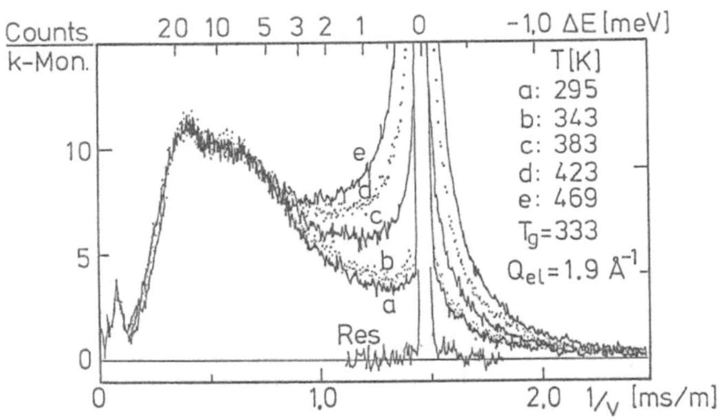

Fig. 2. The directly measured TOF spectra at various temperatures, normalized to each other by division by the Bose factor /5/. The high energy (>5 meV) phonon-like part of the spectra is seen to be independent of the temperature across the glass transition, while at smaller energies there is a clear increase of the spectral density above the conventional glass transition temperature of about 60° C (333 K). The spectrum taken at 200 K (not shown) is identical with that at 295 K.

(The second step can unambiguously be identified with the so-called primary or α-relaxation observed in macroscopic measurements and the first one may be related to high frequency so-called β-relaxation processes). A quantitative analysis of the second relaxation step /7/ reveals that it can very well be reproduced by a simple stretched exponential (Kohlrausch) master function and power law scaling as follows:

$$\Phi(q,t) = (1-h(q))\exp(-(t/\tau)^{\beta}) + h(q)F(q,t), \tag{1}$$

where $h(q)=0.24$ at $q \simeq q_0$ and $\beta=0.58$ are fitted constants, $F(q,t)$ represents the first relaxation step ($F(q,t) \simeq 0$ for $t>10^{-11}$ sec) and the effective relaxation time τ can be represented by a power law

$$\tau = \tau_0(q)(T-T_0)^{-\gamma}. \tag{2}$$

The quality of the scaling behaviour is illustrated in Fig.3 showing all data points observed between 110 and 200° C and plotted against the scaled time t/τ. The continuous line is the master function proportional to the Kohlrausch law as given in (1).

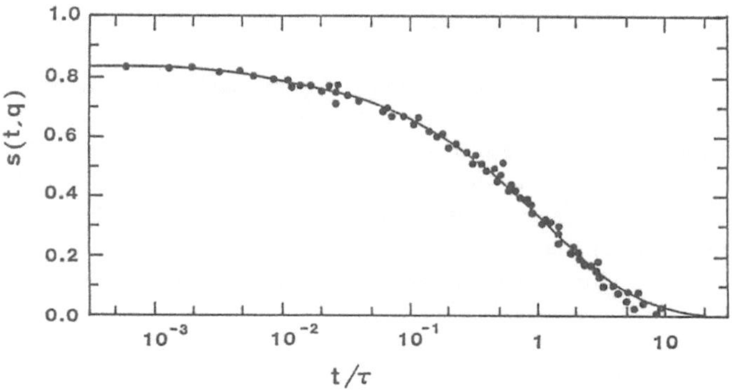

Fig. 3. The decay of the density correlations at $q \simeq 1.9$ Å^{-1} in the time domain of the primary (α) relaxation step plotted against the scaled time /7/. The dots represent all NSE data points (including those in Fig.1) taken at 7 different temperatures between 111 and 196° C. The time scaling parameter τ has been calculated from power law (2) with $\gamma=4.1$ and $\tau_0(q)=0.012$ s for T measured in K. The line is proportional with the stretched exponential (Kohlrausch) function with $\beta=0.58$.

An extension of the experiments to the wider q range of 0.2 to 2.1 Å has shown that $h(q)$ and $\tau_0(q)$ are roughly q independent with smooth oscillatory fluctuations /6/, while β and γ are q independent within an experimental uncertainty of about ±0.1 /8/. Furthermore, the macroscopic relaxation times at q=0 have been observed to agree within a factor of 3 with the τ values found in the neutron scattering experiments /9/. Actually, τ has been found to be proportional to the macroscopic shear viscosity η which is a fundamental signature of the α-relaxation.

However, the power law fit (2) to the measured τ values is not unambiguous. The results obtained in the temperature range of 110° to 200° C are very well reproduced by any choice of T_0 between 80° and 100° C (with γ values changing from 5.0±0.2 to 3.8±0.15, respectively, as a function of T_0.) The dashed lines in Fig.1 show the curves calculated using (1) and (2) with a choice of $T_0=93°$ C, $\gamma=4.1$, $\beta=0.58$, $h(q)=0.24$, and $\tau_0(q)=0.012$ sec if T is measured in K units. ($F(q,t)$ is supposed to be 1 at t=0 and to decay to approximately zero before the time domain of the NSE scans, i.e. $t>4 \cdot 10^{-11}$ sec). The reasons for adopting this particular assumption for T_0 will be discussed below. Note that the range of possible choices of T_0, i.e. 80-100° C in any case lies above the conventional glass transition temperature $T_g \simeq 60°$ C.

The predicted scaling behaviour of the first relaxation step can best be observed in the imaginary part of the frequency dependent susceptibility defined as

$$\chi''(q,\omega) = \omega\, S(q,\omega)/kT \qquad (3)$$

In Fig.4 the susceptibility is shown by the continuous line representing the Fourier transform of the lines (continuous and dotted) of Fig.1, together with directly obtainable TOF data points. The energy ω_ϵ of the minimum of these curves can be considered as a characteristic energy and it is found to be consistent with a power law

$$\omega_\epsilon \propto (T-T_0)^\delta \qquad (4)$$

where δ=2.3±0.3 for the above choice of T_0=93° C. (ω_ϵ is compatible with a power law (4) if 75<T_0<100° C.) Furthermore, the shape of the curves on the logarithmic scales of Fig.4 is also similar around the minimum, i.e. we can talk of a master function tending to power laws on both sides (straight sections), although the range of validity of these power laws is rather limited by the proximity of the microscopic collision regime (E~10 meV) represented by the phonon like part of the spectra, and also present at low temperatures (cf. Fig.2).

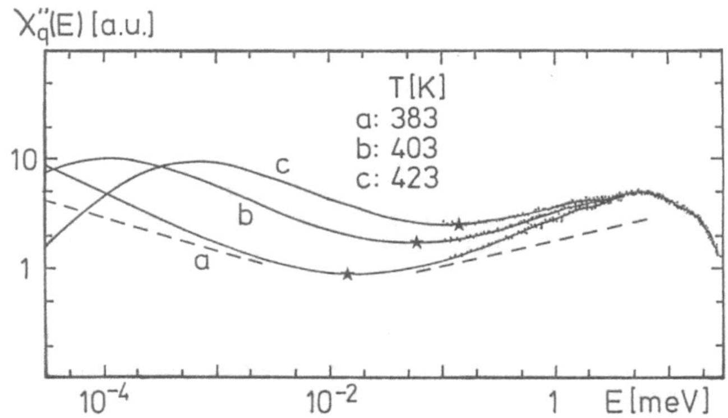

Fig. 4. Frequency (energy) dependence of the susceptibility as deduced from the neutron scattering results /5/. The lines correspond to the Fourier transforms of the spectra shown in Fig.1 as represented by the continuous and dashed lines. In addition, the points show data directly calculated from the TOF spectra. The stars indicate the minima and the dashed lines on the left and right hand sides illustrate the predicted slopes corresponding to power laws (7b) and (7a), respectively, with exponents −0.3 and 0.23.

Thus, we can conclude at this point that the results qualitatively confirm the main predictions of the theory, i.e. that the slowing-down towards non-ergodicity in the liquid phase displays two relaxation steps which are governed by two master functions with two scaling frequencies ω_ϵ and τ^{-1}, respectively, which both follow critical type power laws with a common critical temperature T_0. This amounts to saying that indeed, on cooling the supercooled liquid phase becomes critically unstable against a dynamic ergodic to non-ergodic phase transition. In the next section we shall examine the scaling behaviour in more quantitative terms and discuss the present limits of validity of the theoretical predictions.

3. Comparison with Theory

The general formulation of the mode coupling approach of generalized hydrodynamic theory predicts a few very special relations /11/ between the exponents we were able to identify. Namely, if we represent the exponents γ and δ of (2) and (4) in terms of two positive numbers a and b as

$$\delta = \frac{1}{2a} \tag{5a}$$

$$\gamma = \frac{1}{2a} + \frac{1}{2b} \tag{5b}$$,

a and b have to be the two solutions of the transcendental equation

$$\lambda = \Gamma^2(1+x)/\Gamma(1+2x) \tag{6}$$,

where λ is a free adjustable parameter and Γ is the gamma function. It can easily be verified that with our values δ=2.3 and γ=4.1, i.e. a=0.23, and b=0.30 the fundamental scaling relation (6) is fulfilled. In fact, this is only true if we choose $T_0 \simeq 93°$ C. With other choices we would obtain γ and δ pairs, which would violate (6).

The general theory also predicts that the master function for the shape of the first relaxation step F(q,t) is simply related to a and b. The susceptibility scaling function $\hat{\omega}\tilde{F}(q,\hat{\omega})$, where $\hat{\omega}=\omega/\omega_\epsilon$ and $\tilde{F}(q,\hat{\omega})$ is the Fourier transform of $F(q,t\omega_\epsilon)$, is expected to display the assymptotic behaviours

$$\frac{1}{\sqrt{T-T_0}}\,\hat{\omega}\tilde{F}(q,\hat{\omega})=\hat{\omega}^{\,a} \qquad \text{for} \qquad 1 \ll \hat{\omega} \ll \frac{\Omega}{\omega_\epsilon}, \tag{7a}$$

$$\frac{1}{\sqrt{T-T_0}}\,\hat{\omega}\tilde{F}(q,\hat{\omega})=\hat{\omega}^{-b} \qquad \text{for} \qquad \frac{1}{\tau\omega_\epsilon} \ll \hat{\omega} \ll 1. \tag{7b}$$

The slopes of the straight lines corresponding to these power laws are indicated in Fig.4 with a=0.23 and b=0.30. It is clear that the observed curves show distinctively higher slopes. Furthermore, the value of the minimum drops more like $(T-T_0)$ with T approaching T_0, rather than $(\sqrt{T-T_0})$, as predicted by (7a) and (7b). However, these discrepancies can be due to an ambiguity in the data evaluation process. The energy ω_ϵ (minima in Fig.4) falls just between the scan ranges of the two experimental methods, NSE and TOF, we used. Therefore, the spectra are quite sensitive in this range to the normalization of one set of data with respect to the other. In Ref. /5/ this normalization was obtained by just smoothly joining the Fourier transformed TOF results (continuous lines in Fig.1) to the NSE ones. Smoothness can be achieved by matching the scales of the TOF and NSE signals within a 2-3% window. The exact choice within this window rather drastically influences the depth of the deepest minima in Fig.4. A more probing test of the scaling relations (7a) and (7b) will become feasible after complementary experiments presently in progress have been completed. These experiments are designed to make scans right across this intermediary domain between the NSE and TOF data now available. This is expected to remove most of the present ambiguities.

Note that the general theory does not predict a scaling relation between the fundamental exponents a and b and the effective Kohlrausch exponent β. Such relations could only be numerically established in the framework of schematic models, i.e. by assuming some simple form for the non-linear coupling. Thus, Krieger and Bosse /12/ were able to deduce an effective β value for a given exponent γ within a schematic two component model of ionic liquids. Their result is in perfect agreement with our observation of $\beta \simeq 0.58$ for $\gamma = 4.1$. This is quite remarkable in view of the crude approximation, even if their value of h(q) is clearly incompatible with the experimental data.

The most intriguing open question is what happens between $T_0 \simeq 93°$ C and $T_g \simeq 60°$ C. Fig.5 shows the change of the scattering intensity as a function of temperature as observed in the TOF experiment at various energy transfers. It is clear that there is a change of slope at all energies just around T_g, and the extra increase of intensity above this temperature corresponds to the "first" relaxation step, the scaling behaviour of which above T_0 has just been discussed. Unfortunately, below T_0 we are unable to identify a characteristic energy scale ω_ϵ in view of the lack of any clear signature. Theoretically it has also become clear /13/ that the general hydrodynamic theory cannot be applied in the vicinity of T_g, since it neglects thermally activated atomic processes, such as jump diffusion. A possible interpretation of the difference between T_g and T_0 is that the supercooled liquid is only inherently dynamically stable above T_0,

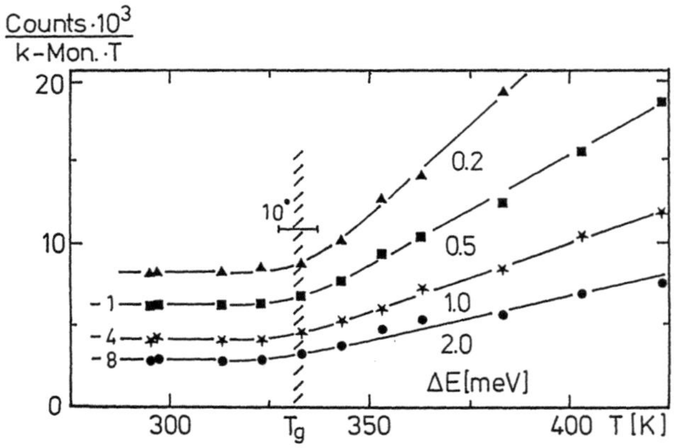

Fig. 5. The temperature dependence of the neutron scattering intensity at various energy transfers ΔE as indicated /5/. The spectra have been normalized by division by the Bose factor in order to remove the trivial part of the temperature dependence. There is a clear, energy independent signature of the conventional glass transition temperature. In order to avoid overlap, some lines are shifted downward in the figure by the amount indicated at their left hand end.

i.e. on approaching from above T_0 there appears to be a singularity. However, in the vicinity of T_0 thermally activated processes stabilize the liquid state for a while, i.e. until T_g is reached. Then the transition at T_g is a smooth, activated phenomenon, without any singularity or scaling.

4. Conclusion

Direct neutron scattering observation of the density-density fluctuations in a typical fragile glassforming system has revealed scaling behaviour and a tendency towards power law singularities with a critical temperature T_0 well above the conventional glass transition temperature T_g . This supports recent theoretical results showing that, on cooling, the supercooled liquid state becomes critically unstable against a dynamic ergodic to non-ergodic phase transition. It is of particular significance that the value of T_0 can be chosen such that the measured exponents fulfill the theoretically predicted scaling relations. This particular choice of T_0 is independently confirmed by the observations that it is compatible with both the predicted sharp drop of f(q) on approaching T_0 from below /6/ and the similar drop of the susceptibility minimum on approaching T_0 from above (cf. Fig. 4) /8/. Thus, we can conclude that in a quite remarkable fashion the observed signatures of a critical dynamic instability of the supercooled liquid state also obey the main scaling relations predicted, and deviation from predicted implications of scaling on the shape of the spectra might either be real or related to experimental ambiguities. A detailed theoretical and experimental understanding of the behaviour in the temperature range between T_0 and T_g and the physical reason for the difference between these two temperatures are the most intriguing outstanding problems.

5. References

1. Geszti, T., J. Phys. C 16, 5805 (1983), E. Leutheusser, Phys. Rev. A29, 2765 (1984), U. Bengtzelius, W. Götze and A. Sjölander, J. Phys. C 17, 5915 (1984), S.P. Das, G.F. Mazenko, S. Ramaswamy and F.F. Toner, Phys. Rev. Lett. 54, 118 (1985).
2. S.A. Edwards and P.W. Anderson, J. Phys. F 5, 965 (1975).
3. Schofield, P. Phys. Rev. Lett. 4, 239 (1960).
4. K.F. Rao, D.B. Helphrey, and C.A. Angell, Phys. Chem. Glasses 14, 26 (1973), L.M. Torell and R.J. Aronsson, Chem. Phys. 78, 1121 (1983).
5. W. Knaak, F. Mezei and B. Farago, Europhys. Lett. 7, 529 (1988).
6. F. Mezei, W. Knaak and B. Farago, Phys. Scr. 19, 363 (1987).
7. F. Mezei, W. Knaak and B. Farago, Phys. Rev. Lett. 58, 571 (1987).
8. F. Mezei (in press).
9. C.A. Angell and L.M. Torell, J. Chem. Phys. 78, 937 (1983).
10. G.M. Glower, A. Matheson, Trans Frad. Soc. 67, 1960 (1971).
11. W. Götze, Z. f. Phys. B 56, 139 (1984); ibid. 60, 195 (1975).
12. U. Krieger, J. Bosse, Phys. Rev. Lett. 59, 1601 (1987).
13. W. Götze and L. Sjögren, J. Phys. C (in press).

Structural Relaxation and Glass Transition of Binary Liquids

J. Bosse

Institut für Atom- und Festkörperphysik, Freie Universität Berlin, D-1000 Berlin 33, Germany

Introduction

Since it was discovered by Leutheußer /1/ and Bengtzelius, Götze, and Sjölander /2/ that the mode-coupling approximation (MCA) in simple classical liquids contains a singularity describing the transition from ergodic to non-ergodic dynamics and their conjecture that this is the liquid to glass transition /3/, there have been many theoretical papers /4-13/ which discuss various implications of MCA for the dynamics of supercooled liquids and the glass transition (see Ref. 9 for a review).

I want to report here on the application of MCA to binary liquids /10-14/ with the aim of investigating the glass transition and related problems. The study of multi-component systems is not only of interest because conventional glass formers are made up of several species of particles, there can also occur interesting additional dynamical phenomena absent in simple (one-component) liquids. While in the latter the glass transition is accompanied with the localization of all particles, it is possible in a binary system that one species of particles retains its mobility, while the other forms a glassy matrix. One could, e.g., think of a glassy structure containing small particles, which undergo a percolation transition as temperature is raised, or of an ionic glass showing an insulator-conductor transition. Such a dynamic transition within the amorphous solid state in addition to the glass transition is, of course, of great interest by itself. Moreover, fascinating new questions arise concerning the influence exerted by one transition on the other.

Non-ergodicity parameters and glass transition

The basic dynamical variables in a mixture of $N = \sum_s N_s$ classical particles in volume V are the Fourier-transformed number densities

$$N_s(\vec{q};\, t) = \sum_{1< j <N_s} \exp(-i\vec{q}\cdot\vec{r}_j^{(s)}(t)) \qquad (1)$$

Springer Proceedings in Physics, Vol. 40 Editors: M. Davidović and A.K. Soper

Static and Dynamic Properties of Liquids

with $\vec{r}_j^{(s)}(t)$ denoting the position of the jth particle of species s at time t. Now, let us assume a homogeneous, isotropic equilibrium state of the system throughout, thus explicitly excluding the possibility of crystallization. This implies the thermal equilibrium averages

$$\langle N_s (\vec{q}) \rangle = N_s \delta_{\vec{q}\vec{o}} \tag{2}$$

to vanish for q ≠ 0 at all temperatures T. Moreover, any small initial density fluctuation of wavevector $\vec{q}$ will relax separately for time t > 0 according to linear response theory:

$$\langle \delta N_s(\vec{q}) \rangle_t = \sum_{\bar{s}\bar{s}'} \Phi(q;t)_{s\bar{s}} \Phi^{-1}(q,o)_{\bar{s}\bar{s}'} \langle \delta N_{s'}(\vec{q}) \rangle_{t=o} . \tag{3}$$

Here $\delta N_s (\vec{q}) = N_s(\vec{q}) - \langle N_s(\vec{q}) \rangle$, brackets with time index $\langle \dots \rangle_t$ denote non-equilibrium averages and

$$\Phi (q;t)_{ss'} = \langle \delta N_s(\vec{q};t)^* \delta N_{s'}(\vec{q}) \rangle /(k_B T) \tag{4}$$

is the classical version of Kubo's relaxation function. At time t = 0 the latter is determined by the static structure factors

$$\Phi (q,o)_{ss'} = S_{ss'}(q) (N_s N_{s'})^{1/2}/(k_B T). \tag{5}$$

In the liquid state, the system will return to equilibrium within a microscopically short time, whenever some small perturbation has created a density fluctuation, i.e. $\Phi (q;t)_{ss'}$ will decay to zero within a few picoseconds. If a liquid is, however, supercooled and driven into the glassy state, it has been suggested and demonstrated for one-component liquids that this transition is accompanied with a breaking of ergodicity signalled by a non-zero long-time limit of the density relaxation function /1-9,15/. This idea is now applied to the liquid mixture. We define the matrix of non-ergodicity parameters /12,14/

$$f (q)_{ss'} = \lim_{t\to\infty} \phi(q,t)_{ss'} \tag{6}$$

with the normalized relaxation functions

$$\phi(q,t)_{ss'} = \Phi(q,t)_{ss'}/[\Phi(q,o)_{ss} \Phi(q,o)_{s's'}]^{1/2},$$

$f(q)_{ss'}$ is a real, symmetric, positive definite matrix obeying

$$o \le f (q)_{ss'} \le 1 \tag{7}$$

$$f (q)^2_{ss'} \le f (q)_{ss} f (q)_{s's'} . \tag{8}$$

If all $f(q)_{ss'}$ vanish identically, Eq.(3) implies that any small initial density fluctuation $\langle \delta N_s(\vec{q}) \rangle_{t=0}$ will smear out into the homogeneous equilibrium density in the course of time. $\langle N_s(\vec{q}) \rangle_{t=\infty}$ equals the unique equilibrium value $\langle N_s(\vec{q}) \rangle$. This behavior-reflecting ergodic dynamics-is, of course, what we expect for the liquid state, i.e. for temperatures higher than some critical temperature T_0. As will be shown below, for $T \leq T_0$ $f(q)_{ss'}$ will become non-zero. Then, according to Eq.(3) any initial deviation from the homogeneous density will result in a density fluctuation which sustains even at long times. Density fluctuations are arrested, and there will be no unique equilibrium state of the system for $T \leq T_0$. The density $\langle N_s(\vec{q}) \rangle_{t=\infty}$ will be inhomogeneous and it will depend on the initial value of the density fluctuation, i.e. the present state of the system will depend on its remote past. This behavior of density fluctuations appears to be the most outstanding feature of the glassy state of the mixture. Therefore, for $T \leq T_0$ the system is called a glass. A remark seems necessary here. If we take the long-time limit literally as $t = \infty$, we must stop here and think anew about the statistical mechanics of the glassy state, in which density fluctuations are arrested in the way described above. Our theoretical description would then break down at T_0, and the quantities $f(q)_{ss'}$ for $T \leq T_0$ would have no physical relevance. If, however, we may interpret "$t = \infty$" as some large (but finite) multiple of the experimental time scale (~ 1 ps for neutrons) admitting that any density fluctuation will - at still later times - eventually decay to zero due to a residual mobility of the particles in the glass, then the same theoretical framework may be used on both sides of the "transition". This interpretation is supported by the fact that the static structure does not change considerably when passing through T_0. Then, the matrix of non-ergodicity parameters (determining the heights of extremely long-lasting plateau values in the relaxation functions) is of direct experimental relevance: it determines the cross-section for (quasi-) elastic coherent scattering, thus generalizing to a multicomponent glass the concept of Debye-Waller Factor (DWF). Indeed, the scattering cross-section in Born approximation for energy- and momentum transfer, $\hbar\omega$ and $\hbar\vec{q}$, resp. is

$$\frac{d^2\sigma}{d\Omega d\omega} \propto \int_{-\infty}^{+\infty} dt\ e^{it\omega}\{ \sum_{ss'} (b_s^{coh})^* \langle N_s(-\vec{q};t)^* N_{s'}(-\vec{q}) \rangle b_{s'}^{coh} + \sum_s N_s\ |b_s^{inc}|^2 \psi^{(s)}(\vec{q};t)\phi_s(q;t)\} \tag{9}$$

with b_s^{coh} and b_s^{inc} the coherent and incoherent scattering lengths, $\psi^{(s)}(\vec{q},\omega) = \int dt e^{it\omega}\ \psi^{(s)}\ (\vec{q},t)$ the excitation spectrum of s-particles' internal degrees of freedom, and $\phi_s(q,t) = \langle \exp(-i\,\vec{q}\cdot\vec{r}_o^{(s)}(t))\exp(i\vec{q}\cdot\vec{r}_o^{(s)}) \rangle$ the normalized tagged-particle density relaxation function. Thus, in a glass ($T < T_0$) the coherent elastic scattering contribution for $q \neq 0$ is determined by the long-time limit of $\Phi\,(q,t)_{ss'}$,

$$\left(\frac{d^2\sigma}{d\Omega d\omega}\right)_{elastic}^{coherent} \propto \sum_{ss'} (b_s^{coh})^{*} \sqrt{N_s S_{ss}(q)}\; f(q)_{ss'}\; \sqrt{N_{s'} S_{s's'}(q)}\; b_{s'}^{coh}\; \delta(\omega), \quad (10)$$

while the long-time limit of the tagged-particle relaxation functions, $f_s(q) = \lim_{t\to\infty} \phi_s\,(q,t)$, determines the incoherent recoilless scattering contribution

$$\left(\frac{d^2\sigma}{d\Omega d\omega}\right)_{recoilless}^{incoherent} \propto \sum_s N_s\, |b_s^{inc}|^2\, \psi^{(s)}(\vec{q};\omega)\; f_s(q) \quad (11)$$

i.e. $f_s\,(q)$ is the Lamb-Mößbauer Factor (LMF) of species s, while $f\,(q)_{ss}$ is its DWF.

Mode Coupling Approximation (MCA)

The determination of the non-ergodicity parameters as well as the dynamical behavior close to the glass transition is achieved by numerical solution of approximate equations of motion. These are derived from the exact Mori-Zwanzig equations for a set of relaxation functions /16/

$$\phi(t)_{\mu\nu} = \langle\, \delta A_\mu(t)^{*} \delta A_\nu(o)\, \rangle/(k_B T) \equiv (A_\mu|e^{-it\mathcal{L}}|A_\nu), \quad (12)$$

which read

$$\dot{\phi}(t)_{\mu\nu} + i \sum_\sigma \omega_{\mu\sigma}\, \phi(t)_{\sigma\nu} + \sum_\sigma \int_0^t dt'\; m(t-t')_{\mu\sigma}\, \phi(t')_{\sigma\nu} = o \quad (13)$$

with initial conditions /17/ $\phi\,(t = 0)_{\mu\nu} = \delta_{\mu\nu}$ and the abbreviations

$$\omega_{\mu\nu} = (A_\mu|\mathcal{L}|A_\nu) \quad (14)$$

$$m\,(t)_{\mu\nu} = (\tilde{A}_\mu|e^{-i\mathcal{L}_A}|\tilde{A}_\nu),\; \mathcal{L}_A = Q_A\, \mathcal{L}\, Q_A,\; Q_A = 1-\sum_\sigma |A_\sigma)(A_\sigma| \quad (15)$$

$$|\tilde{A}_\mu) = Q_A\, \mathcal{L}|A), \quad (16)$$

where Mori's scalar-product notation /16/ has been introduced in Eq.(12). The coupling of variables $|A_\mu)$ to the product variables $|A_\sigma A_\lambda^+)$ is now taken into account explicitly in Eq.(15) by making use of the identity Eq.(A3) of Ref. 18 with the orthonormalized set

$$|B_{\sigma\lambda}\,) = \sum_{\sigma'\lambda'} Q_A |A_{\sigma'} A_{\lambda'}^+)\; (AA^+|Q_A|AA^+)^{-1/2}_{\sigma'\lambda';\sigma\lambda}\;. \quad (17)$$

This leads to a decomposition of the memory functions

$$m(t)_{\mu\nu} = m(t)_{\mu\nu}^{MC} + \tilde{m}(t)_{\mu\nu} \tag{18}$$

into the mode-coupling contribution and a remainder. The former can be shown to have the form

$$m(t)_{\mu\nu}^{MC} = \sum_{\sigma\lambda}\sum_{\sigma'\lambda'} (\tilde{A}_\mu|B_{\sigma\lambda})\, \phi^{(o)}(t)_{\sigma\sigma'} \phi^{(o)}(t)^*_{\lambda\lambda'} (B_{\sigma'\lambda'}|\tilde{A}_\nu) + o(\tilde{A}^3), \tag{19}$$

while the latter is given by

$$\tilde{m}(t)_{\mu\nu} = (\tilde{A}_\mu|Q_{AB}\, e^{-it\mathcal{L}_{AB}}\, Q_{AB}|\tilde{A}_\nu) \quad . \tag{20}$$

Here $Q_{AB} = Q_A - \sum_{\sigma\lambda} |B_{\sigma\lambda})(B_{\sigma\lambda}|$ projects onto the subspace orthogonal to $\{A_\mu, B_{\sigma\lambda}\}$, $\mathcal{L}_{AB} = Q_{AB}\, \mathcal{L}\, Q_{AB}$, $\phi^{(0)}(t)_{\lambda\lambda'}$ is the solution of Eq.(13) for $m(t)_{\mu\nu} = 0$, and $|B_{\sigma\lambda}) = Q_{AB}\, \mathcal{L}|B_{\sigma\lambda}) = o(\tilde{A})$. The relative importance of the mode-coupling contribution and the remainder depends, of course, on the choice for the set $\{A_\mu\}$. If we chose for $\{A_\mu\}$ the <u>complete</u> set of one-particle phase-space functions, the remainder would <u>vanish</u> for systems with two-body forces: $\tilde{m}(t) \equiv 0$. Our choice will be a subset of the phase-space densities, namely the particle number densities and current densities which must be properly orthonormalized

$$\{A_\mu\} \to \{\hat{N}_s(\vec{q}), \hat{J}_s(\vec{q}), \hat{N}_{so}(\vec{q}), \hat{J}_{so}(\vec{q})\};\ \mu \to (s,\vec{q},n)$$

where n =1 to 4 enumerates the number density and the 3 components of current density variables and n = 5 to 8 denotes the corresponding variables of a tagged particle. With fixed choice for the $\{A_\mu\}$ one might expect an increase of the mode-coupling contribution relative to the remainder if one enlarges the set of variables in Eq.(17). Systematically, one could consider the coupling to higher order products, $C_{\mu\nu\lambda} \sim Q_{AB}|A_\mu A_\nu A_\lambda)$, etc. Their incorporation into $m(t)_{\mu\nu}^{MC}$ is straightforward. However, their contribution is not expected to be large in view of the above and thus has been taken into account neither in the theory of liquids nor in the following. Moreover, it has been demonstrated /4/ that coupling to triple products will not change the results qualitatively at the glass transition in one-component liquids.

The mode-coupling approximation (MCA) now consists of replacing the exact expression for $m(t)^{MC}$, Eq.(19), by

$$m(t)_{\mu\nu}^{MC} \cong \sum_{\sigma\lambda\sigma'\lambda'}' (\tilde{A}_\mu|B_{\sigma\lambda})\, \phi(t)_{\sigma\sigma'}\, \phi(t)^*_{\lambda\lambda'}\, (B_{\sigma'\lambda'}|\tilde{A}_\nu) \tag{21}$$

and

$$m(t)_{\mu\nu} \simeq 2\,\gamma_{\mu\nu}\,\delta(t) + m(t)_{\mu\nu}^{MC} \quad , \tag{22}$$

assuming the <u>full</u> relaxation functions in Eq.(21) will adequately represent the higher order terms in $\tilde{A}$ which were not shown explicitly in Eq.(19). For the long-time dynamics close to the glass transition, only the coupling to products of the non-ergodic density variables is relevant. This is indicated by the prime on the summation sign in Eq.(21). Terms containing products of current densities in Eq.(19) will decay within microscopic times. These contributions together with the remainder will make up the regular part of the memory functions which does not determine the asymptotic long-time behavior. In Eq.(22) these contributions are approximated by damping constants. Taking into account the symmetries of the system, it is now straightforward to show that Eqs.(13,14) and (21,22) lead to:

a) a closed set of equations for the density relaxation functions

$$\ddot{\phi}(q;t)_{ss'} + \sum_{\bar{s}} \Omega^2(q)_{s\bar{s}}\, \phi(q,t)_{\bar{s}s'} + \sum_{\bar{s}} \int_0^t dt' K_L(q;t-t')_{s\bar{s}}\, \dot{\phi}(q;t')_{\bar{s}s'} = 0 \qquad (23)$$

with $\phi(q,0)_{ss'} = S_{ss'}(q)/(S_{ss}(q)S_{s's'}(q))^{1/2}$ and $\dot{\phi}(q,0) = 0$. $K_L(q,t)_{ss'}$ is the longitudinal part of the friction tensor with MC contribution

$$K_{\alpha\beta}(\vec{q};t)^{MC}_{ss'} = \sum_{\substack{\vec{k} \\ \lambda\lambda'\sigma\sigma'}} V_{\alpha\beta}(\vec{q},\vec{k})^{\lambda\lambda'\sigma\sigma'}_{ss'}\, \phi(k,t)_{\lambda\lambda'} \cdot \phi(|\vec{q}-\vec{k}|,t)_{\sigma\sigma'} \qquad (24)$$

where the vertex function can be expressed in terms of static structure factors only, if proper factorization approximations are applied for three- and four-particle static correlations;

b) equations for the transverse current relaxation functions $\phi_T(q,t)_{ss'}$

$$\dot{\phi}_T(q;t)_{ss'} + \sum_{\bar{s}} \int_0^t dt'\, K_T(q;t-t')_{s\bar{s}}\, \phi_T(q;t')_{\bar{s}s'} = 0 \qquad (25)$$

with $\phi_T(q,0)_{ss'} = \delta_{ss'}$ and the transverse part $K_T(q,t)_{ss'}$ of the friction kernel from above. (Note that this equation is trivial in the sense that there is no feedback: K_T is not a functional of ϕ_T. The long-time behavior of the transverse currents is completely ruled by the long-time properties of the density relaxation function.) and finally one finds:

c) nontrivial equations for the tagged-particle density relaxation functions

$$\ddot{\phi}_s(q,t) + v_s^2 q^2\, \phi_s(q,t) + \int_0^t dt'\, K_s(q;t-t')\, \dot{\phi}_s(q;t') = 0 \qquad (26)$$

with $\phi_s(q,0) = 1$, $\dot{\phi}_s(q,0) = 0$, $v_s^2 = k_B T/m_s$ and the friction kernels' MC contribution

$$K_s(q;t)^{MC} = \sum_{\vec{k}\lambda\lambda'} V_s(\vec{q};\vec{k})^{\lambda\lambda'} \phi(k,t)_{\lambda\lambda'} \phi_s(|\vec{q}-\vec{k}|,t) \quad . \tag{27}$$

Explicit expressions for the vertex functions in terms of static structure factors are given in Ref. 12,14.

Applications:

(i) Schematic Model for Molten Salt

Eqs.(23,24) have not yet been solved for a binary mixture. However, we investigated the following schematic model /10,11/ for two coupled relaxation functions

$$\ddot{\phi}_{\pm}(t) + \Omega_{\pm}^2 \phi_{\pm}(t) + \int_0^t dt' \, K_{\pm}(t-t') \, \dot{\phi}_{\pm}(t') = 0$$

$$K_+(t)^{MC} = \Omega_+^2 \lambda \left[\alpha \, \phi_+(t)^2 + (1-\alpha) \, \phi_-(t)^2\right] \tag{28}$$

$$K_-(t)^{MC} = \Omega_-^2 r\lambda \, \phi_+(t) \, \phi_-(t),$$

which reflects the symmetries of mass (+) and charge (-) relaxation functions in a simple molten salt. This model shows a glass transition. The non-ergodicity parameters $f_{\pm} = \lim_{t\to\infty} \phi_{\pm}(t)$ which obey (cf. (28) for $t\to\infty$)

$$f_+ = \lambda \left[\alpha \, f_+^2 + (1-\alpha) \, f_-^2\right] (f_+-1),$$

$$f_- = r\lambda \, f_+ f_- \, (f_- - 1) \tag{29}$$

take on non-zero values /19/ for $\lambda > \lambda_c$ (and fixed α, r). Near to the transition the model exhibits critical dynamical behavior with Kohlrausch-type long-time relaxation (see Fig.3 of Ref. 11 and Fig.1 below) in the liquid phase as well as

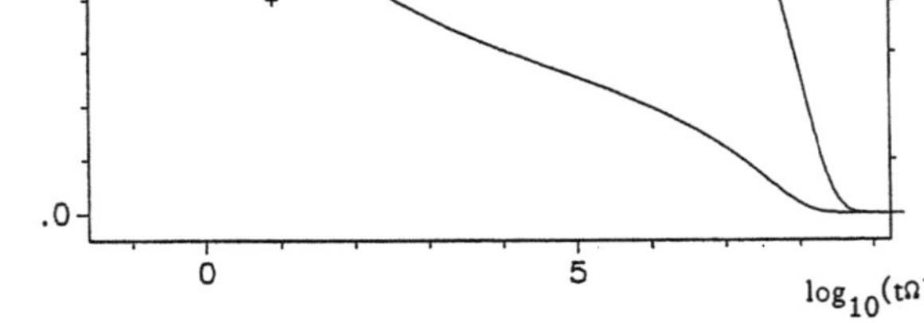

Fig.1: Solutions of Eqs.(28) for $\epsilon=1-\lambda/\lambda_c=0.025$ and $r=6.35$, $\alpha=0.932$, $\gamma_+=5\Omega_+$, $\Omega_-^2=\Omega_+^2/r$.

scaling behavior. For application to the supercooled melt $K_{0.6}Ca_{0.4}(NO_3)_{1.4}$ the parameters in Eq.(28) were determined by fitting the measured temperature-dependence of the viscosity to a power-law implied by Eqs.(23-25) /13/. The resulting slow time-dependence of the relaxation functions and their scaling behavior are in excellent agreement with recent neutron spin echo experiments (see Fig.3 of Ref.13).

Note that the above model, in which $\phi_+(t)$ and $\phi_-(t)$ are coupled only via the mode-coupling kernels, may be regarded as a special case of a one-component liquid with two relevant wavenumbers. Therefore, the general predictions for wavenumber-dependent one-component liquids /9/ apply to this model. Our numerical solution of Eq.(28) confirms these predictions.

(ii) Delocalization of Small Particles in a Glassy Matrix:

From Eqs.(23,24) one finds for the matrix of non-ergodicity parameters by letting $t \to \infty$

$$\sum_{\bar{s}} \Omega^2(q)_{s\bar{s}}\, f(q)_{\bar{s}s'} + \sum_{\bar{s}} K_L(q,\infty)_{s\bar{s}} \left[f(q)_{\bar{s}s'} - \phi(q,o)_{\bar{s}s'}\right] = o\,,$$

$$K_L(q,\infty)_{s\bar{s}} = \sum_{\substack{\vec{k}\\ \lambda\lambda'\sigma\sigma'}} V_L(\vec{q};\vec{k})_{s\bar{s}}^{\lambda\lambda'\sigma\sigma'}\, f(k)_{\lambda\lambda'}\, f(|\vec{q}-\vec{k}|)_{\sigma\sigma'} \qquad (30)$$

and from Eqs.(26,27)

$$v_s^2 q^2\, f_s(q) + K_s(q,\infty)\left[f_s(q)-1\right] = o\,,$$

$$K_s(q,\infty) = \sum_{\vec{k}\lambda\lambda'} V_s(\vec{q};\vec{k})^{\lambda\lambda'}\, f(k)_{\lambda\lambda'}\, f_s(|\vec{q}-\vec{k}|) \quad . \qquad (31)$$

These nonlinear integral equations for DWF and LMF have been solved for a binary mixture of hard spheres /12,14/ for which the static structure factors $S_{ss'}(q)$ are available from the analytical solution of Percus-Yevick integral equations /20/. The mixture shows a glass transition at a total packing fraction $\eta_B \simeq 0.52$ similar to a one-component hard sphere system /2/. For sufficiently small diameter ratio $\delta = \sigma_1/\sigma_2$, however, only the large particles become localized at η_B, while the small particles form a liquid within the voids of the glassy matrix. This phase is characterized by non-zero DWF's and a non-zero LMF of the large particles, while the LMF of small particles is zero /12/. For still higher densities, the small particles will also become localized at a critical $\eta_A > \eta_B$. The LMF of the small particles shows characteristic deviations from the conventionally assumed Gaussian behavior (Fig.2) due to the strong increase of the localization length l_1 as the delocalization transition is approached. This

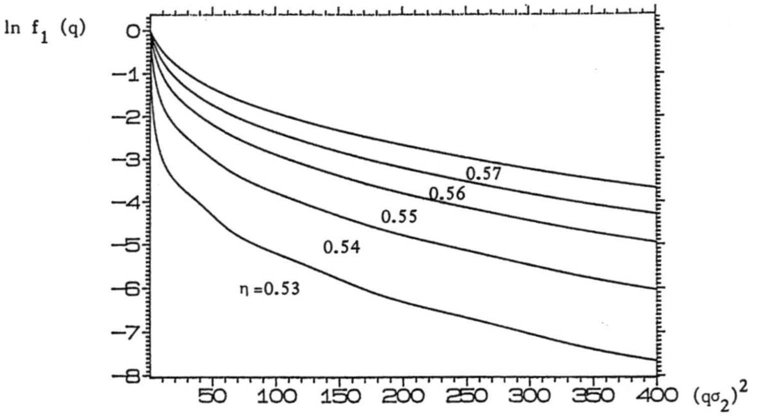

Fig.2: Small particles' LMF in a binary glass of hard spheres for $\eta > \eta_A \simeq 0.525$. ($\sigma_1=\sigma_2/5, N_1=9N_2$)

phenomenon appears to have been observed in the intensity of quasielastic incoherent neutron scattering on hydrogen in metals /21/, which is determined by $f_1(q)$. The experiments show the same typical increase in localization length (the initial slope in Fig.2 equals $-l_1^2/2$) with increasing temperature. However, the explanation offered so far /21/ differs from the above. Let me emphasize that the above theory for the first time clearly reveals a qualitative difference between DWF and LMF of one species of particles in an amorphous solid /12/. Since these two quantities become identical at low temperatures (harmonic solid approximation), one often does not distinguish between the two.

References

1. E. Leutheußer, Phys. Rev. A29, 2765 (1984)
2. U. Bengtzelius, W. Götze, and A. Sjölander, J. Phys. C17, 5915 (1984)
3. Earlier suggestions along similar lines of thought are due to T. Geszti, J. Phys. C16, 5805 (1983), P. Papon, P. Meier, Physica 101A, 477 (1980)
4. W. Götze, Z. Phys. B56, 139 (1984), ibid. B60, 195 (1985), H. De Raedt, W. Götze, J. Phys. C19, 2607 (1986)
5. S. P. Das, G. F. Mazenko, S. Ramaswamij, and J. J. Toner, Phys. Rev. Lett. 54, 118 (1985)
6. T. R. Kirkpatrick, Phys. Rev. A31, 939 (1985)
7. U. Bengtzelius, Phys. Rev. A34, 5059 (1986)
8. W. Götze and L. Sjögren, J. Phys. C20, 879 (1987), Z. Phys. B65, 415 (1987)
9. W. Götze, Liquid-Glass Transition, Troisieme cycle de la Physique en Suisse romande, Lecture Notes, Neuchatel 1987)
10. U. Krieger, J. Bosse, in "Amorphous and Liquid Materials", E. Lüscher et. al. eds., Martinus Nijhoff Publishers, p.34 (1987)
11. J. Bosse, U. Krieger, J. Phys. C19, L609 (1986)
12. J. Bosse, J. S. Thakur, Phys. Rev. Lett. 59, 998 (1987)
13. U. Krieger, J. Bosse, Phys. Rev. Lett. 59, 1601 (1987)
14. J. S. Thakur, J. Bosse, to be published

15. R. G. Palmer, D. L. Stein, E. Abrahams, and P. W. Anderson, Phys. Rev. Lett. 53, 958 (1984); R. G. Palmer, Advances in Phys. 31, 669 (1982)
16. H. Mori; Prog. Theor. Phys. 33, 429 (1965); 34, 399 (1965); R. Zwanzig, in "Lectures in Theoretical Physics", eds. W. Brittin and L. Dunham (Wiley-Interscience, New York, 1961), Vol.3, p. 135
17. Note that the isothermal susceptibilities $(\alpha_\mu|\alpha_\nu)$ of a set of linearly independent variables $\{|\alpha_\nu)\}$ must be known in order to form the orthonormal set $\{|A_\mu)\}$.
18. J. Bosse, T. Munakata, Phys. Rev. A25, 2763 (1982)
19. An interesting discussion of the complete phase diagram of this model has been given by W. Götze and R. Haussmann, preprint March 1988
20. J. L. Lebowitz; Phys. Rev. 133, A895 (1964), gives Ornstein-Zernike direct correlation functions
21. N. Wakabayashi, B. Alefeld, K. W. Kehr, T. Springer; Sol. State Commun. 15, 503 (1974)

Structural Relaxation in Glass-Forming Liquids

M. Elmroth, L. Börjesson, and L.M. Torell

Department of Physics, Chalmers University of Technology,
S-41296 Gothenburg, Sweden

We present experimental results on the structural relaxation time of a glass-forming ionic hydrate melt, $Ca(NO_3)_2 \cdot 8H_2O$, over a wide time range from times typical of the very fluid state (~10^{-11}s) to the long relaxation times (~10s) close to the glass transition. The data are compared with previous results for ionic glassformers and discussed in terms of a recent classification of glass-forming liquids on a scale ranging from "strong" to "fragile". It is found that the relaxation times of the ionic hydrate melt scale with those of non-network ionic glassformers using Tg as a scaling parameter. As for the ionic liquids, crossover from non-exponential to exponential decay of the relaxation function is observed in the time range ~10^{-11}s. The results imply that the hydrate melt is extremely fragile which makes it a model material for comparisons with recent theories of structural relaxation in glassy liquids. The relaxation times can be well fitted to a Vogel-Tammann-Fulcher (VTF) equation and with a characteristic temperature T_o related to the glass transition. Within the experimental error, they can equally well be fitted to a power law $(T-T_x)^{\mu}$ with μ = -1.8 as predicted from the mode coupling theory. The VTF-fitting holds for the whole time / temperature range studied while the power-law fitting is only applicable in a temperature region which is far above the range normally associated with the glass transition. Crossover from power-law correlation to a VTF-like relaxation region is observed at a temperature $T_x \approx 1.3$Tg in accordance with recently reported crossover temperatures for some other glass-forming liquids.

1. Introduction

The present study concerns structural relaxation in supercooled viscous liquids with focus on simple glass-forming liquids. A relaxation process is often described by the average structural relaxation time $<\tau>$ and from its temperature dependence important information is provided about the mechanism of viscous flow in the fluid and the nature of the glass transition. Recently it has been suggested that the $<\tau>$ vs T plot for various glass-forming liquids can be used as a basis for a classification of the type of glass-former on a scale ranging from "strong" to "fragile"/1/. In an attempt to determine if the characteristics proposed for such a classification are general we investigate some extreme cases of this scheme, namely some ionic and ionic hydrate melts of which some belong to the most fragile systems observed so far. The general form of $<\tau>$ vs T for

Springer Proceedings in Physics, Vol. 40
Editors: M. Davidović and A.K. Soper

Static and Dynamic Properties of Liquids

the various liquids will also be compared with predictions from the mode coupling theory/2,3/ for which "fragile" systems are suggested to be model systems. The data presented are in part from previous observations in this laboratory of the structural dynamics of ionic melts of the non-network glass-forming type such as $Ca(NO_3)_2$ - KNO_3/4/ and KCl-2$BiCl_3$/5/, and of network materials such as $ZnCl_2$/6/, and in part from a recent study of an aqueous ionic glassformer, namely $Ca(NO_3)_2 \cdot 8H_2O$/7/.

Structural relaxation of a liquid is the process in which the fluid rearranges towards an equilibrium state after a perturbation has been introduced. It can be characterized by the structural relaxation time, τ, which is the time needed for the system to reach equilibrium. When the time required for structural change is much longer than the time of observation the fluid becomes a rigid glass on the experimental time scale. The material then *appears* solid even if it flows on a longer time scale. Using short period times of the perturbations (i.e high probe frequencies) it is then possible to "freeze" the liquid at temperatures well above the glass transition temperature and by varying the probe frequency the structural relaxation can be followed over a wide temperature range.

In the present work we combine high frequency / high temperature relaxation data from light scattering in various ionic and ionic hydrate melts, obtained in this laboratory using the Brillouin scattering technique, with results from ultrasonic and viscosity measurements at lower frequencies / temperatures. The combined data result in a characterization of the relaxation process from that representing the very fast dynamics of a fluid liquid ($\sim 10^{-11}$s) to the slow processes (~10 s) close to the glass transition. The corresponding relaxation times are thereby obtained over the wide time range of 12 decades. This allows for a determination of the two general characteristics of the structural relaxation time, i.e. the temperature dependence of its average value and the frequency dependence of its distribution function. It is these properties which are taken as the bases in the "strong" to "fragile" classification scheme/1/ as discussed in the following.

Strong glassformers are generally represented by an Arrhenius temperature dependence of $\langle\tau\rangle$ and a narrow time distribution function. Typical examples are network materials such as the vast majority of commercial glasses of various silicates. In the latter the cations are four-coordinated and the structure consists of a three-dimensional random network of linked tetrahedra which change very little upon passing T_g. The similar structures of the supercooled and glassy states of a strong glassformer explain the small change in heat capacity at the glass transition and the slowly changing (Arrhenius) viscosity in the supercooled state.

At the other extreme, the fragile liquids demonstrate the opposite behaviour. These materials have generally larger coordination numbers, which are fluctuating from ion to ion, and there is no network structure. Examples are simple molecular and ionic liquids. Large heat capacity changes at the glass transition reflect the "disorder" of the fragile systems, which can be attributed to the multiplicity of configurations of the ground states. The rapid breakdown of the short range order as temperature increases is also demonstrated by the rapidly decreasing viscosity. In an Arrhenius plot, the viscosity η, as the closely related average structural relaxation time $<\tau>$, will then show significant curvature, which is increasing in magnitude with increasing fragility. The form of these bent curves can be characterized by two activation energies E_A; a high-E_A region just above Tg followed by a crossover to a low-E_A / high temperature region. The crossover occurs at smaller η and $<\tau>$ values for increasing fragility. However, it is not only the general form of $<\tau>$ which changes at the crossover. We have reported from light scattering studies of various ionic melts that the broad distribution of relaxation times (non-exponential decay), which characterizes the low temperature / high-E_A region, sharpens up to become single (exponential decay) in the high temperature / low-E_A region/4-6,8/. In the present work we will investigate if these observations are general by including the case of an ionic hydrate melt. Brillouin scattering will be used since it provides the short time domain ($\sim 10^{-11}$s) expected for crossover of $<\tau>$ and of its distribution function.

Another kind of transition is predicted to occur in the viscosity plot from mode coupling theories/2-3/. The theories are based on the Zwanzig-Mori memory function approach for a dynamical description of dense liquids and are later extended for the discussion of the liquid glass transition. They suggest the existence of a transition temperature T_x at which the viscosity η diverges as $(T-T_x)^{\mu}$. In the simplest models, considering a single viscous liquid, an exponent of $\mu = -1.8$ is predicted. For a more complex two-component system μ can vary over a considerable range /9/. However, in all cases comparisons with experimental results show that power-law behaviour is only adequate for the fluid liquid and in the high temperature range of the supercooled state. At lower temperatures, instead of viscosity divergence at T_x, η continues to increase continuously for $T<T_x$ /1,10/. It should then be noted that although T_x is not a sharp transition and does not coincide with Tg, it seems to control the temperature dependence of the viscosity at $T>T_x$. The viscosity range probed by structural relaxation in Brillouin scattering is the fluid liquid range in which the power-law description of η and the mode coupling theory have shown to be successful. We will discuss our Brillouin data in this context to test the power-law prediction and to estimate the crossover temperature from structural relaxation time data for the ionic and ionic hydrate systems.

2. Brillouin Scattering Technique

Brillouin scattering in a fluid is the scattering of light by long wavelength thermal density fluctuations or sound waves in the material. The Brillouin spectrum consists of the Rayleigh line centered at the frequency of the incident light and a frequency shifted doublet, i.e the Brillouin components. In glassforming systems there are two of these doublets because of the

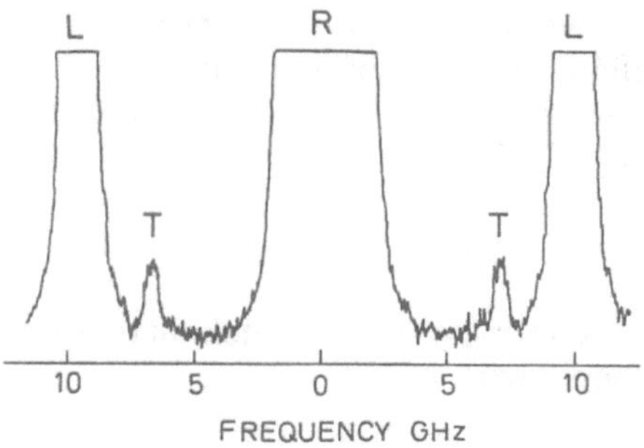

Fig.1. Brillouin spectrum of $Ca(NO_3)_2 \cdot 8H_2O$; R, L, and T are Rayleigh, longitudinal Brilllouin and transverse Brillouin components, respectively.

presence of both longitudinal and transverse (shear) waves in highly viscous systems, see Fig.1. The transverse modes are usually weak in intensity, about 10 to 100 times smaller than the longitudinal peaks, which in turn are orders-of-magnitudes weaker than the central line, whose intensity is governed by static inhomogeneities in the medium. The frequency shifts ν_L and ν_T are related to the longitudinal and transverse sound velocities v_L and v_T according to

$$v_\mu = \nu_\mu(\mathbf{q}) / q, \qquad \mu = L,T. \tag{1}$$

The magnitude of the wave-vector of the observed sound waves, q, is then determined from the geometry of the experiment;

$$q = 2\, nk_i \sin(\theta/2). \tag{2}$$

Here n is the refractive index of the material, k_i is the wavevector of the incident light and θ the scattering angle.

The melt is characterized by the ability to be compressed and sheared as longitudinal and transverse sound waves move through the fluid. At sufficiently low frequencies, the ionic configurations of the liquid structure can change during the passage of the propagating pressure wave and the liquid flows and obeys Newton's viscosity law. At higher frequencies, when the time necessary to reach the new equilibrium state is comparable to the time period of the perturbation, energy will be lost in the medium and relaxation phenomena will occur. This will lead to an attenuation of the longitudinal sound waves which is manifested in a broadening of the Brillouin components. The half-width Γ_B (half-width at half height) of the Brillouin lines are related to the absorption coefficient α according to

$$\Gamma_B = \alpha v_L / 2\pi. \tag{3}$$

Maximum absorption will then be observed at a Brillouin frequency $\nu_{\alpha max}$ such that the peak condition, i.e.

$$\omega \langle\tau\rangle \approx 1, \tag{4}$$

can be applied for $\omega = 2\pi\nu_{\alpha max}$, $<\tau>$ being the average structural relaxation time of the liquid. Generally, a single relaxation time is not sufficient to describe the relaxation process. It has been shown that a relaxation function of the Kohlrausch-Williams-Walts (KWW) form/11/, i.e.

$$\phi(t) = \exp(-(t/\tau)^{\beta}) , \tag{5}$$

closely fits the data for a broad variety of glassforming liquids including ionic systems/4/. The average relaxation time is then given by

$$<\tau> = (\tau/\beta)\,\Gamma(1/\beta) \tag{6}$$

where $\Gamma(x)$ is the gamma function. The relaxation characteristics can then be obtained from the behaviour of the imaginary part of the relaxing complex longitudinal modulus M". Writing M" in a reduced form as

$$N''(\omega) = \frac{M''}{M_\infty - M_o} = \frac{2\,\alpha\, v^3}{\omega(v_\infty^2 - v_o^2)} \tag{7}$$

it is, at a particular temperature, directly related to the sound velocity v and absorption α, which can be directly obtained from the Brillouin spectrum. v_∞ and v_o are the low- and high- frequency limits of the velocity, usually determined from the low and high temperature side of the relaxation range, respectively. Analytically N" can be expressed by

$$N''(\omega') = \omega' \int_0^\infty dt' \cos(\omega' t')\, \phi(t') \tag{8}$$

where $t' = t/\tau$ and $\omega' = \omega\tau$. By comparing experimental results (7) and the analytical equation (8) the relaxation function $\phi(t')$ can be obtained.

The experimental set up in the present study includes a single mode Ar^+ laser operating at 488.0 nm, a triple-passed piezoelectrically scanned Fabry-Perot interferometer, and photon-counting detection techniques. Details of the specific arrangements used for investigating the various liquids are given elsewhere, together with descriptions of the various routes for sample preparations/4-7/.

3. Results and Discussion

Brillouin spectra of the ionic melts $Ca(NO_3)_2$ - KNO_3, KCl - $2BiCl_3$ and $ZnCl_2$, and of an ionic hydrate melt $Ca(NO_3)_2 \cdot 8H_2O$ have been recorded all the way from the glass transition over the relaxation range to the high temperature region of the relaxed fluid liquids. Detailed results from recordings of the various systems are given elsewhere/4-8/. Here we concentrate on comparing some of the results with the recent data for the ionic hydrate liquid $Ca(NO_3)_2 \cdot 8H_2O$/7/, with the general characteristics of the "strong-fragile" classification scheme/1/ and the mode coupling theory/2,3/ in mind.

Strong relaxation effects were observed in all the investigated glassformers both from the data of the Brillouin frequency shifts and from the Brillouin line broadening. In the low temperature range close to the glass transition where the viscosity is large, weak transverse modes due to propagating shear waves could be observed, see for example Fig.1. As the temperature increases, the shear modes disappear and the longitudinal modes start to broaden due to the relaxation effects.

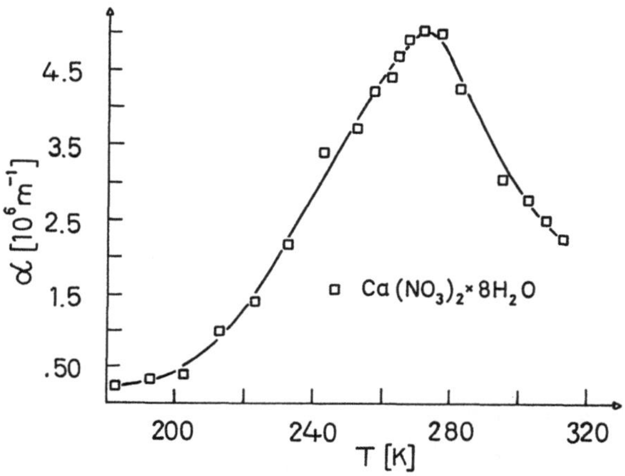

Fig.2. Temperature dependence of hypersonic attenuation in $Ca(NO_3)_2 \cdot 8H_2O$.

In Fig.2 the absorption coefficient for $Ca(NO_3)_2 \cdot 8H_2O$, calculated from the observed longitudinal line width, is plotted versus temperature. Since the absorption maximum, α_{max}, is recorded, a value for the structural relaxation time, τ, can directly be obtained from the peak condition (4). The peak is observed at T ≈ 273 K for a Brillouin frequency shift of 8.64 GHz which then corresponds to a structural relaxation time $\tau \approx 1.84 \times 10^{-11}$s. At lower temperatures, where the liquid structure changes too slowly to follow the perturbation and therefore exhibits the elastic properties of a glassy solid, the shear relaxation time $<\tau_s>$ can be calculated from the expression

$$<\tau> = \eta_s / G_\infty \quad (9)$$

where G_∞ is the limiting high frequency shear modulus. G_∞ is determined from the shear velocity according to

$$G_\infty = \rho \, v_T^2. \quad (10)$$

Using the frequency shifts of the observed transverse modes in combination with (1), (2), (10) and (9) shear relaxation times in the range 10^{-6} -1s were obtained.

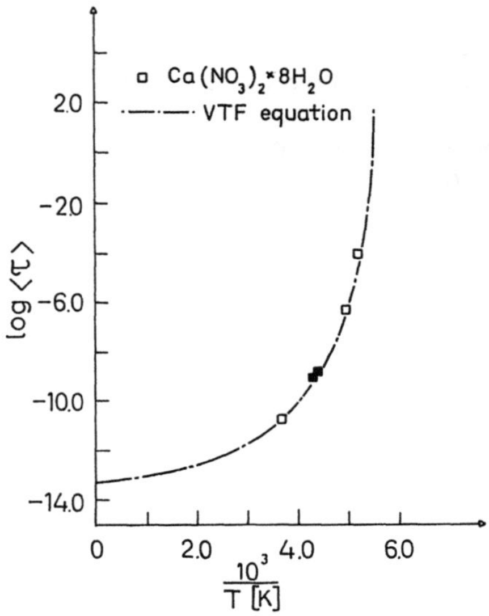

Fig.3. Arrhenius plot of the structural relaxation time in $Ca(NO_3)_2 \cdot 8H_2O$ from Brillouin scattering (□) and ultrasonic measurements (■, /13/). Dot-dashed line represents a Vogel-Tamman-Fulcher fit to the data.

In Fig.3 we show results for both the structural and shear relaxation times for $Ca(NO_3)_2 \cdot 8H_2O$ in an Arrhenius representation. In the calculations, viscosity and density data were taken from AMBRUS et al. /11,12/. We also include in Fig.3 some relaxation times estimated from ultrasonic measurements of sound absorption maxima by using the peak condition of (4). The solid line in Fig.3 represents a Vogel-Tamman-Fulcher fit to the data and best correlation was obtained for the following expression

$$<\tau> = 4.83 \cdot 10^{-14} \exp[658/(T-162)] \ . \quad (11)$$

Figure 3 demonstrates that an extrapolation to $1/T = 0$ carries τ directly into the region of typical inverse quasilattice frequencies/4/ which thus determine the lower limit on τ. At the other end, the low temperature limit is given by 162 K which is close to the value 161 K obtained for a VTF fit of viscosity data/12/ and it may be taken as the ideal glass transition temperature T_o. The calorimetric determined glass transition temperature Tg is for this system reported to be larger, Tg = 183 K, as is generally observed. Figure 3 thus demonstrates an outstanding feature of the relaxation time of $Ca(NO_3)_2 \cdot 8H_2O$, where it can be seen that one single equation can describe the temperature dependence over a very broad time range. Moreover, in the high and low temperature limits it correlates with the behaviour of other physical properties of this system/4,12/.

To compare the structural relaxation data of the present ionic hydrate melt with those of the ionic systems previously obtained, we show the various results in a scaled Arrhenius presentation in Fig.4 using Tg as a scaling parameter. This is the graphical form which has proved fruitful to demonstrate correlations between structural properties and the structural relaxation time and which is used in the classification scheme of glass-

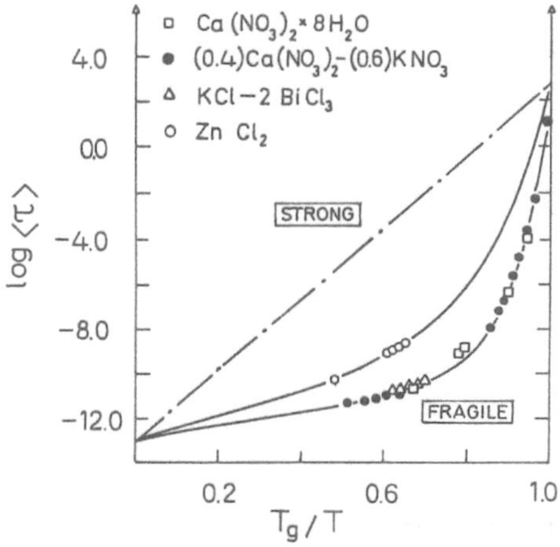

Fig.4. Arrhenius plot of the structural relaxation times of various ionic melts using Tg as a scaling parameter. Non-network materials superimpose on the fragile extreme. The weak network-former $ZnCl_2$ falls in the intermediate range.

formers discussed above/1/. In Fig.4 the linear (Arrhenius) temperature dependence of typical strong networks is drawn as a reference. Departure from the linear form signals weaker networks with a curvature which is increasing with increasing "fragility". We found that the present data for the ionic hydrate melt, after scaling by Tg, superimpose on the relaxation time data of the ionic systems $Ca(NO_3)_2$ - KNO_3 and KCl - $2BiCl_3$, whereas the data for $ZnCl_2$ demonstrate a different behaviour. $Ca(NO_3)_2 \cdot 8H_2O$ can therefore be classified in the group of extremely fragile liquids of which the molten salts of the non-network type are good examples. In $ZnCl_2$, on the other hand, the tetrahedral network, even though weak, introduces a short range order which shifts the structural relaxation time towards the "strong" edge and places $ZnCl_2$ close to the center in the classification scheme.

Next we will examine the relaxation time distribution to see whether the correlation of fragility with relaxation time width seen for the ionic systems/8/ will hold for the case of an ionic hydrate melt. For the ionic systems of the non-network kind, crossover to single relaxation time behaviour has been observed in the time range typical of Brillouin scattering, i.e. $\sim 10^{-11}$s. In the case of $ZnCl_2$ the crossover was found to occur already at the lower temperatures / lower frequencies probed by ultrasonics/14/ with typical times $\sim 10^{-8}$s. From the graph in Fig.4 it then seems that the characteristic crossover times mark the bend of the structural relaxation time curve at which it curls over from the rapid changes in the low temperature range to the more slowly decreasing times in the high temperature region. We therefore expect the crossover in case of $Ca(NO_3)_2 \cdot 8H_2O$ to occur in the time range accessible by the Brillouin technique. To analyze the relaxation time distribution we present the data for the imaginary part of the relaxing modulus (using the reduced form N") versus $\log(\omega<\tau>)$ in Fig.5. N" was calculated from the absorption and velocity data (7), for details see ELMROTH et al. /7/, and values for $<\tau>$ are taken from the VTF-relation (11). In Fig.5 we also include results for a non-network ionic melt, $Ca(NO_3)_2$ - KNO_3, together with data for the network $ZnCl_2$ system. A

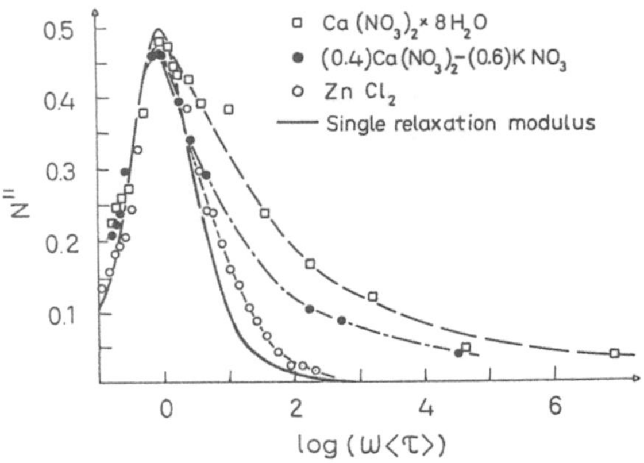

Fig.5. Imaginary part of the normalized longitudinal modulus, N", for various ionic melts. Solid line represents single relaxation time behaviour.

theoretical curve representing single relaxation time behaviour is also shown in the figure for clarity.

From Fig.5 it can be seen that the relaxation data for the three systems coincide within the experimental error on the high temperature side of the maximum, i.e for $\log(\omega<\tau>) < 0$. The relaxation function (5) is in this range described by a single relaxation time ($\beta=1$), i.e. exponential decay. On the low temperature side systematic differences appear which imply increasing non-exponentiality of the relaxation processes in the order ZnCl < $Ca(NO_3)_2$ - KNO_3 < $Ca(NO_3)_2 \cdot 8H_2O$, characterized by decreasing β-values. It should then be noted that the observed coupling between the longitudinal propagating modes and the relaxing internal modes of the systems are obtained for a more or less fixed probe frequency in the GHz range since the Brillouin frequency changes very little upon temperature changes of the liquid. Departures from single relaxation time behaviour on the low temperature side therefore describe on a temperature scale the relaxation distribution function at very short times ($\sim 10^{-11}$s).

In the case of $ZnCl_2$, it can be seen from Fig.5 that a single relaxation time closely describes the process also at lower temperatures and therefore at lower probe frequencies than those of Brillouin scattering, as in fact the ultrasonic study already showed/14/. For $Ca(NO_3)_2$ - KNO_3 the data in Fig.5 suggest that a single time can only describe the relaxation at temperatures close to and above the observed hypersonic absorption maximum. For $Ca(NO_3)_2 \cdot 8H_2O$ this behaviour is even more restricted and can only be seen on the high temperature side of the maximum, i.e the single relaxation time range has just been reached at hypersonic frequencies. The close fits to the single relaxation time curve on the short time / high temperature side and the departure on the long time / low temperature side for $Ca(NO_3)_2 \cdot 8H_2O$

and $Ca(NO_3)_2$ - KNO_3 then demonstrate that the Brillouin range (10^{-11}s) in fact corresponds to the cross-over region for the non-network systems in accordance with the proposed characteristics of the classification scheme discussed above. However, since the latter two liquids are both of about the same fragility, Fig.5 makes clear that there can be no simple correlation of non-Arrhenius with non-exponential behaviour as originally suggested/[15]/. Rather it seems likely that for liquids of equal fragility, those with more complex atomic level structures (ions + solvation shells) will relax in the most complex way, i.e. will show the largest departure from $\beta=1$.

Finally we analyze the structural relaxation data of the various ionic liquids in the context of the existence of a characteristic temperature T_x for viscosity divergence. According to the mode coupling theory/[2,3]/ the viscosity will follow a power-law

$$\eta = \eta_o\ (T/T_x - 1)^{\mu} \qquad (12)$$

with $\mu = -1.8$ in case of a simple fluid. This implies a glass transition at a temperature T_x. Experimental data for a wide variety of liquids show the success of the power-law correlation, however, in a temperature region far above the normal glass transition/[10,15]/. We have found that the structural relaxation time, which is closely related to the viscosity, in case of $Ca(NO_3)_2 \cdot 8H_2O$ can be fitted to a VTF-equation (11). To test the power-law correlation we present the results in Fig.6 in a $\tau^{-1/1.8}$ plot. In the figure we use a scaled temperature, T/Tg, to be able to incorporate data for $Ca(NO_3)_2$ - KNO_3 and $ZnCl_2$ in the same graph. For the network systems the figure shows that the data points in fact fall on a straight line in the high temperature region. For comparison a straight line with a temperature intercept at $T \approx 1.3$ Tg is shown in Fig.6, since this is the temperature of viscosity divergence recently reported for some other liquids/[15]/. It can be seen that, within the accuracy of the experiment, VTF and power-law behaviour equally well describe the structural relaxation at $T>T_x$. The extrapolated singularity temperature lies in the supercooled metastable liquid regime and seems to have the common value of $T_x \approx 1.3$ Tg. However, there is no temperature at which $<\tau>$ (or the viscosity) really diverges, and T_x therefore signals the crossover to another regime. A recent neutron spin echo study of $Ca(NO_3)_2$ - KNO_3 also reports on an effective transition at a temperature considerably above Tg/[16]/. In the latter, the observation of two distinct slow relaxation processes are in qualitative agreement with mode coupling predictions for a value of $T_x \approx 1.1$ Tg. In the low temperature region $Tg<T<T_x$ we find that a VTF-relation fits the data of $Ca(NO_3)_2 \cdot 8H_2O$ while the $Ca(NO_3)_2$ - KNO_3 results are better described by an Arrhenius dependence/[4]/. Crossover to this region occurs just in the viscosity range $\eta \approx 1$ Pas in which activated processes have been suggested to be of importance/[15]/. At temperatures $T<T_x$ such activated processes between various configurational states, characterized by potential barriers of various heights, seem to become a more dominant mechanism than the cooperative behaviour described in the mode coupling theory. The rapid change of $<\tau>$ in the fragile glassformers in the low temperature range can, in this picture, be attributed to the high degeneracy of configurational states in the energy landscape of these systems/[15]/.

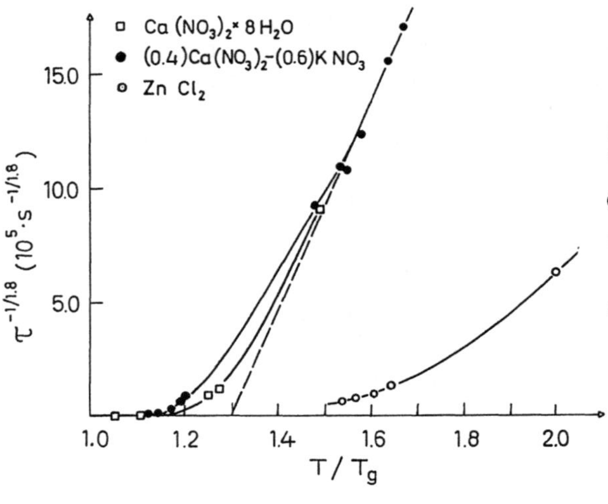

Fig.6. $\tau^{-1/1.8}$ as a function of temperature for various ionic systems using Tg as a scaling parameter. Solid lines are best Vogel-Tamman-Fulcher fits to the experimental data. Dashed line shows high temperature power-law behaviour for the fragile systems of Fig.5.

In case of the network material, i.e. $ZnCl_2$, the structural relaxation time changes more slowly with temperature and Fig.6 shows that the short time domain for the power-law correlation is hardly reached. Since $ZnCl_2$ corresponds to an intermediate glassformer, this behaviour will be even more pronounced for strong glass-formers. Materials for which the dynamics on supercooling can be described by mode coupling theories are therefore restricted to the liquids of the fragile type, i.e weakly bonded liquids with only van der Waals, Coulomb or dipole-dipole interactions.

In conclusion it is found that the structural relaxation time for an ionic hydrate melt, $Ca(NO_3)_2 \cdot 8H_2O$ is, with regard to a scaled temperature dependence, identical to those of non-network ionic glassformers. The relaxation data imply a classification of these systems as being extremely fragile. The correlation of fragility with the width and the form of the average relaxation time, as seen for the ionic liquids, also holds for the case of an ionic hydrate melt. The average relaxation time is well represented by a Vogel-Tammann-Fulcher form with a characteristic temperature which correlates with the glass transition temperature Tg. Within the accuracy of the measurements the data in the high temperature range can equally well be represented by a power-law, with an exponent μ = -1.8 in accordance with the mode coupling theory. The characteristic time T_x is then much higher than the glass transition temperature and it can be extrapolated to $T_x \approx 1.3$ Tg as recently reported for other fragile liquids. The common T_x value therefore seems to define the structural relaxation in the high temperature range of fairly low viscosities, irrespective of detailed structural differences. In the high viscosity / low temperature range $T_x<T<Tg$ a more complete theory than mode coupling in its present form is needed to account for phenomena close to the glass transition.

References

1. C.A. Angell: In Relaxations in Complex Systems, ed. by K. Ngai and G.B. Wright, National Technical Information Service, U.S. Department of Commerce, Springfield, VA 22161, 1 (1985); J. Non-Cryst. Solids 73, 1 (1985).
2. E. Leutheusser: Phys. Rev. A 29, 2765 (1984).
3. U. Bengtzelius, W. Goetze and A. Sjölander: J. Phys. Chem. 17, 5915 (1984); U. Bengtzelius and A. Sjölander, Ann. N.Y. Acad. Sci. 484, 229 (1986).
4. L.M. Torell: J. Chem. Phys. 76, 3467 (1982); C.A. Angell and L.M. Torell, J. Chem. Phys. 78, 937 (1983).
5. L.M. Torell, D.C. Ziegler and C.A. Angell: J. Chem. Phys. 81, 5053 (1984).
6. H.E.G. Knape: J. Chem. Phys. 80, 4788 (1984).
7. M. Elmroth, L. Börjesson and L.M. Torell (to be published).
8. L.M. Torell: In Molten Salt Chemistry, ed. by G. Mamantov and R. Marassi (Reidel 1987), p 161-180.
9. J. Bosse and U. Krieger: J. Phys. C 19. L 609 (1986).
10. P. Taborek, R.N. Kleiman and D.J. Bishop: Phys. Rev. B34, 1835 (1986).
11. G. Williams and D.C. Walts: Trans. Faraday, Soc. 66, 80 (1970).
12. J.H. Ambrus, C.T. Moynihan and P.B. Macedo: J. Electrochem. Soc. 119,192 (1972).
13. J.H. Ambrus, H. Dardy and C.T. Moynihan: J. Phys. Chem. 76, 23 (1972).
14. G.J. Gruber and T.A. Litovitz: J. Chem. Phys. 40, 13 (1964).
15. C.A. Angell: J. Phys. Chem. Solids (in press).
16. F. Mezei, W. Knaak and B. Farago: Physica Scripta T19, 363 (1987).

Aspects of Atomic Dynamics in Liquid Rubidium Above and Below the Melting Temperature

T. Gaskell[1], *U. Balucani*[2], *and R. Vallauri*[2]

[1]Department of Physics, The University, Sheffield S37RH, UK
[2]Istituto di Elettronica Quantistica del Consiglio Nazionale delle Ricerche, I-50127 Firenze, Italy

Abstract: We concentrate on two features of microscopic dynamics together with the associated transport coefficients. These are (i) transverse current correlations and shear viscosity and (ii) single particle motion and the self-diffusion coefficient.

From molecular dynamics data for the transverse current correlation function in liquid rubidium, just above the melting point, the memory function, $K_T(q,t)$, and the generalized viscosity, $\tilde{\eta}(q)$, have recently been obtained. We discuss the limitations of using a 2-exponential model to fit $K_T(q,t)$, and then apply mode-coupling theory in an attempt to explain the results. Both approaches are successful at small q, but for the wavevector range where well-defined shear waves are supported by the liquid, mode-coupling theory completely fails to describe the details revealed by the computer data.

Information about individual particle dynamics in supercooled states of liquid rubidium is also available from computer simulation studies of the velocity autocorrelation function. The most striking feature is the development of a slowly decaying tail in the memory function, whose amplitude increases significantly with the degree of supercooling and consequent decrease of the self-diffusion coefficient. We argue that this effect is largely due to the coupling of single particle motion with the transverse modes, and demonstrate explicitly that it correlates with the appearance of an extensive tail in $K_T(q,t)$ at small wavevectors.

1. Introduction

From computer simulation data of the transverse current autocorrelation function, covering a range of wavevectors, the associated memory functions have recently been derived for a liquid metal-like system near its melting point /1/. An attempt was made to analyze the results by means of a fit to a 2-exponential model. At small wavevectors, $q\sigma \lesssim 1$, where σ is a measure of the core diameter in the effective pair-potential, this proved to be successful. For intermediate wavevectors, where the liquid can support well-defined propagating shear wave modes, this is no longer the case. A characteristic feature in this regime is the appearance of a plateau or shoulder in the memory function which cannot be described by this simple model. We have, instead, applied a mode-coupling theory result for the memory function. This too fails to give an adequate explanation of the details, except within the small wavevector range quoted above. Indeed, beyond this range, the viscoelastic model is superior.

Finally, we discuss some recent molecular dynamics data on single particle motion in a liquid rubidium model, at and below the melting temperature. A feature of the data is the development of an extensive tail in the velocity autocorrelation memory function as the liquid is supercooled. The effect is investigated using a microscopic velocity field approach to momentum transfer and velocity correlation. This describes single particle dynamics through its coupling to the longitudinal and transverse momentum current density fluctuations. It is demonstrated that the tail in the memory function can be correlated with the development of a similar effect in the transverse current memory function at small wavevectors.

Springer Proceedings in Physics, Vol. 40
Editors: M. Davidović and A.K. Soper

Static and Dynamic Properties of Liquids

2. Transverse current: memory function

The longitudinal and transverse current correlation functions, $C_L(q,t)$ and $C_T(q,t)$, are defined through the equation

$$\langle \underline{j}(\underline{q},o).\underline{j}(-\underline{q},t)\rangle/N = \langle \Sigma_{i,j}\, m\underline{v}_i(0).\, m\underline{v}_j(t)\, \exp[i\underline{q}.(\underline{r}_i(0)-\underline{r}_j(t))]\rangle/N$$

$$\equiv C_L(q,t) + 2C_T(q,t), \qquad (1)$$

where L and T refer to the velocity components with respect to the direction of $\underline{q}$. The memory function, $K_T(q,t)$, associated with the transverse component is defined as usual through the equation

$$dC_T(q,t)/dt + \int_0^t ds\, K_T(q,s)C_T(q,t-s) = 0 \qquad (2)$$

Hence, in terms of the respective Laplace transforms

$$\hat{C}_T(q,z) = C_T(q,t=0)/[z + \hat{K}_T(q,z)]\,. \qquad (3)$$

It is convenient to introduce the normalized memory function, $n_T(q,z)$, and the generalized wavevector- and frequency-dependent viscosity, $\eta(q,z)$, through the equations

$$\hat{K}_T(q,z) \equiv (q^2/nm)G(q)\hat{n}_T(q,z) \equiv (q^2/nm)\hat{\eta}(q,z),$$

$G(q)$ being the rigidity modulus and n the number density. The structure of the memory function, for a range of wavevectors, has been obtained quite recently from computer data of $C_T(q,t)$. Details of the computer experiment are given in the reference /1/.

Within the viscoelastic approximation $n_T(q,t)$ is given a simple exponential time decay, with a wavevector-dependent relaxation time. A prescription for the latter /2/, which is based on an extrapolation scheme between the hydrodynamic and ideal gas limits, is quite successful at small q. However, it was shown by BALUCANI et al./1/ that in this regime ($q\sigma < 1$) a 2-exponential model for $n_T(q,t)$, namely,

$$n_T(q,t) = [1 - \alpha(q)]\exp(-t/\tau_S(q) + \alpha(q)\exp(-t/\tau_L(q)), \qquad (4)$$

produces an improved fit of the data. The physical picture which supports this model, assumes that τ_S is a short relaxation time predominantly associated with binary collisions, and τ_L with cooperative effects in which correlated sequences of collisions play an important role. There is an obvious difficulty at small t where the exponential models produced a finite initial slope. This apart, excellent overall agreement was obtained. As the magnitude of q increases, however, this model becomes less satisfactory. The memory function no longer has a monotonous time decay, but instead begins to display a structure which for some wavevectors develops into a broad maximum. Note, also, that this effect produces a distinct shoulder in the generalized viscosity $\tilde{\eta}(q) \equiv \hat{\eta}(q,z=0)$, for $4 \lesssim q\sigma \lesssim 8$ /1/. These features are clearly inconsistent with the simple 2-relaxation time picture, and a more sophisticated approach is required.

With this in mind, it was decided to apply a mode-coupling expression for the memory function. This is derived by projecting the so-called random force onto the subspace of bilinear products of an appropriate set of conserved variables. They are coupled to the random force through vertex functions which determine the modes responsible for the decay mechanism. In a dense liquid, for small enough wavevectors, we may neglect the kinetic term in the random force in comparison to the potential contribution. It then follows that the vertex functions arising from the momentum current density modes vanish. The surviving contribution is due to coupling to the density fluctuations, and this leads to the result /3,4/

$$K_T(q,t) = (q^2/nm)\eta(q,t)$$

$$= (k_BT/16\pi^3 nm) \int_{k,|\underline{k}-\underline{q}| \le k_c} d\underline{k}\ [k^x]^2 [(S(|\underline{k}-\underline{q}|)-S(k))/S(|\underline{k}-\underline{q}|)S(k)]^2\ F(|\underline{k}-\underline{q}|)F(k,t)\ . \quad (5)$$

In the above equation S(k) and F(k,t) are the static structure factor and the intermediate scattering function respectively. This result may be expected to apply at intermediate and long times, the integral cut-off emphasizing that the theory can only be justified when the expression is dominated by a limited range of wavevectors. A different treatment of the dynamics is required at sufficiently small values of t, but its application to any system other than rigid spheres has proved difficult. Here, we use a simple extrapolation scheme, which involves choosing k_c so that the correct initial value for K(q,t) is produced. This has no serious justification, but, in a calculation of the stress autocorrelation function and shear viscosity coefficient on the basis of eq.(5), it has been shown to be a surprisingly successful extrapolation procedure /5/. We find that $k_c\sigma$ = 11.1, the choice in the stress calculation, reproduces the initial values to within a few percent. In figures 1 and 2 we compare some results from eq.(5) with those from computer data for the thermodynamic state of the liquid rubidium model, T = 332K and n = 0.01058 x $10^{24}cm^{-3}$. A viscoelastic model of F(k,t) is used in the calculation. Once again the agreement is satisfactory at the smaller wavevector. For $q\sigma$ = 1.53 it is less so, and the discrepancy rapidly becomes more significant as q increases. For $q\sigma \gtrsim 3$, the theory is no longer even qualitatively correct.

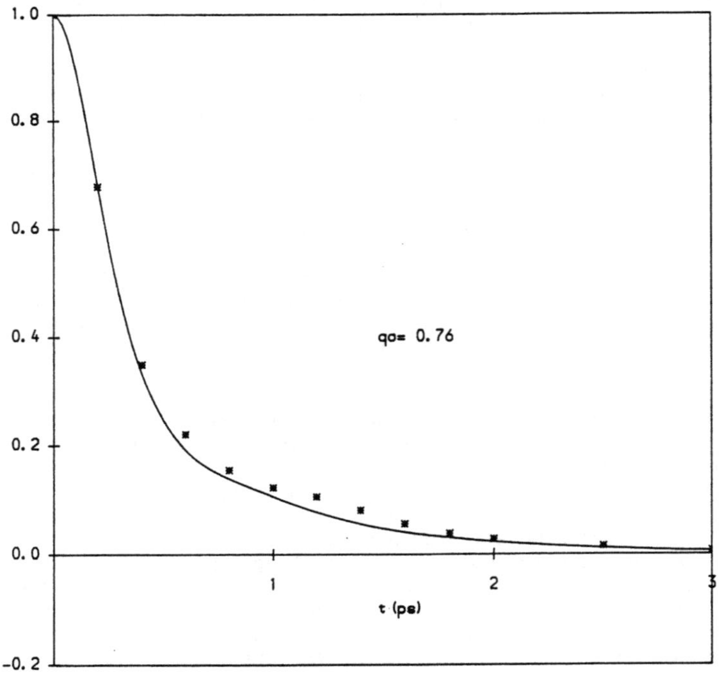

Figure 1. Normalized memory function of the transverse current. The sold curve represents the mode-coupling result, and the crosses are obtained from computer data.

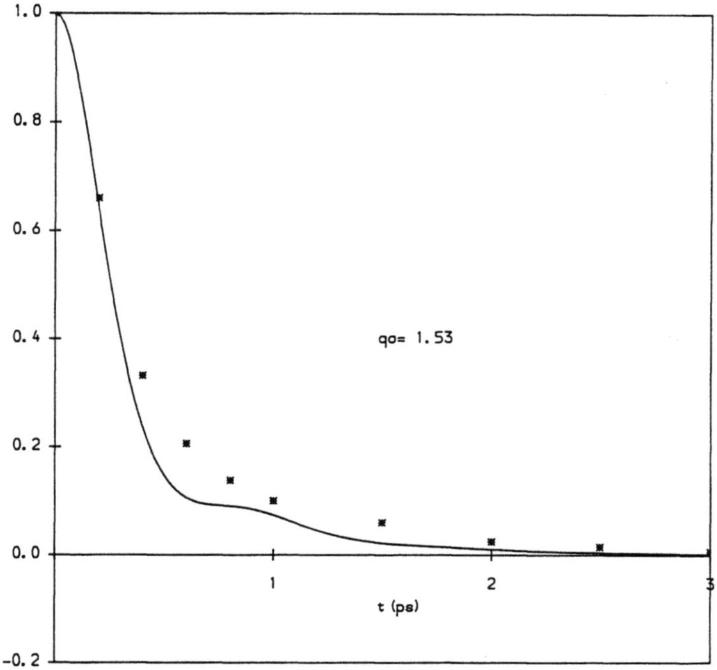

Figure 2. Normalized memory function of the transverse current. The sold curve represents the mode-coupling result, and the crosses are obtained from computer data.

3. Single particle dynamics: the effects of supercooling.

The velocity autocorrelation function is used to discuss single particle motion, within a theoretical framework which involves the introduction of a microscopic velocity field, $\underline{v}(\underline{r},t)$. The latter takes the form $\underline{v}(\underline{r},t) = \Sigma_i \underline{v}_i(t) f(|\underline{r}-\underline{r}_i(t)|)$, where $f(r)$ is essentially a step function whose width a is determined by the macroscopic requirement $\int d\underline{r}\, f(r) = n^{-1}$. It leads to the result /6/

$$\psi(t) = (1/3mk_BT)(2\pi)^{-3} \int d\underline{q}\, \tilde{f}(q)\,[C_L(q,t) + 2C_T(q,t)]F_S(q,t) \qquad (6)$$

with $\tilde{f}(q)$ and $F_S(q,t)$ the Fourier transforms of $f(r)$ and the self-correlation function, $G_S(r,t)$, respectively. Results from this expression have been discussed before. Here, we focus attention on the memory function, $K(t)$, and the way it changes as the liquid is supercooled. Our earlier discussion of the transverse current will be particularly relevant in this context.

In figure 3 we show the details for a state close to the melting point (312K), when both $C_L(q,t)$ and $C_T(q,t)$ are based on simple viscoelastic models, and compare them with a result derived from computer data for $\psi(t)$ obtained by RAHMAN (private communication; see also ref. /6/). Now according to eq.(6) the value of the self-diffusion coefficient, D, depends almost entirely on the contribution from $C_T(q,t)$. Indeed, on putting $F_S(q,t) = 1$ (which has little effect on $\psi(t)$), D is completely determined by the coupling to the transverse current. In this case /7/

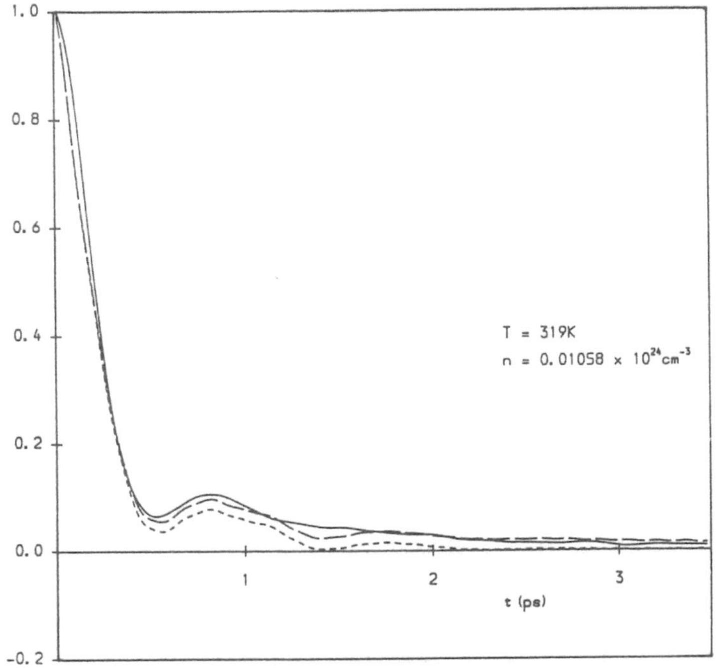

Figure 3. Normalized memory function of the velocity autocorrelation function. The solid curve is derived from computer data. The dashed curve is a theoretical result from eq. (6), with viscoelastic models for both $C_L(q,t)$ and $C_T(q,t)$, the broken one is obtained when the modified viscoelastic model is used for $C_T(q,t)$.

$$\begin{aligned} D = (K_BT/m)\int_0^\infty dt\psi(t) &\equiv (K_BT/m)\left[\int_0^\infty dt\ K(t)\right]^{-1} \\ &= (nK_BT/3\pi^2)\int_0^\infty dq\tilde{f}(q)/G(q)\hat{n}_T(q,z=0) \\ &= (nK_BT/3\pi^2)\int_0^\infty dq\tilde{f}(q)/\tilde{\eta}(q) \qquad (7) \end{aligned}$$

is clearly overestimated by the theory. This is consistent with our previous conclusion that a single exponential choice for $n_T(q,t)$ is inadequate. Although the overall shape of the memory function is approximately correct, the theory fails to reproduce properly the structure at relatively small times ($0.6 \leq t \leq 1.5$ps), nor does it account for the rather extensive tail. We believe that the absence of the rather broad maximum in the model we use for $n_T(q,t)$, which was mentioned in section 2, is responsible for the first of these defects. However, the slow decay at large t, which,on supercooling the liquid becomes the more important of the two, as the data below shows, can be explained by introducing two different relaxation times into the model of $n_T(q,t)$. Note that when $C_T(q,t)$ is derived on the basis of this memory function it will be referred to as the modified viscoelastic model. In figure 3 we demonstrate the improvement in the result for K(t).

The theory requires a quasi-continuous range of values for the q-dependent parameters in eq.(4) in order to perform the wavevector integral, and we rely on phenomenological arguments to obtain them rather than trying to interpolate from a few fitted values. The steps are as follows. (i) An estimate of the decay time $\tau_S(q=0)$ is obtained as $\tau_S(q=0) = \eta_E/G(q=0)$, where η_E is the Enskog viscosity of a rigid sphere fluid at the packing fraction of the liquid. Values at finite q are obtained by using the extrapolation formula of AKCASU and DANIELS /2/. (ii) Obtain the relaxation time τ_L, which is the most difficult to assess. In the absence of a theory, we are forced to use, as an initial value, the result obtained from the fitting procedure discussed in /1/, for the smallest wavevector available i.e. $\tau = 2.29$ps at $q\sigma = 0.76$. Mode-coupling theory, which correlates the relaxation of K(t) with F(q,t), implies that τ_L is connected with structural relaxation effects. Since the time-scale of the latter is linked to diffusive motion of the atoms we suggest that $\tau_L \sim D^{-1}$. We therefore predict τ_L at another state point by scaling the above result with the ratio of the two diffusion coefficients. This oversimplifies the situation of course, but the results below suggest that it is not unreasonable. Having obtained an estimate for $\tau_L(q=0)$ in this way, we make the further approximation of keeping this value for all q. (iii) Use the hydrodynamic relation

$$(1 - \alpha(0))\tau_S(0) + \alpha(0)\tau_L(0) = \eta/G(0) \qquad (8)$$

to fix the the parameter $\alpha(0)$, η being the shear viscosity coefficient. The fitting procedure referred to earlier also indicates that $\alpha(q)$ decays rapidly as q increases. We select $\alpha(q) = \alpha(0)\exp[-q/q_m]$, and when q_m is determined the transverse current memory function is completely specified. Clearly, the value for D predicted by means of eq.(7) should be consistent with the one we use to obtain τ_L. The parameter q_m is chosen to achieve this self-consistency. At a temperature just above the melting point, and also for the two supercooled states discussed below, we find that $q_m\sigma \approx 3.1$ fulfills this requirement. The parameter $\alpha(0)$ also is relatively insensitive, and varies from 0.131 at T = 319K to 0.140 at T = 234.4K. Over the same temperature range τ_L changes from 2.52 to 7.32ps.

In figures 4 and 5, a comparison is made of the theoretical prediction for the memory function and the result obtained for K(t) from computer data for two supercooled states of rubidium. The computer study has been reported by KINELL and LOVESEY /8/. The modified viscoelastic model has been used for the transverse current and the usual viscoelastic framework for longitudinal current. The latter requires details of the static structure factor, and since we have no comprehensive data for the supercooled states we have had to assume that it remains essentially unchanged. In addition, the replacement of the single exponential memory function by the 2-exponential model may be appropriate also in the longitudinal current framework, especially in supercooled states. However, it is more difficult to implement in this case. We no longer have the advantage of information on the longer relaxation time, because a fit of longitudinal current data is not available. Nor is there a self-consistency condition to satisfy as a means of selecting another of the parameters. A relatively crude test of the effects of a 2-exponential model has been made for T = 273.2K, with the conclusion that it tends to increase the amplitude of the oscillations in the longitudinal contribution to $\psi(t)$. It must be borne in mind that the area under the memory function curve is not changed by any modification to $C_L(q,t)$; the latter can only change the shape. In our test the overall shape was not significantly altered and, in particular, the effect on the slowly decaying tail was small.

4. Conclusions.

We have applied a form of mode-coupling theory to the memory function, $K_T(q,t)$, of the transverse current in a liquid rubidium model, in an attempt to explain the details revealed by computer simulation data. Although successful at long wavelengths, it fails to describe the structure of the memory function for wavelengths of the order of the interparticle separation. A comprehensive theory of $K_T(q,t)$ and hence the generalized viscosity, $\eta(q)$, remains an outstanding problem.

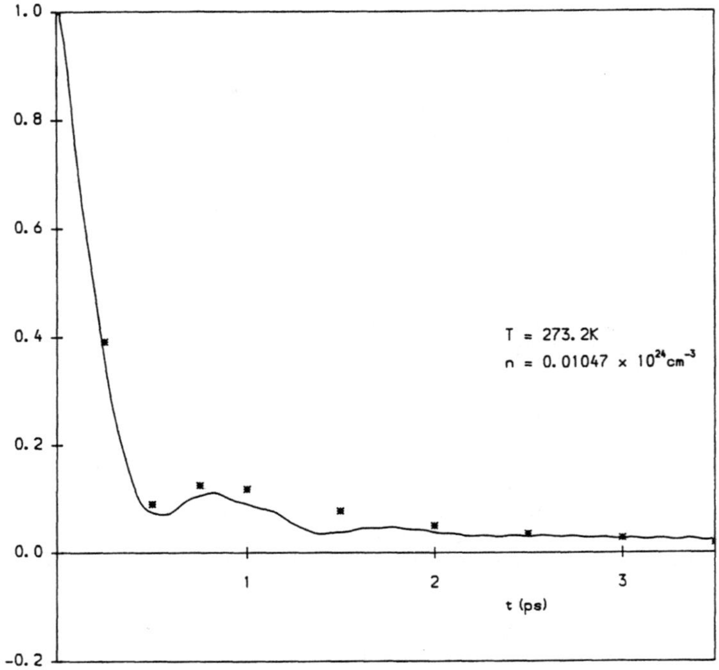

Figure 4. Normalized memory function of the velocity autocorrelation function. The solid curve is a theoretical result using viscoelastic models for the current correlation functions in eq. (6), but with the modification to $C_T(q,t)$ discussed in the text. The crosses are predicted values from computer data.

Individual particle dynamics in a supercooled state of the liquid has also been investigated via the velocity autocorrelation function $\psi(t)$. The theoretical framework stresses the coupling of single particle motion with the collective modes which the liquid can sustain. Consequently, this part of the work too involves the transverse current modes and hence $K_T(q,t)$. The importance of the latter, as the degree of supercooling is increased, has been stressed. Because of the limited success of mode-coupling theory, we have use a 2-exponential form for this memory function in the calculations, since it has a wider application as far as the wavevector range is concerned. Given the acknowledged defects of this model, the agreement in figures 4 and 5 is satisfactory. If we fail to include the longer relaxation time, the substantial tail which develops in the memory function associated with $\psi(t)$, as the system is supercooled, is absent. Our work, therefore, strongly suggests (a) that this effect correlates with the development of a marked reduction in the the rate of decay of the stress autocorrelation function, and (b) that assuming the relaxation time involved to be inversely proportional to the self-diffusion coefficient is not inconsistent with the limited information available from computer data.

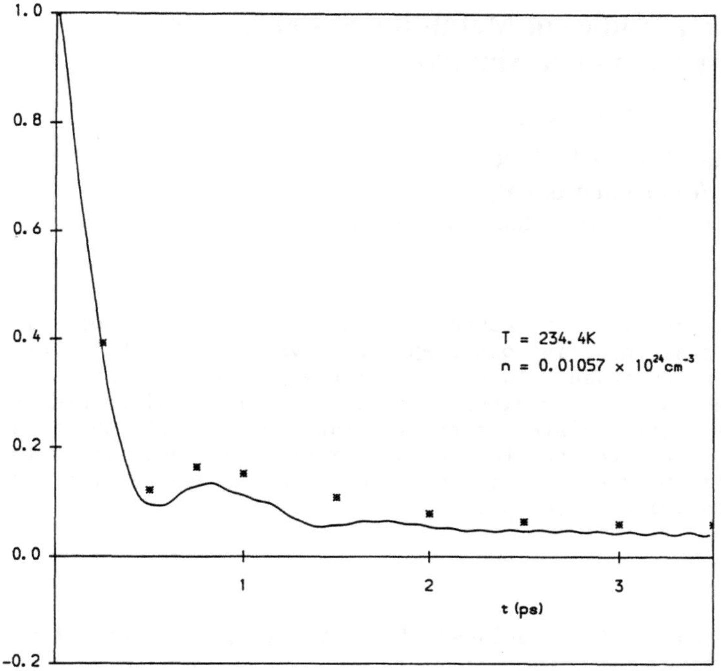

Figure 5. Normalized memory function of the velocity autocorrelation function. The solid curve is a theoretical result using viscoelastic models for the current correlation functions in eq. (6), but with the modification to $C_T(q,t)$ discussed in the text. The crosses are predicted values from computer data.

References

/1/ U. Balucani, R. Vallauri and T. Gaskell, Phys. Rev. A35, 4263 (1987)
/2/ A.Z. Akcasu and E. Daniels, Phys. Rev. A2, 962 (1970)
/3/ T. Munakata and A. Igarashi, Prog. Theor. Phys. 60, 45 (1978)
/4/ U. Bengtzelius, W. Götze and A. Sjölander, J. Phys. C17, 591 (1984)
/5/ U. Balucani, R. Vallauri and T. Gaskell, Phys. Rev. A37, 3386 (1988)
/6/ T. Gaskell and S. Miller, J. Phys. C11, 3749 (1978)
/7/ U. Balucani, R. Vallauri, T.Gaskell and M. Gori, J. Phys. C18, 3133 (1985)
/8/ T. Kinell and S.W. Lovesey, J. Phys. C19, L791 (1986)

Dynamical Correlation Studies of Metal-like Systems in the Supercooled Phase by Computer Simulation

U. Balucani[1], *R. Vallauri*[1], *and T. Gaskell*[2]

[1]Istituto di Elettronica Quantistica del C.N.R., Via Panciatichi 56/30, I-50127 Firenze, Italy

[2]Department of Physics, The University, Sheffield S3 7RH, UK

Abstract: Some recent results of computer simulation studies of time dependent correlation functions, in one component systems, in the supercooled phase will be reviewed. In particular the process of structural arrest (which is a precursor of the glass transition) will be discussed by the analysis of the intermediate scattering function $F(q,t)$), which is directly accessible to a neutron scattering experiment. Its connection with the stress autocorrelation function will be emphasized and new results for a metal-like system (rubidium) reported.

1. Introduction

Computer Simulation (CS) techniques for the study of the glass transition have extensively been used in recent years. In particular simple model fluids have been investigated, despite the lack of experimental results to be compared with, due to the intrinsic difficulty of achieving the fast quenching rates required to bring a monatomic system in a supercooled phase and eventually in a glassy state. Somehow CS is exploited to look for those peculiar features of thermodynamic quantities (discontinuities of specific heat, enthalpy etc..), which occur in laboratory systems with more complicated interactions between the constituents; the understanding of the behaviour of such quantities in systems which can easily be dealt with by a simulation, will hopefully give a closer realistic insight into the processes which are responsible for the appearance of long lived metastable states of matter. From this point of view it is natural to consider firstly systems modelled by Lennard-Jones (LJ), truncated LJ and Soft-Sphere interaction potential. The changes due to the use of different potential models are "per se" an interesting field of analysis. For an extensive review of the studies up to 1980-81, one can usefully refer to the papers by ANGELL, CLARKE and WOODCOCK [1] and FOX and ANDERSEN [2].

More recently the theoretical analysis of the glass transition performed by different authors [3,4], which makes use of mode-coupling concepts, has stimulated more CS [5,6] and even experimental [7] work,

Springer Proceedings in Physics, Vol. 40
Editors: M. Davidović and A.K. Soper

Static and Dynamic Properties of Liquids

aimed to verify the reliability of such an approach. In essence, the theory assumes the glass transition to be ruled by a dynamical process by which a rapid decrease of the temperature or an increase of the density induces a slowing down of the structural relaxation , which in turn yields an increase of the viscosity of the system. A feedback mechanism switched on by the direct connection between the diffusion coefficient D and the viscosity η (e.g. through a Stokes-Einstein relation), leads naturally to the appearance of a transition at a temperature where $D \rightarrow 0$ and $\eta \rightarrow \infty$ (see Ref. 4). An important feature of the mode-coupling analysis is that it can describe, in principle, the long time behaviour of the relaxation of a physical variable and therefore it appears fully consistent with a realistic description of the problem under investigation. In fact the presence of a long time tail of the stress autocorrelation function is of overwhelming importance in determining the value of the viscosity coefficient as soon as the system is brought into a supercooled (metastable) state.

2. Outline of the mode-coupling results

It can be easily shown that if one defines the stress tensor component

$$\sigma^{\alpha\beta}(t) = \sum_i [mv_i^{\alpha}(t)v_i^{\beta}(t) + r_i^{\alpha}(t)F_i^{\beta}(t)] \qquad (1)$$

where α and β indicate cartesian components, $\vec{v}_i$ the velocity, $\vec{r}_i$ the position and $\vec{F}_i$ the total force acting on particle i, the viscosity coefficient η turns out to be

$$\eta = \frac{1}{K_B TV} \int_0^{\infty} dt \, \langle \sigma^{\alpha\beta}(0) \cdot \sigma^{\alpha\beta}(t) \rangle \qquad (2)$$

where $\alpha \neq \beta$ and V is the volume occupied by the system. The stress autocorrelation function which appears in (2), breaks naturally into three terms (kinetic, potential and cross), derived from the kinetic and potential contributions to the stress tensor. In particular the potential component reads

$$\eta_{PP}(t) = \frac{1}{K_B TV} \langle \sum_{i,j} r_i^{\alpha}(0)F_i^{\beta}(0) \cdot r_j^{\alpha}(t)F_j^{\beta}(t) \rangle \qquad (3)$$

and it is found to give the overwhelming contribution to the total stress auto correlation function in the liquid phase. By using mode-coupling arguments it can be written as [4,8]

$$\eta_{pp}(t) = \frac{K_B T n^2}{60\Pi^2} \int_0^\infty dq\, q^4 \left[c'(q) \right]^2 F^2(q,t) \qquad (4)$$

where $c'(q)$ represents the derivative of the direct correlation function $c(q) = 1/n\,[1 - 1/S(q)]$ (with $n=N/V$, number density), with respect to the wavevector q, and $S(q)$ is the structure factor. In (4), $F(q,t)$ represents the intermediate scattering function, which is predicted to develop a long time tail as soon as a structural relaxation slowing down takes place, eventually leading to a structural arrest (with a flat long time tail) at the glass transition [4].

Whereas these predictions for $F(q,t)$ have been confirmed both by CS [5] and in a real experiment [7], their consequences on the stress autocorrelation function, as the temperature of the system is rapidly quenched, have not yet fully been analyzed. To our knowledge the only results available in the literature are those by AMAR and MOUNTAIN [9], who studied the behaviour of various transport coefficients in going from the liquid to the supercooled phase, for a Soft-Sphere potential model. The aim of this paper is to report some results on the behaviour of both the stress and intermediate scattering autocorrelation functions in supercooled rubidium.

The choice of such a particular system, for which an effective interaction potential has been developed [10] and successfully tested over various properties [11], has been dictated by two main reasons. The first one is that a comparison between mode-coupling theory and CS results has already been performed in the liquid phase [12], and very good agreement obtained. Since mode-coupling arguments emphasize the role of slowly-varying variables, their applicability is a priori expected to be appropriate when approaching the glass transition, where effects at intermediate and long times are particularly important. The second reason is the possible higher relevance of finite-wavevectors effects in metal-like systems, as far as the relation between D and η is concerned. Indeed, already in the liquid phase, we have shown that the actual value of D can only be accounted for by a generalized version of the Stokes-Einstein relation [13]. We believe this feature to be quantitatively relevant on the behaviour of D near the glass transition, but we do not attempt such a study in the present contribution.

On the other hand the possibility of obtaining rubidium in a glassy state has extensively been explored by several authors [14], so that a better guidance to the way the system has to be treated can be at hand. A major problem for our study is the easiness of nucleation as the system is quenched from an equilibrium liquid configuration. In fact, according to our experience, in evaluating the stress autocorrelation function runs as long as 10^{-9} s are necessary, in order to achieve a reasonable statistical accuracy at long times. Of course during this lapse of time the system has to remain in the metastable state and no nucleation should occur.

3. The CS "experiment"

We start from liquid configurations, with a system of N = 250 particles at $n^* = n\sigma^3 = 0.9045$ and T = 318 K, where $\sigma = 4.405$ A is the distance where the potential turns out to be zero for the first time. The temperature of the system is rapidly lowered by scaling the velocity by a factor μ at each time step, thus reaching the state points indicated in Fig.1, where the total energy of the system is reported as a function of the temperature. In order to check the influence of the quenching rate over the evaluated quantities and the overall properties of the system, we have used two values of μ, namely 0.9999 and 0.999, which correspond to two different quenching rates, i.e. $1.26 \cdot 10^{12}$ K/s and $1.23 \cdot 10^{13}$ K/s respectively. It appears that as far as the energy is concerned no difference in the final state point can be noticed. For comparison, the corresponding curve relative to a bcc crystal is also reported; it is evident that by fast quenching non-equilibrium states are attained.

Various quantities are then monitored on a time interval of 4000 time steps (here the time step $\Delta t = 10^{-14}$ s), namely, the pair distribution function g(r), the mean square displacement $\langle|\vec{r}_i(0)-\vec{r}_i(t)|^2\rangle$ and the distribution of

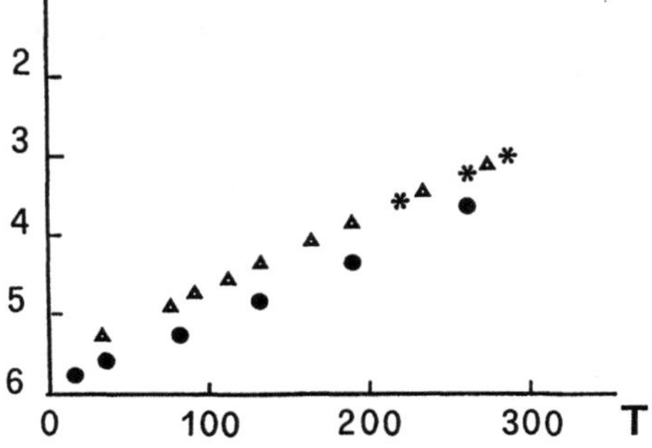

Figure 1. Energy per particle vs temperature at $n^* = 0.9045$. Stars and triangles correspond to the lowest and fastest quenching rates, μ = 0.9999 and 0.999 respectively. Dots represent the energy of the bcc crystal. Energies are in units of the well depth $\varepsilon = 5.52 \cdot 10^{-14}$ erg, of the model potential for rubidium

angular bonds P(cos(θ)) for the first shell of neighbours. The last quantity gives the distribution of the cosine of the angle between the vectors joining a central position particle with that of two neighbours, chosen to be within a distance $R_{MAX} = 1.5\sigma$ from the central particle. Both g(r) and P(cos(θ)) show the typical liquid like behaviour. The diffusion coefficient decreases rapidly with the temperature decrease, as shown by the behaviour of the mean square displacement, reported in Fig.2.

The corresponding values of D are reported in Table 1 for temperature in the range of 318 K (liquid phase) to 160 K (deeply supercooled state).

After this, a run of 10,000 time steps was started to evaluate the stress autocorrelation and intermediate scattering functions. For the former quantity the three independent components of the stress tensor have been evaluated, in order to check for the isotropy of the system; moreover the kinetic, potential and cross contributions have been calculated separa-

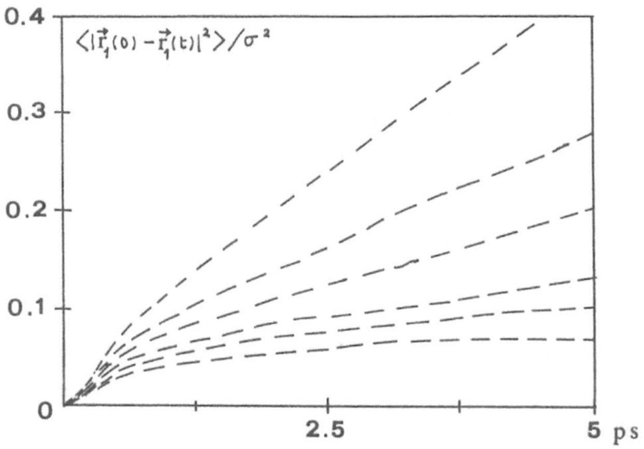

Figure 2. Mean square displacement at different temperatures, in units of σ. From top to bottom T = 318, 271, 230, 197, 160 and 141 K.

Table 1. The diffusion coefficient D at different temperatures derived from the slope of the mean square displacement.

T [K]	D [cm²/s]
318	$2.57 \cdot 10^{-5}$
270	1.50 "
238	1.03 "
197	0.32 "
160	0.08 "

tely, and confirm that the potential part is by far the most important. As far as F(q,t) is concerned, the wavevectors q have been chosen to lie along the cartesian axes with a magnitude $q = (2\Pi/L)l$ where $l = 1 \div 10$, and an average over the three independent components have been performed. The box length L turns out to be equal to 28.69 A, so that the minimum accessible wavevector is $q_{min} \approx 0.22\ A^{-1}$.

4. Results and discussion

For all the state points, except the one at T ≈ 270 K, there is evidence that nucleation is taking place during the run. This is signalled by an abrupt increase of the temperature (and an equally sharp drop of the pressure). Such an effect turns out to be independent of the way the system had been brought into the supercooled state. A very sensitive indicator of the nucleation occurrence is the P(cos(θ)) distribution. In Fig.3 we show such a quantity evaluated before and after the nucleation, starting from T ≈ 197 K and ending at T ≈ 260 K; for comparison the result for a bcc crystal at the same temperature is also reported. The sharp peak around cos(θ) = - 1, the shift of the second peak at a value ≈ -0.56 and the appearance of a peak at cos(θ) = 0, are the distinct marks of a bcc crystal, with respect to a liquid-like distribution pertaining to the supercooled state. The state point at T ≈ 270 K seems on the other hand more "stable" against nucleation and a detailed analysis of the required quantities has been carried out.

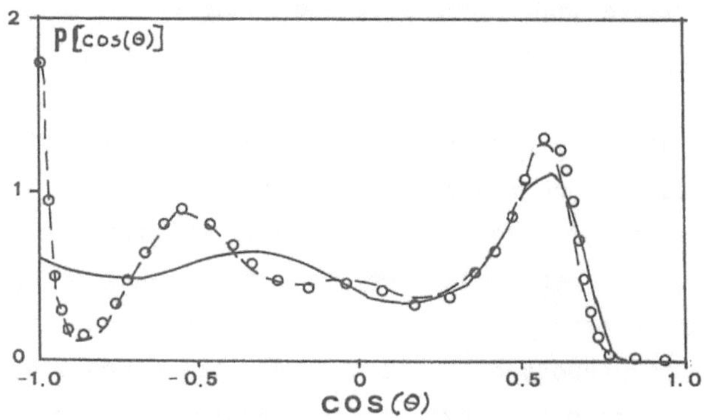

Figure 3. The distribution of angular bonds P(cos(θ)) before (full curve, T = 197 K) and after (dashed line, T = 260 K) nucleation. The open circles represent the same quantity evaluated for the bcc crystal at the same temperature

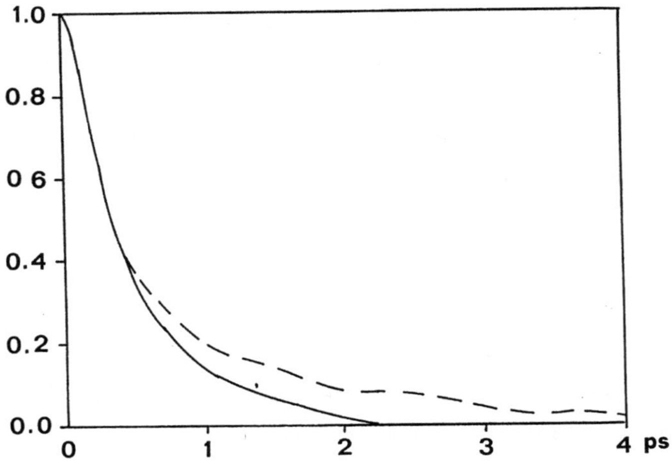

Figure 4. Normalized stress autocorrelation functions in the liquid phase (full curve, T = 318 K) and in the supecooled state (dashed curve, T = 270 K).

The normalized stress autocorrelation function is reported in Fig.4 and compared with the corresponding one at T $\approx$ 318 K. The reported curve at T $\approx$ 270 K has been obtained by averaging the results of a run of 60,000 time steps, with the system brought into the supercooled region by the lowest quenching rate, and a run of 40,000 time steps after the fastest quenching had been applied. It turns out that the results are essentially the same for both quenching rates. An increase of the long time tail, even if not so spectacular, is apparent. The viscosity coefficient changes from $\eta(T\approx318K)$ = 6.10 mpoise to $\eta(T\approx270K)$ = 7.26 mpoise (we want to recall that the value at the state point where the mode-coupling theory has been checked in Ref.12, is $\eta(T\approx335K)$ = 5.50 mpoise).

We wish to point out that by assuming the validity of a simple Stokes-Einstein relation, i.e. $D\eta/T$ = const, one would predict a value of η = 6.87 mpoise at T = 270 K. As a matter of fact the slowing down of structural relaxation becomes of some relevance, as indicated by the behaviour of F(q,t) even in this slightly supercooled state. The intermediate scattering functions for four values of l, are reported in Fig.5, where the comparison with the results in the liquid phase is shown. The wavevectors have been chosen to be around the first maximum of S(q), which from the data turns out to be at $q\sigma$ = 6.75 (i.e. l = 7) and reaches a value of 3.0 and 3.8 at T $\approx$ 318 K and 270 K respectively. First of all it is worth noticing that a

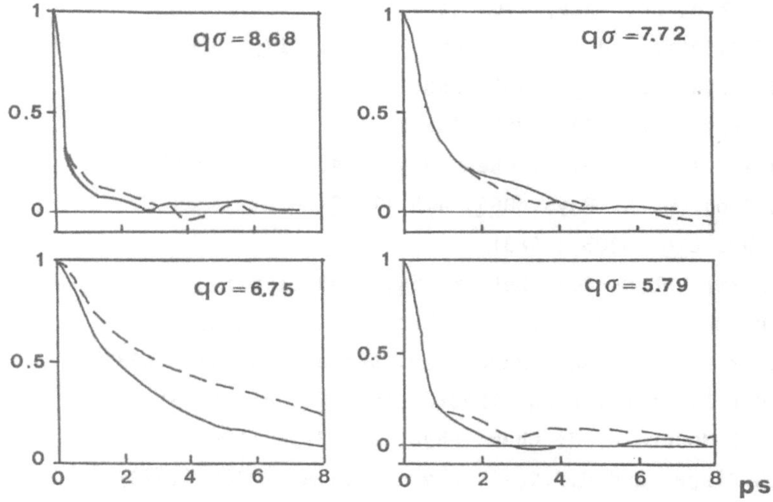

Figure 5. Normalized intermediate scattering functions at various wavevectors, in the liquid phase (full curves, T = 318 K) and in the supercooled state (dashed curves, T = 270 K).

slowing down occurs when approaching the maximum of S(q) (typical of the so called "de Gennes" narrowing of the spectrum). More interesting is to note that a slowing down effect in going from the liquid to the supercooled phase is only appreciable at the value corresponding to the maximum. Except at small t, the slowing down effect cannot be accounted for simply by the increase of S(q) at the position of the principal peak. A precursor of a structural arrest can be assumed to be responsible for this behaviour and related to the increase of the long time tail of the stress autocorrelation function.

Acknowledgements : We wish to thank Prof. S.W. Lovesey for stimulating discussions; one of us (RV) is indebted to him for the kind hospitality at Rutherford and Appleton Laboratory (U.K.), where part of the programs has been implemented. A grant by The British Council (BC), under a convention between Consiglio Nazionale delle Ricerche (CNR) and BC, is gratefully acknowledged.

References

1. C.A.Angell, J.H.R. Clarke and L.V. Woodcock, Adv. Chem. Phys. 48, 397 (1980).

2. J.R.Fox and H.C.Andersen, Ann. N.Y. Acad. Sci. 378, 123 (1981)

3. E.Leutheusser, Phys. Rev. A29, 2765 (1984)
U.Bengtzelius, W.Götze and A.Sjölander, J.Phys. C17, 5915 (1984)
U.Bengtzelius, Phys. Rev A33, 3433 (1986)
U.Bengtzelius, Phys. Rev A34, 5059 (1986)
U.Bengtzelius and A.Sjögren, J. Chem. Phys. 84, 1744 (1986)
W.Götze and L.Sjögren, Z. Phys. B65, 415 (1987)
4. T.Geszti, J. Phys C16, 5805 (1983)
5. J.J.Ullo and S.Yip, Phys. Rev. Letters 54, 1509 (1985)
6. B.Bernu, J.P.Hansen, Y.Hiwatari and G.Pastore (preprint)
7. F.Mezei, W.Knaak and B.Farago, Phys. Rev. Letters 58, 571 (1987)
8. T.Munakata and A.Igarashi, Prog. Theor. Phys. 60, 45 (1978)
9. J.G.Amar and R.D.Mountain, J. Chem. Phys. 86, 2236 (1987)
10. D.L.Price, K.S.Singwi and M.P.Tosi, Phys. Rev. B2, 2983 (1970)
11. A.Rahman, Phys. Rev. Letters 32, 52 (1974); Phys. Rev. A9, 1667 (1974)
12. U.Balucani, R.Vallauri and T.Gaskell, Phys. Rev. A37, 3386 (1988)
13. U.Balucani, R.Vallauri and T.Gaskell, Phys. Rev. A35, 4263 (1987)
14. C.S.Hsu and A.Rahman, J. Chem. Phys. 70, 5234 (1979); ibidem 71, 4974 (1979)
R.D.Mountain, Phys. Rev. B26, 2859 (1982)
R.D.Mountain and P.K.Basu, Phys. Rev. A28, 370 (1983)
R.D.Mountain and P.K.Basu, J. Chem. Phys. 78, 7318 (1983)

Electrical Properties of Liquid Quenched Alloys

H. Leitz[1], *P.-G. Häberle*[2], *C. Politis*[2,3], *and I. Wolf*[2]

[1]Fa. Edmund Bühler, P.O. Box 1126, D-7454 Bodelshausen, Fed. Rep. of Germany
[2]Kernforschungszentrum Karlsruhe, Institut für ElettNukleare Festkörperphysik, D-7500 Karlsruhe, Fed. Rep. of Germany
[3]University of California, San Diego, USA

Amorphous alloys combining early transition elements like Ti, Zr and Hf with Cu or Ni have been prepared by splat cooling and in some cases by melt spinning. The amorphousness has been confirmed by conventional x-ray diffraction and by differential thermal analysis (DTA), which gives information on the crystallization of the amorphous samples. The temperature dependence of the electrical resistivity has been measured in the range from 80 K to 300 K using a standard 4-probe (dc) technique. The temperature coefficient 1(/ρ) dρ/dT is negative for the amorphous samples and depends on temperature itself.

I. Introduction

Amorphous metals or metallic glasses are still of great interest because of their outstanding physical properties or unusual combinations of them. Besides this "economic" interest in metallic glasses, a deeper theoretical understanding of these physical properties is still a challenge for fundamental science, since the non - crystalline state of matter is the rule rather than the exception in nature.

The advantage of reducing the description of crystals to their basic units is lost in the case of amorphous materials. These objects have static fluctuations in their basic units which do not allow a periodic arrangement of atoms. In contrast to the crystalline state, the amorphous state is a metastable state. Nevertheless there is a quite good similarity between amorphous solid alloys and their liquid counterparts which are stable states in the sense of thermodynamics. So the liquid state can serve as a reference system for the corresponding amorphous state. This has increasingly turned out to be the case in the past /1/. Especially in the case of metallic glasses this is rather plausible since they are produced mainly by ultrarapid quenching techniques directly from the liquid or the vapor. The atomic structure of the liquid alloys seems to be "frozen-in" /2/. From this point of view it has been tempting to explain the electrical properties, especially the temperature coefficients of resistivity, within the framework of an extended Faber-Ziman theory /3,4/, which has been used successfully in explaining experimental results of liquid alloys /5/. Recent results obtained by resistivity measurements on metallic glasses have shown temperature coefficients which are inconsistent with predictions of the Faber-Ziman formalism. The aim of this paper is to test the reproducibility of some Ti, Zr, Hf and Cu, Ni amorphous alloys, to measure temperature coefficients and to discuss them.

Springer Proceedings in Physics, Vol. 40 Static and Dynamic Properties of Liquids
Editors: M. Davidović and A.K. Soper

II. Preparation and Characterisation

The starting alloys were mixed by weighing amounts of the pure elements and melting them in high purity argon using a modern arc-melting apparatus /6/. Then small (150 - 200 mg) spherical samples of these alloys were used to prepare the ultra rapidly quenched "splats" in our splat cooling equipment /6/. First the sample is molten in argon atmosphere (~ 800 mbar) using an inductive levitation technique (see top and middle parts of Fig. 1). This crucible-free method avoids contamination of the melt and ensures a homogeneous temperature of the molten alloy. Switching off the rf power supply of the levitation coil allows the molten alloy to drop down and be pressed in free fall by two high-energy pistons to a thin (30 - 50 µm) splat of 20 - 30 mm diameter.

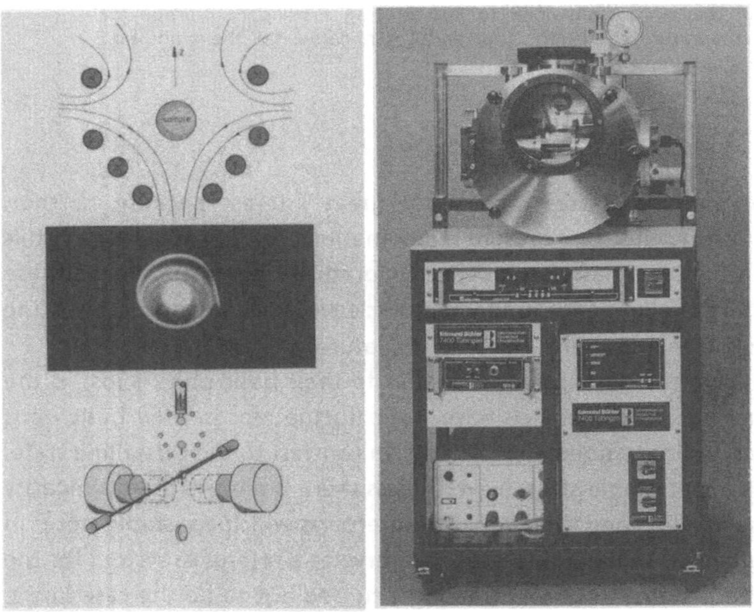

Fig. 1: Levitation melting and splat cooling. Schematic drawing and picture of the splat cooling apparatus.

Some NiZr alloys have been produced in the form of amorphous ribbons (thickness: 25 µm) by melt spinning. To avoid reactions of these molten alloys with the crucible we used an argon arc for melting these samples (approx. 3 g) in a cold crucible (water cooled copper) having a suitable bore-hole for ejecting a fine liquid jet onto the copper roller of the melt spinner /6/.

All rapidly quenched alloys have been checked by X-ray investigation in a θ-2θ goniometer using Cu-K_α radiation. The typical amorphous diffraction patterns are

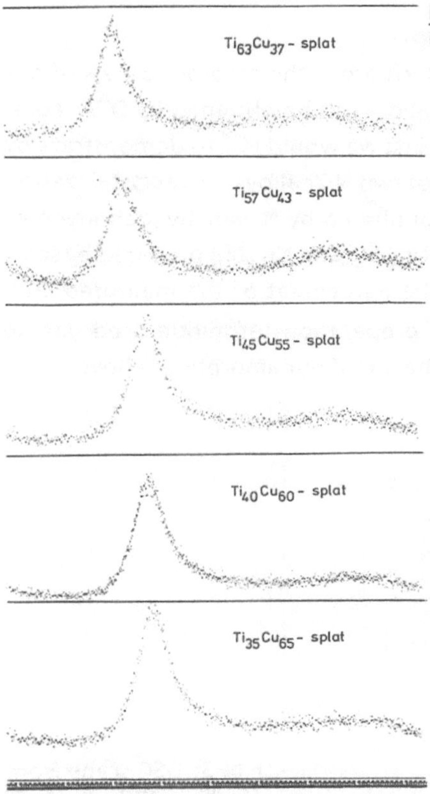

Fig. 2: X-ray diagrams of some amorphous alloys produced by splat cooling.

shown in Fig. 2 for all CuTi alloys being prepared. The X-ray profiles of a splat and a ribbon having the same composition show, within the data´s scatter, a very good agreement, which is also true for two other compositions. In the next step we investigated all samples by differential thermal analysis (DTA). This has been done using a commercial apparatus /7/ in a helium atmosphere.

The resistivity measurements were carried out using a conventional dc 4-probe method. The samples - cut pieces of 10 x 1 mm from splats or ribbons - were pressed onto gold wires between two sapphire plates. This arrangement was in good thermal contact with the sample holder (copper block) which also carries a Pt 100 (4-lead technique) for temperature measurement in the range 80 K to 300 K. This sample holder can be locked into the sample chamber of a small liquid nitrogen cryostat. By filling this chamber with gaseous helium or evacuating it, a thermal coupling or decoupling between sample and liquid nitrogen can be achieved. By this means and with the aid of a heater fixed on the sample holder the temperature can be varied in a controlled manner.

III. Results

In Sect. II we showed some X-ray measurements which prove the amorphousness of our samples. More information about the amorphous alloys can be obtained by DTA measurements which show the crystallization behavior. First we would like to demonstrate by Fig. 3 the very nice reproducibility in the case of two NiZr alloys. The crystallization peaks measured by us agree very well with those published by H. van Swijgenhoven et al. /8/. DTA measurements for three alloys $Ni_{24}Zr_{76}$, $Ni_{36}Zr_{64}$ and $Ni_{64}Zr_{36}$ prepared by splat cooling and by melt spinning have been made. The agreement of the measured DTA traces is perfect in all cases. This means that the preparation technique used has no influence on the stability and the crystallization behavior of our amorphous alloys.

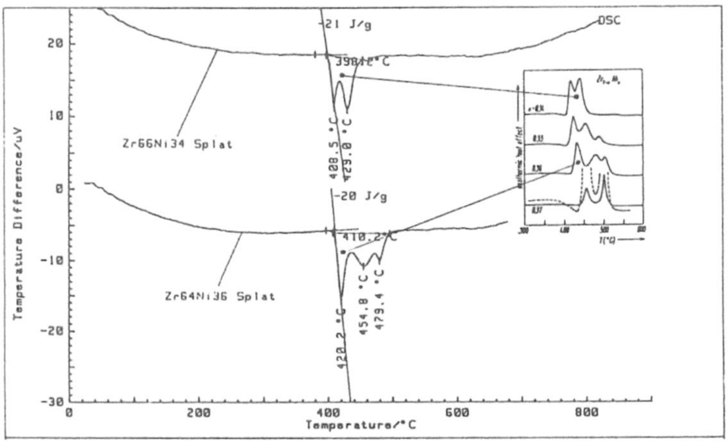

Fig. 3: DSC scans from a $Zr_{66}Ni_{34}$ and a $Zr_{64}Ni_{36}$ splat. The insert shows results after /8/ obtained from the same nominal compositions.

The same kind of measurement has been done in the case of amorphous CuTi and NiHf alloys. The onsets and peak positions of the measured DTA traces are plotted in Fig. 4. In the case of CuTi alloys the crystallization temperature does not change very much when the Ti content is changed. The smooth dip in the crystallization temperature at 50 at. % Ti may be attributed to the crystalline phase CuTi, which has a slightly increased melting point with respect to neighboring compositions (see phase diagrams inserted into the figures). A relatively strongly increasing crystallization temperature with increasing Ni content is observed in NiZr and NiHf alloys. Both systems exhibit extremely high melting intermetallic compounds NiZr and NiHf as shown in the phase diagrams (insets in the figures). At the given quenching rate of approx. 10^6 K/sec it was not possible to stabilize the amorphous phase in this range around 50 at. % Zr, or Hf. This behavior reflects the well-known fact that in contrast to deep eutectics the existence of high melting point compounds is not a good presupposition for obtaining metallic glasses.

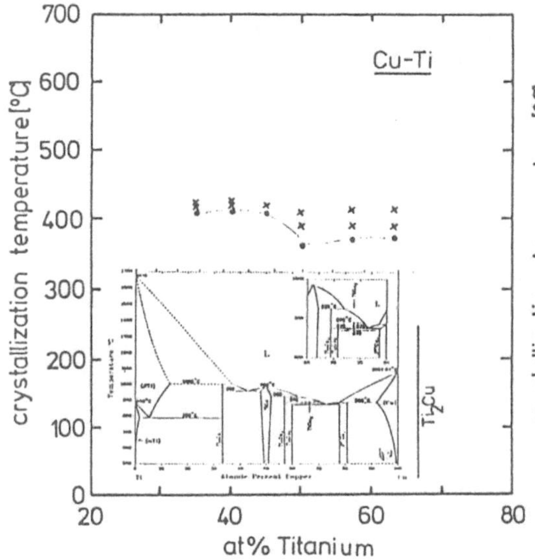

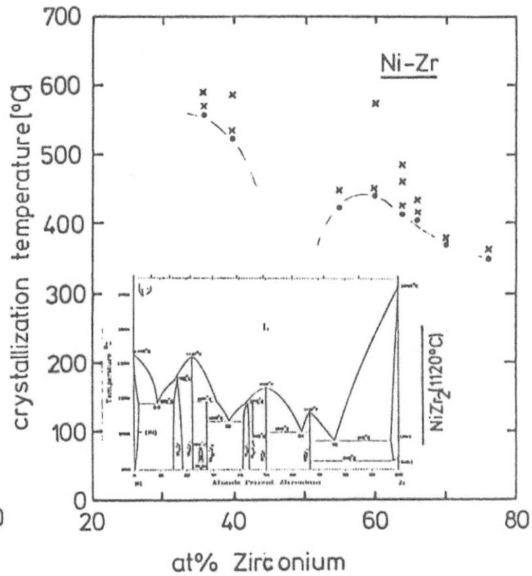

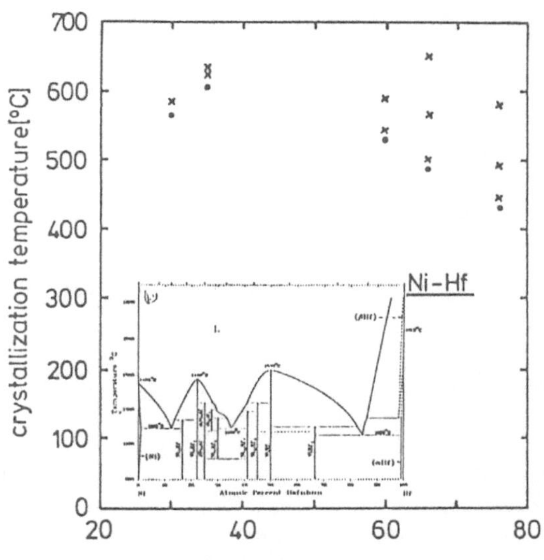

Fig. 4: Crystallization temperatures of some amorphous CuTi, NiZr and NiHf alloys. Crosses indicate peak positions in the DSC scans. The onset of crystallization is given by full circles. The phase diagrams are inserted.

In Fig. 5 the variation of resistivity with temperature is shown for an amorphous $Ni_{40}Zr_{60}$ splat (X-ray profile inserted). The resistivity, being approx. 200 Ωμcm (error ± 10 %), increases with decreasing temperature. The reversible changes in resistivity are reproducible within 0.1 %. From Fig. 5 it is quite obvious that the negative temperature coefficient (NTC) is not a constant value, but increases with decreasing temperature. The results of all resistivity measurements are given in Fig. 6. They show the normalized resistivity ρ(T)/ρ(300 K) versus temperature for some amorphous CuTi, NiZr and NiHf

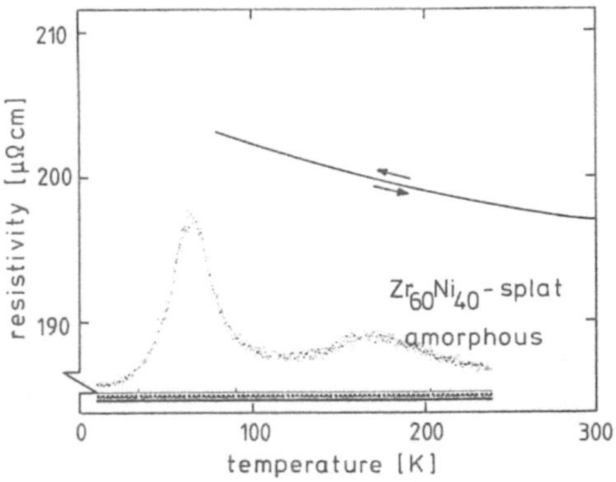

Fig. 5: Resistivity of an amorphous $Zr_{60}Ni_{40}$ splat in the range 80 - 300 K.

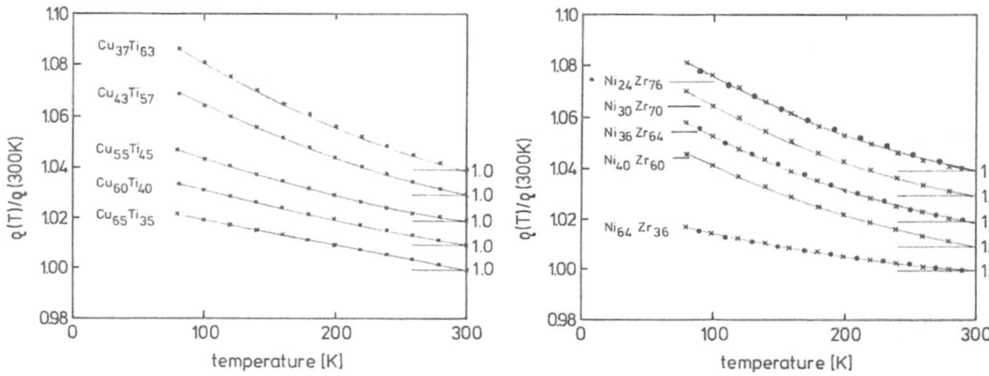

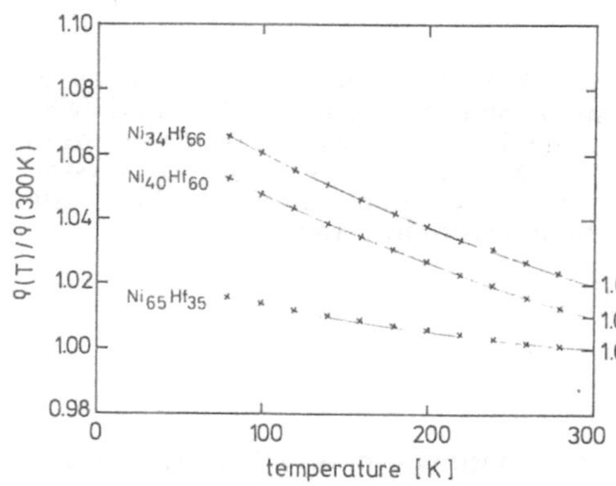

Fig. 6: Normalized reversible resistivity of amorphous CuTi, NiZr and NiHf alloys. In the case of NiZr alloys results obtained from the ribbons (full circles) are included.

alloys. Here again in the case of NiZr alloys the results obtained from some ribbons fit excellently into those obtained from splats. They all show a NTC increasing with decreasing temperature with the general trend of a decreasing NTC with increasing contents of the late transition metals. Only in the case of $Cu_{65}Ti_{35}$ did we find a linear variation of ρ(T) corresponding to a constant NTC.

IV. Discussion

Though the explanation of TC´s (especially NTC´s) with aid of an extended Faber-Ziman theory seemed to be obvious, as mentioned in Sect. I, there are some aspects which have been discussed critically in the past. One major concern about using the Ziman theory is that it may not be appropriate for strong scattering systems, which most amorphous metals are. A simple estimate for metals having $\rho \approx 200\ \Omega\mu cm$ leads to mean free paths for the conduction electrons of the order of one interatomic distance which is almost two orders of magnitude lower than a typical mean free path of a simple liquid. Furthermore, these types of amorphous metals (this work) do have significant contributions of d-electrons at the Fermi level, so the wave number k_F has only the meaning of a fitting parameter. Treating all these problems in a theoretically correct way, Meisel and Cote /3/ found that above the Debye temperature Θ_D, $\rho(T)$ should be proportional to $\pm$ T, or, in other words, TC should be a constant. In contrast to this there are experimental results which show - like ours - that TC, especially NTC, increases with decreasing temperature /9-12/. Tsuei /9/ has found that a Kondo-type scattering mechanism can fit his results on amorphous Nb_3Ge above Θ_D. Mizutani and coworkers /10, 11/ have shown that the expression $\rho(T)/\rho(300\ K) = A + B \exp(-T/\Delta)$ fits their data. Furthermore, localization effects on $\rho(T)$ have been discussed by Howson /12/. A dependence $\rho(T) \sim \sqrt{T}$ can fit some experimental data.

The experimental results, including our own measurements, show on the one hand that any simple diffraction model for the resistivity alone is not able to explain the observed behavior. On the other hand there are different attempts based on quite different physical effects which can fit the experimental data. But it seems to us from the study of this literature that these fits cannot prove the validity of one physical picture. So more work-experimentally and theoretically - must be done to gain better insight into this problem.

References

1. Hafner, J. : in Topics in Appl.Phys. : Glassy Metals I, eds. H.J.Güntherodt and H.Beck (Springer-Verlag, Berlin 1981)
2. Leitz, H. : Z.Phys. B40, 65 (1980)
3. Meisel, L.V., Cote, P.J. : Phys.Rev. B17, 4652 (1978)
4. Wochner, P., Jäckle, J. : Z.Phys. B44, 293 (1981)
5. Ziman, J. M. : Philos.Mag. 6, 1013 (1961)
6. "Bühler News", Fa. E. Bühler, 7454 Bodelshausen, FRG
7. Fa. Netzsch: DTA 404
8. Van Swijgenhoven, H., Stals, L. M., Buschow, K. H. J.: phys. stat. sol. (a) 72, 153 (1982)
9. Tsuei C. C.: Sol. St. Com., 27, 691 (1978)
10. Mizutani, U.: Rapidly Quenched Metals, 977 (1985) eds. S. Steeb, H. Warlimont
11. Yamada, Y., Itoh, Y., Matsuda, T. and Mizutani, U.: J. Phys. F: 17, 2313 (1987)
12. Howson, M. A.: J. Phys. F: 14, L 25 - L 31 (1984)

A Molecular Dynamics Study of the Liquid-Glass Transition in Lennard-Jonesium

M.J.D. Brakkee and S.W. de Leeuw

Laboratory for Physical Chemistry, University of Amsterdam,
Nieuwe Achtergracht 127, NL-1018 WS Amsterdam, The Netherlands

1. Introduction

In recent years considerable attention has been paid to relaxation phenomena in undercooled liquids and to the liquid-glass transition. Theoretical models, based on mode-coupling principles, have been proposed which provide an explanation of this transition. One of the first models is due to LEUTHEUSSER [1]. More recently, GOETZE [2] and BENGTZELIUS and SJOEGREN [3] worked out these ideas in a more rigorous way. BENGTZELIUS [4,5] made some calculations specifically for the Lennard-Jones system.

In these theories the intermediate scattering function plays a central role. A non-linear feedback is derived for this function, leading to non-decaying solutions in a certain range of density and temperature. This behaviour is called structural arrest. In this picture it is the main feature of the glassy state. The persistent part of the intermediate scattering function is regarded as a dynamical order-parameter for the glass transition. Structural arrest also causes the singular behaviour of thermodynamic quantities at the transition. In the limit of infinitely small cooling rates there is a well defined transition temperature T_0. For any finite experimental duration, however, the "thermodynamic" transition temperature will be higher than T_0.

Recent MD-simulations on the LJ-system are by ULLO and YIP [6]. They limited the LJ-potential to its repulsive part. We present here the results of MD-simulations on the (almost) full Lennard-Jones system with the potential truncated at 2.5 σ. Where possible we compare these results with BENGTZELIUS' [4,5] calculations. Throughout this paper we use reduced units as customary, except for the unit of time $\tau^2 = m\,\sigma^2/24\,\varepsilon$. Our unit of viscosity is $\varepsilon\tau/\sigma^3$.

2. Results from MD-simulations

We performed a series of MD-simulations in the undercooled LJ-liquid. Our system consisted of 958 particles and we used NOSE [7] constant (p,T)-dynamics. We started off in the stable liquid at a reduced temperature of T = 1.0 and reduced pressure p = 3.20, and subsequently cooled it along the isobar stepwise down to T = 0.30 with steps of $\Delta T = 0.1$. At each temperature a simulation of 10^4 timesteps was carried out. With a timestep of 0.02 τ and an equilibration period of 1000 timesteps this means an average cooling rate of 4.5 10^{-4} ε/τ which is equivalent to 1.2 10^{11} K/s for Argon. At low temperatures, time averaging becomes less and less effective: the wiggles in the tails of the time-correlation-functions are no longer a good indication of the error. We therefore performed several quenches in the same way. Often nucleation occurred and the quench had to be abandoned. The fluctuations between the remaining quenches appear small enough not to invalidate our conclusions. In this respect our results are only preliminary and more study is needed.

We observed many of the characteristic features of the glass transition. First of all, the thermal expansion coefficient

$$\alpha_T = \frac{1}{V}\left(\frac{\partial V}{\partial T}\right)_p$$

and the specific heat

Springer Proceedings in Physics, Vol. 40
Editors: M. Davidović and A.K. Soper

Static and Dynamic Properties of Liquids

$$C_p = \left(\frac{\partial H}{\partial T}\right)_p$$

show a rapid change in a comparatively small temperature region around a certain temperature T_g. This phenomenon provides the empirical definition of an observational glass-transition temperature T_g which we found to be $T_g \approx 0.45$. We plotted volume per particle v and enthalpy per particle h in such a way as to make this clearly visible (Fig. 1).

The stress-tensor correlation function (Fig. 2)

$$C_\sigma = < \sigma^{xy}(t)\ \sigma^{xy}(0) >$$

and intermediate scattering function

$$F(k,t) = \frac{1}{N} < \sum_i e^{ikr_i(t)} \sum_j e^{-ikr_j(0)} >$$

develop non-decaying tails which is evidence for structural arrest. In the spectrum of the transverse current correlation function an elastic peak shows up (Fig. 3); a clear indication of solid-like behaviour with transverse sound at long wavelengths. The shear viscosity coefficient η grows beyond measure and the self diffusion coefficient D tends to zero. These divergences, as far as they can be observed, occur at still lower temperatures than the "thermodynamical" T_g.

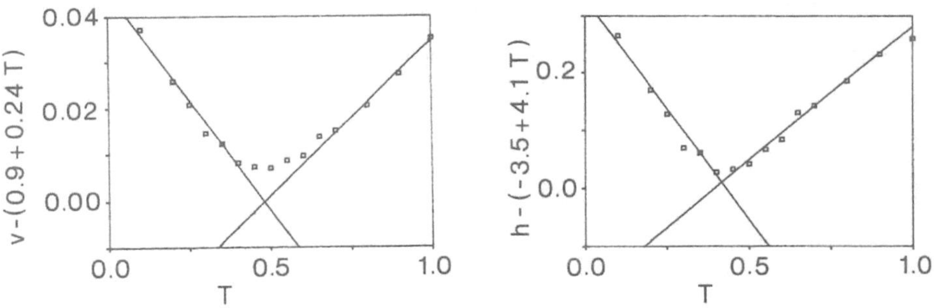

Fig. 1: Volume per particle v and enthalpy per particle h versus temperature; a straight line is subtracted in each case to magnify the discontinuity in slope.

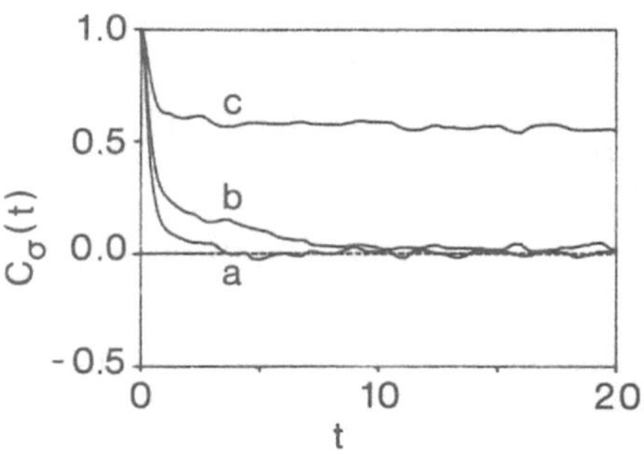

Fig. 2: Normalized stress-tensor correlation function as a function of time;
a: T = 1.00
b: T = 0.60
c: T = 0.30

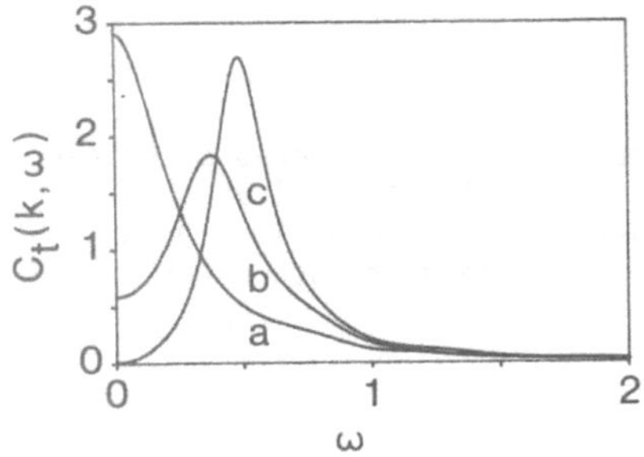

Fig. 3: Spectrum of transverse current correlation function normalized to unit area, $k \approx 0.6$;
a: T = 1.00
b: T = 0.60
c: T = 0.30

We compared the observed behaviour of η with the relation given by BENGTZELIUS [4] that it should diverge as $\eta \cong |T-T_0|^{-\alpha}$ with α = 1.765. We were able to fit our data with a comparatively wide range of T_0's and α's, but with a strong correltion between α and T_0. If we demand that $\alpha \approx 1.8$ we obtain $T_0 \approx 0.28$ (Fig. 4). However, an Arrhenius fit to the data is also possible.

BENGTZELIUS [4] calculated the limiting values

$$f_k = \lim_{t\to\infty} F(k,t)$$

of the intermediate scattering function on the glass transition point. We compare our results for T = 0.30 with these values (Fig. 5). The comparison seems to be rather poor except for the point $k = k_{max}$, that value of k where S(k) has its maximum, where the agreement is remarkably good.

We find a glass transition point at a lower temperature and higher density than calculated by BENGTZELIUS [5] and in agreement with simulations of ULLO and YIP [6].

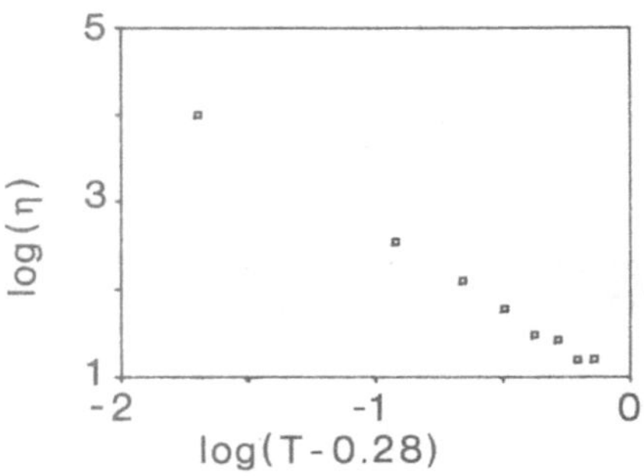

Fig. 4: Viscosity versus temperature showing a possible power law dependence near T_0 = 0.28.

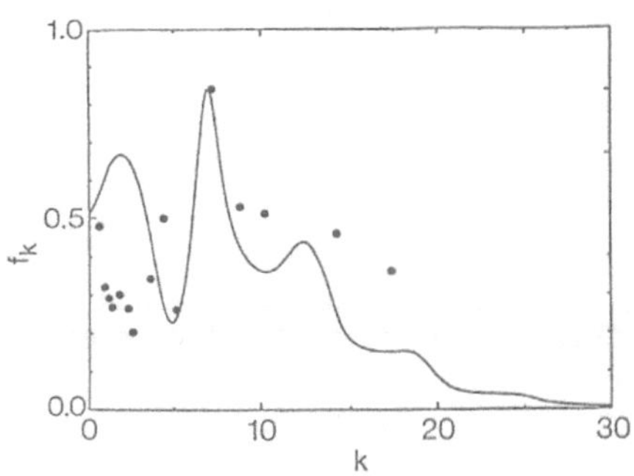

Fig. 5: Persistent part f_k of the intermediate scattering functions on the glass transition; the drawn line is the calculation by BENGTZELIUS [4];the dots represent the MD-data.

3. Conclusions

The LJ-system shows a change in thermodynamical and dynamical properties at low temperatures in the supercooled liquid region. Structural arrest is apparent from the growing structural relaxation times. The persistent part of the intermediate scattering function agrees well with theory at $k = k_{max}$ (where mode-coupling probably works best), but the agreement is rather poor at other values of k. The appearance of transverse sound at long wavelengths is equivalent to elastic response on applied shear, which is the most common property of a solid. The divergence of the viscosity does not seem to point to a certain temperature unless we assume the theoretical relation to be valid, in which case a best value is $T_g \approx 0.28$. Our data are not yet accurate enough to discriminate between this relation and, for example, Arrhenius behaviour. In any case, if a finite T_g is associated with these phenomena it is certainly considerably lower than the theoretical predictions.

REFERENCES

1. E. Leutheusser, Phys. Rev. A 29, 2765 (1984)
2. W. Götze, Z. Phys. B - Condensed Matter 60, 195 (1985)
3. U. Bengtzelius and L. Sjögren, J. Chem. Phys. 84, 1744 (1986)
4. U. Bengtzelius, Phys. Rev. A 34, 5059 (1986)
5. U. Bengtzelius, Phys. Rev. A 33, 3433 (1986)
6. J.J. Ullo and S. Yip, Phys. Rev. Lett. 54, 1509 (1985)
7. S. Nosé, J. Chem. Phys. 81, 511 (1984)

Part V

Structure of Liquids, Effective Pair Potentials and Three-Body Forces

The Electronic Structure of Liquid Metals

*J. Hafner, W. Jank, and M. Tegze**

Institut für Theoretische Physik, Technische Universität Wien,
Karlsplatz 13, A-1040 Wien, Austria
*Permanent address: Central Research Institute for Physics,
Hungarian Academy of Sciences, Budapest

We present an investigation of the interplay of the atomic and the electronic structure of liquid metals, based on interatomic forces derived from ab initio pseudopotentials, molecular dynamics simulations of the atomic structure, and supercell-LMTO calculations of the electronic structure. Trends in the electronic structure of the liquid polyvalent metals and in liquid and amorphous simple-metal alloys are discussed.

1. Introduction

Over the last two decades the electronic theory of crystalline solids has advanced at a rapid pace. With the advent of modern theoretical concepts such as density functional theory and linearized band-structure methods the state-of-the-art has progressed from a mere qualitative prediction of the electron states in bulk solids to a quantitative prediction of materials' properties and to a detailed discussion of the electron states at defects, surfaces, interfaces, etc[1].

For the electronic theory of liquids the situation is entirely different. Until very recently the traditional belief that the electronic structure of many liquid simple metals is very close to the free-electron model had remained unchallenged by both experimentalists and theorists. Because of the tremendous experimental difficulties associated with photoelectron- or X-ray-spectroscopic studies of liquid metals only a small number of investigations had been performed, and very often with photon energies so low that only the upper part of the valence band could be studied. The recent work of the Basel group[2,3] provides us for the first time with a coherent picture of the systematic variations in the electronic structure of the liquid polyvalent metals (Fig.1). The total electronic density of states shows an increasing tendency to a separation of the partial s and p contributions with increasing atomic number within each group and with increasing valence within each row of the periodic table. Hence there is increasing deviation from a free-electron behaviour whose explanation is a challenge to theorists. For binary systems, the recent work on glassy alloys[4] (i.e. on frozen liquid alloys) complements the rather narrow data base on the electronic structure.

The reason why the electron theory of liquids is in a much poorer state of development than the theory of crystalline solids is that the ions are arranged periodically in the crystal but are spatially disordered in the liquid. The translational invariance of the crystalline lattice provides the analytic framework of the theory (Blochs theorem) and it is the lack of a comparably powerful concept which has left the electronic theory of liquids in a relatively primitive state .

Another difficult point is the following. In principle, even a pure metal is a two-component system, being composed of electrons

Springer Proceedings in Physics, Vol. 40
Editors: M. Davidović and A.K. Soper

Static and Dynamic Properties of Liquids

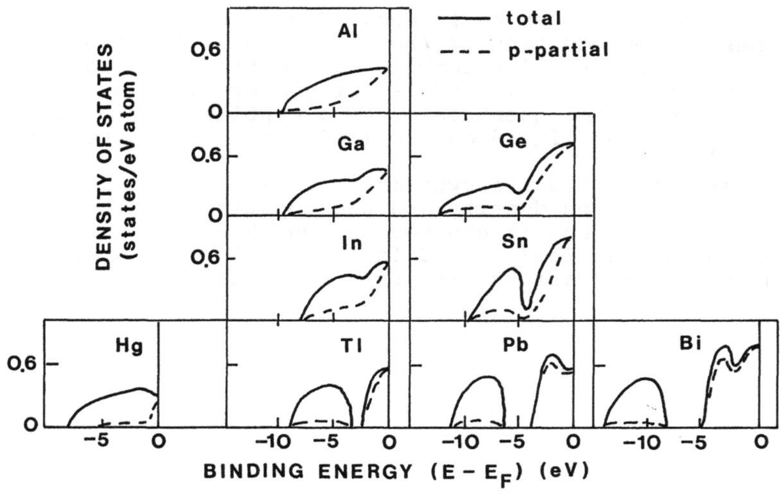

Fig.1: Photoemission density of states of the liquid polyvalent metals from UPS and XPS investigations. After Indlekofer et al.[2,3].

and ions. For a crystal the ionic coordinates are known from a diffraction experiment, and as long as the adiabatic approximation holds we do not have to worry about the ionic degrees of freedom. For a liquid, a diffraction experiment yields only statistical information on the distances between pairs of atoms (the pair distribution function) and this is much less information than necessary for an accurate calculation of the electronic structure. Therefore any attempt to calculate the electronic structure of a liquid metal has to start with a calculation of the atomic structure. This provides us with a set of ionic coordinates which serve as the basis for the calculation of the electronic structure. Still we have to solve the problem of calculating the electronic states in a condensed material without the use of Bloch's theorem - or to develop approximations allowing us to continue to use this powerful concept, even though translational periodicity has disappeared.

2. Interatomic Forces and Atomic Structure

Evidently the starting point of our investigations has to be the determination of the interatomic forces. A possible way to decompose the total energy of a metal into a set of volume, pair, and many-body forces is based on the response of the conduction electrons to the electron-ion potential[5,6]. However, the application of reponse theory is restricted to systems whose ions scatter the electrons only weakly and are describable by pseudopotentials. This excludes the strong-scattering d-electron systems immediately. The contributions of the d-electrons to the interatomic forces may be calculated within a mixed tight-binding framework using a momentum expansion of the electronic density of states to arrive at the desired decomposition of the total energy[7,8].

Once the interatomic potentials are known, the structure of the liquid may be determined straightforwardly by a molecular dynamics simulation. We have used a highly effective (N,V,E) molecular dynamics routine[9]. Parallel simulations for large ensembles (up to 1372 atoms in a cubic box with periodic boundary conditions) and small models (about 60 atoms) have been performed. The simulations on the large ensembles permit reproduction of the available diffraction data with good accuracy; the small ensembles create the basis for the electronic structure calculations. Models for glassy systems may be produced by molecular-dynamics quenches or by a mapping of the atomic configurations of the liquid on local minima in the potential energy[10,11].

For the pure simple metals the systematic variations in the interatomic forces with electron density and pseudopotential have been thoroughly investigated[12,6] and it has been shown that the characteristic trends in the crystalline[2,12] and in the liquid[6,13] structures arise from these variations in the interatomic potentials (Fig.2).

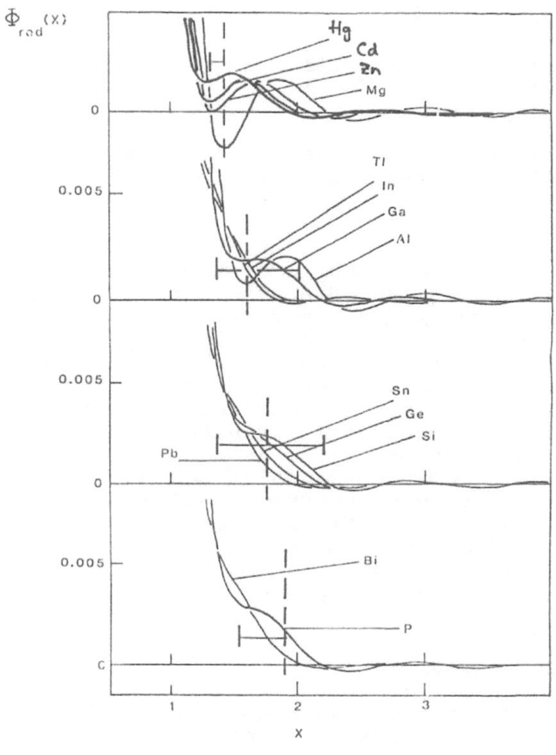

Fig.2: Variation of the reduced interatomic potentials $\Phi_{red}(r)$ ($\Phi(R) = Z^2/R \cdot \Phi_{red}(R)$) for the polyvalent B-group elements. Distances are given in units of the Friedel wavelength, $x=R/\lambda_F$. After [6,12].

Put most simply, one finds that the basic question is the compatibility of the geometric and the electronic factors influencing the liquid structures. At low electron densities and large core radii both are compatible with hard-sphere-like structures. As the electron density increases and/or the core radius decreases, this constructive interference disappears and the oscillations in the potential are shifted relative to the repulsive core. The diameters of the repulsive core and the Friedel wavelength, which gives the periodicity of the long-range oscillations in the potential, now set two different characteristic distances. The complex structures of Zn, Cd, Ga, In, Si, Ge, and Sn result from the interplay of these different length scales. However, at high atomic numbers with a further reduction of the effective core radius (let us note at once that the reduction of the core radius with increasing atomic number accounts for both the existence of a d-electron core and for relativistic bonding effects[12]) the node of the pseudopotential approaches $2.k_F$, hence the oscillations in $\Phi(R)$ are strongly damped and only the influence of the geometrical factors prevails - this explains the return to hard-sphere-like structures in Tl and Pb.

For binary alloys the situation is more complicated. In systems with large differences in the electron densities and/or the electronegativities of the constituents, large composition-dependent changes in the interatomic forces are predicted. The origin of these changes is twofold: (a) Because of the reduction of the conduction electron density, the screening of an ion in an alloy with a lower electron density than in the pure metal (e.g. Pb in K-Pb) will be much weaker. In our example this results in a strongly enhanced Pb-Pb interaction, see Fig.3a. (b) The second point is intimately related to the quantum-mechanical nature of the pseudpotential. The repulsive part of the pseudopotential arises from the fact that on penetrating the ionic core the valence electrons acquire a high kinetic energy because of the orthogonality constraint of the Pauli principle. Formally, this additional kinetic energy may be treated as a repulsive potential. In an alloy, where an ion sees a lower electron density than in the pure metal, this orthogonalization effect will be weaker and hence the ionic pseudopotential will be more attractive than in the pure metal, compensating for some of the reduced screening effect (see Fig.3b). We find that both effects together are important for obtaining a realistic form of the pair interactions compatible with the interatomic distances in the crystalline intermetallic compound and the average distances in the liquid alloy (Fig.3).

The magnitude of this effect depends strongly on the angular momentum - the s-component of the valence-electron states will be more affected because the s-partial wave has larger amplitudes in the core region. Of course we should expect that these effects are reflected in the density of states.

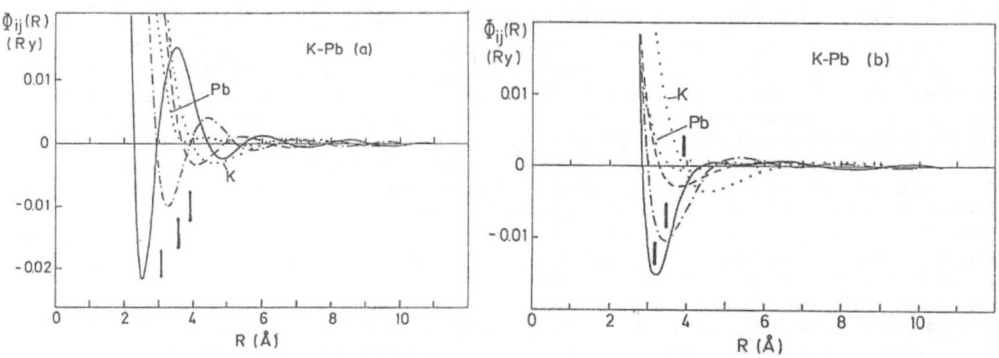

Fig.3: Interatomic potentials in a $K_{0.5}Pb_{0.5}$ alloy . (a) Calculated with concentration-independent empty-core model potentials - only the screening effect is reflected in the interatomic potentials. (b) Calculated with first-principles pseudopotentials optimized for the composition of the alloy, the pair potentials reflect the change in the screening and the variation of the ionic pseudopotentials. Full lines : Pb-Pb interactions, dot-dashed lines: K-Pb, and dashed lines: K-K interactions. The dotted lines show the pure-metal pair interactions, the vertical lines mark the nearest-neighbour distances in the crystalline compound KPb; they are very close to the average nearest-neighbour distances in the liquid alloy[14].

3. The Electronic Structure of Liquids

Many attempts have been made to calculate the electronic density of states of liquid metals. They may be classified in three groups: (a) effective medium approximations (EMA), (b) real-space techniques such as the recursion and equation-of-motion techniques, and (c) cluster and supercell techniques. In the EMA[15,16] the DOS is expressed in terms of configurationally averaged quantities related to the scattering-path operator. The configuration averages are expressed in terms of pair-correlation and approximate multi-particle correlation functions. The solution of the integral equations defining the configuration-averaged scattering path operator is a very extensive computational task. At present it is impossible to achieve electronic self-consistency. The recursion[17] and equation-of-motion[18] techniques offer very efficient routines for the calculation of the DOS based on the full structural information contained in the coordinates of a large structural model (1000 atoms and more). Up to now, self-consistency has been achieved only for an average atom[19]. The advantage of the cluster[20] and supercell techniques[21] (a supercell calculation is a conventional band structure calculation for a periodically repeated cell, it may be performed with any available technique) is the possibility of achieving electronic self-consistency for each atom individually. This allows one to observe the influence of local fluctuations in the atomic structure on the electronic DOS. The disadvantage is the unknown influence of the boundary conditions, although it is clear that the periodic boundary conditions of a supercell are a less stringent restriction than the free surface of a free cluster.

We have recently presented a number of supercell calculations[22,23] for amorphous metallic alloys based on the linearized muffin tin orbital (LMTO) method[24] in the atomic sphere approximation. This method (which has been adopted for its computational efficiency) is applied here for the first time to liquids. Given the limitations of our computer capacity, we can handle up to about 60 atoms per supercell with a full s-p-d basis set. Usually supercell calculations are done for the Γ-point (i.e. in practice $k=0.01\cdot 2\pi/L$) of the Brillouin zone of the supercell only, and a continuous DOS is derived from the discrete spectrum of the Γ-point eigenvalues by applying an appropriate Gaussian broadening. For the rather narrow bands of amorphous transition metal alloys this is an acceptable approximation, but for the much wider bands of simple metal systems this yields unrealistic results. A possible way to simulate a much larger cell is to do a more extensive $\vec{k}$-space sampling - a four-point calculation essentially simulates a cubic cluster with twice the cell-diameter (but ignores the effects of disorder for distances larger than the cell-diameter). In order to test whether this produces a realistic DOS we have recently performed in cooperation with S.K.Bose and O.K.Andersen[25] a comparative study of the recursion and supercell techniques. The average self-consistent potentials of a 60-atom supercell calculation for an amorphous Ca_7Al_3 alloy were used as the basis for a tight-binding- LMTO[26] recursion calculation on an 800-atom model. This allows us to test the influence of the periodic boundary conditions and of the $\vec{k}$-space sampling in the supercell calculations. The results for the total and the partial DOS are very similar, both methods require a comparable computational effort. This shows that even with these relatively small models the supercell method produces reliable results for the electronic DOS of disordered simple-metal systems. Because of the advantage of giving a locally self-consistent DOS we think that the supercell method deserves preference over the other techniques.

3.1 The Electronic Structure of Crystalline and Liquid Germanium

In contrast to most other solids, the elements Si and Ge shrink when they melt. In the crystalline form stable at normal pressures these materials are semiconducting, they have an open structure with coordination number four. Under pressure a series of polymorphic transitions to a metallic form, first with the β-tin structure (coordination number six) and then to a simple hexagonal structure, are observed. Melting is accompanied by a semiconductor/metal transition and diffraction experiments show that the number of nearest neighbours in the melt is about 6 to 7. The metallic liquid may be quenched to a semiconducting amorphous form, which may again be metallized by applying pressure[27]. The semiconducting and amorphous forms of Si and Ge have recently been the subject of extensive theoretical studies[10,28-30]. Most of these investigations are based on empirical pair and many-body forces fitted to a large base of theoretical and experimental data. In contrast to these studies we have started a systematic investigation of the group IV metals based on pseudopotential-derived interatomic forces[6,27,31]. We find that the atomic structures of liquid Si, Ge, Sn, and Pb are all well described in terms of the interplay of volume- and density-dependent pair forces - the physical mechanism stabilizing the atomic arrangement is precisely the one described in Sect.2. As an example we show in Fig.4 the comparison of our molecular-dynamics simulations with the available diffraction data. However, it is also necessary to go beyond the pair correlations. The simplest thing is to have a look at the bond-angle distributions (Fig.5).

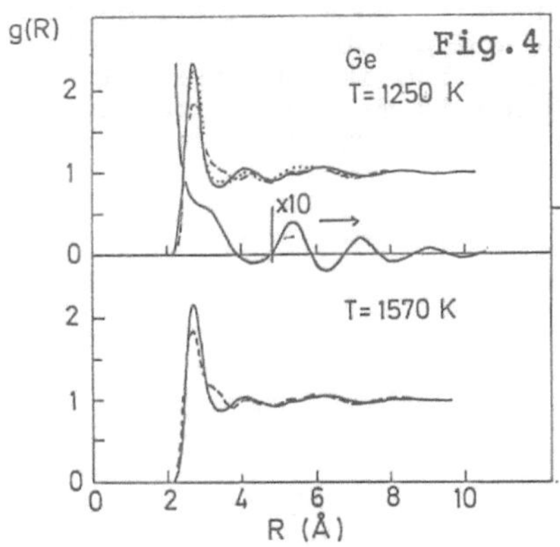

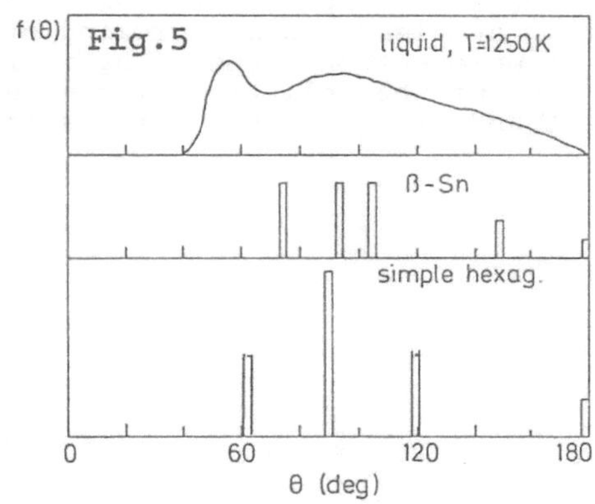

Fig.4: Pair potential Φ(R) and pair correlation function g(R) for liquid Ge. Full line - MD simulation, dotted and dashed lines - experiment. After [27].

Fig.5: Bond-angle distribution f(θ) in liquid and supercooled liquid Ge. The distribution of the bond-angles in the crystalline high-pressure polymorphs is given for comparison. After [27].

We find that although there is some similarity between the distribution of the bond angles in liquid Ge and the bonds in the crystalline metallic high-pressure polymorphs, there are also characteristic differences: the peak near $\theta \approx 60°$ indicates the existence of triplets forming nearly equilateral triangles which are absent in the β-tin structure and exist only in the simple hexagonal form of Ge which is formed at very high pressure. In simulations based on empirical pair and triplet forces this 60° peak is sometimes present, sometimes missing. We think that it reflects a genuine difference in the local atomic arrangement, which is also reflected in the electronic structure.

The electronic structure of α-Ge shows the characteristic signature of the sp^3-hybrids stabilizing the tetrahedral coordination in the diamond structure (Fig.6a). The local arrangement in β-Ge shows a flattened diamond grouping with four neighbours along the distorted tetrahedral axes and two more only slightly further away along the c-axis. The retention of part of the local order is reflected in the electronic DOS (Fig.6b). In simple hexagonal Ge the remaining directional character of the bonds is lost and the electronic DOS is very close to a free-electron parabola, with a weak structure near E_F (not shown in the figure). As in any free-electron metal we observe a large overlap of the s-and p-partial DOS. It was therefore quite surprising when the photoemission experiments[2,3] showed such a sharp structure in the DOS of liquid Ge (see Fig.1). The calculated electronic DOS of liquid Ge and of a hypothetical metallic amorphous form (obtained by projecting liquid configurations on local potential-energy minima) show the deep minimum about 4 eV below the Fermi level which was observed in the UPS[32]. We find that the s,p hybridization of the valence band is completely lost, there is little overlap between the s- and the p-bands. We think that there are two reasons for the dehybridization of the electron states: (a) The large disorder in the directionality of the bonds forbids the formation of hybrid orbitals associated with a certain preferred nearest-neighbour geometry. (b) The relatively low coordination number ($N_c \approx 6$-7 compared to $N_c \approx 10$-11 in "normal" metals) reduces the overlap and leads to a narrowing of the s- and p-subbands, so that the "free-electron" type of hybridization is also impossible. In Si and in Ge the splitting of the valence band is mainly a structural effect and we think that our results for the electronic structure

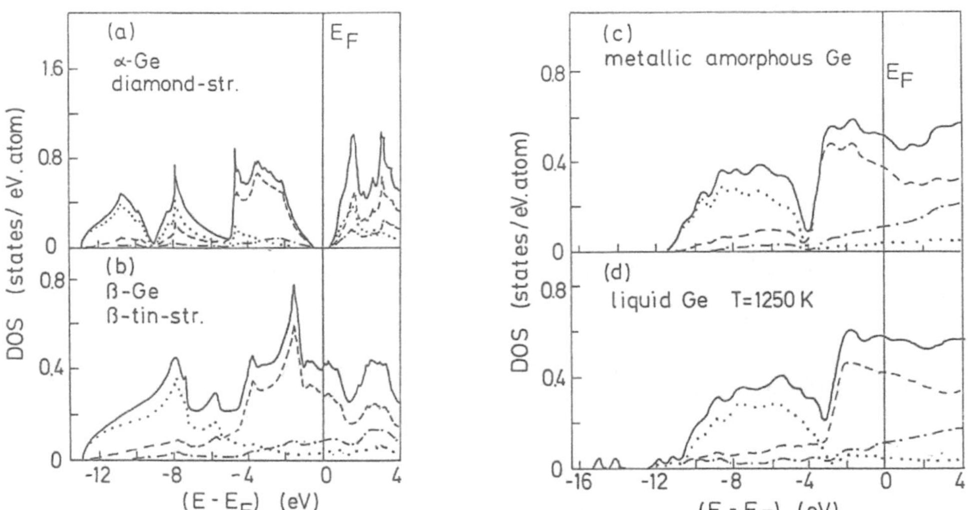

Fig. 6(a-d): Electronic density of states for crystalline, liquid, and quenched liquid (≈metallic amorphous) Ge. Full line - total DOS, dotted - s-, dashed - p-, dot-dashed d-electron contribution .

corroborate the arguments[6,13,31] we had put forward on the basis of our studies of the atomic structure alone. In the heavier elements the splitting of the band is reinforced by relativistic effects, which leads to a stronger binding of the s-electrons compared to the p-electrons. Remember that this is precisely the effect which dominates the trend to purely repulsive interatomic forces and the return to close-packed structures. The importance of relativistic effects for the crystal structure of lead has also been stressed by Christensen[33].

Finally we want to make a brief comment on the differences in the electronic structure of the liquid and the metallic amorphous phases. In the atomic structure the difference is mainly in much sharper structures in the correlation functions of the quenched phase. In the electronic structure the larger spread of the nearest-neighbour distances results in resonant bound states at high binding energies. These states are created and destroyed at the time-scale of the local density fluctuations. If a configuration average were taken, they would produce a long, flat tail at the bottom of the conduction band. These states are missing in the DOS of the band, in the case of liquid Ge mainly in the p-band. This leads to a narrowing of the p-band. Similar effects have been observed in the electronic DOS of liquid and amorphous Au-Si alloys[2].

3.2 The Electronic Structure of Amorphous Simple-Metal Alloys

Very few systematic investigations for the electronic DOS of liquid simple-metal alloys are known. On the other hand there has been considerable interest in the electronic structure of amorphous alloys. In the example of Ge we have seen that the difference in the electronic structure of a liquid and an amorphous metal is small. The main difference is the existence of a small number of short-lived localized resonance states at low energies. These localized resonances arise from short interatomic distances due to the thermal fluctuations. These states are missing in the band and lead to a slight narrowing of the p-band. With these differences in mind, we can safely discuss the electronic structure of liquid and amorphous alloys together.

The first point we want to investigate is of long-standing interest. More then ten years ago Nagel and Tauc[34] suggested that in analogy to the Mott-Jones interpretation of the Hume-Rothery rules, the sharp first peak should induce a minimum in the electronic DOS, and if this minimum falls at the Fermi level, i.e. if $Q_P \approx 2k_F$, this would contribute to a stabilization of the amorphous alloy. The search for this structure-induced DOS minimum turned out to be a frustrating task, both experimentally and theoretically[35]. This was mainly due to the fact that in the amorphous Hume-Rothery phases formed by noble metals and polyvalent metals, where this effect should be strongest, it is difficult to distinguish the structure-induced effect from a DOS minimum induced by s-d hybridization.

Therefore we have concentrated our efforts in the prototype free-electron glass Mg-Zn[23]. In Mg-Zn we have $Q_P/2k_F = 1.08$, so we would expect a DOS minimum at E_F. The DOS of the stable crystalline intermetallic compound $MgZn_2$ shows a broad s-p band overlapping at its bottom with a narrow Zn-3d band. The s-p band is strongly structured by the van Hove singularities of the complex C14 structure (hexagonal Laves phase). The essential features of the DOS are preserved in the glassy phase, but the structure-induced features have nearly completely disappeared with the breakdown of translational periodicity. The DOS is now very close to a free-electron parabola, except for the weak structure-induced minimum at E_F (Fig.7). We calculate a Mott factor of $g=n(E_F)/n_{free}(E_F)= 0.90$; electronic specific

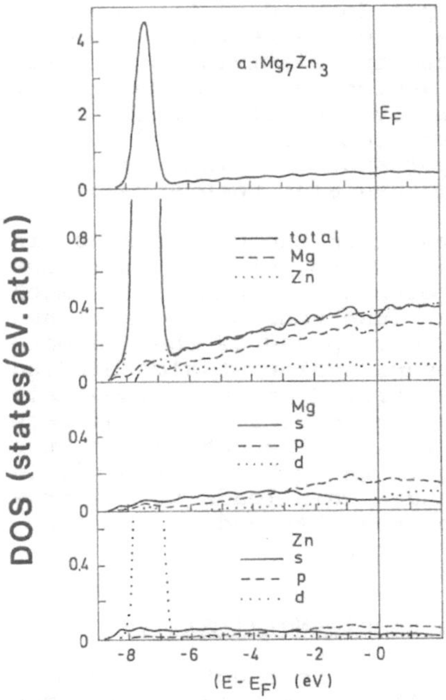

Fig.7 Total and partial electronic DOS of an amorphous Mg_7Zn_3 alloy. Note the small structure-induced DOS minimum at E_F, see text. After [23].

heat measurements yield g≈ 0.77-0.97. The existence of the DOS minimum is also well comfirmed by photoemission experiments. Thus we conclude that the interplay of the atomic and the electronic structures suggested by Nagel and Tauc exists, but is rather weak in this free-electron system.

Stronger cross-correlations between the atomic and the electronic structures (but of a somewhat different nature) are expected in Ca-based glasses. Ca-Mg and Ca-Al glasses show interesting differences in the structural, thermodynamic, electronic, and electronic transport properties (see [22] for details). The calculation of the interatomic potentials shows that the changes in the screening and in the pseudopotentials lead to a strong compression of the effective diameter of the Ca ions on alloying with Al, and to much stronger Al-Al interactions, compared to the pure metals[6]. These effects are much weaker in Ca-Mg alloys. The difference shows up in the atomic structure. The bond-angle distribution for Ca-Mg glasses (Fig.8) shows peaks at the icosahedral bond angles, those for Ca-Al glasses at $\theta \approx 100°$ and $\theta \approx 150°$ originating from the Ca sites. This suggests that there are considerable differences in the local environment of Ca in Ca-Mg and Ca-Al, Ca-Al glasses deviate more strongly from a predominantly icosahedral short-range order (see [22] for a more detailed discussion). We also find distinct differences in the electronic DOS (Fig.9). Ca-Mg glass has a parabolic s-p band at higher binding energies, and a strong onset of a Ca-3d DOS at E_F. We find g = 1.69, in good agreement with photoemission data[36] (g = 1.74). In Ca-Al on the other hand a DOS minimum about 3 eV below the Fermi level separates the Al-3s band from the Al-3p band; only the Al-3p states interact with the Ca states.

Rather surprisingly g = 1.58 is lower in Ca_7Al_3 than in Ca_7Mg_3, but again this is well confirmed by photoemission experiments (g ≈ 1.34)[37]. The narrowing of the Al-3p band in the amorphous alloys compared to the Al-rich intermetallic compound $CaAl_2$ and to pure Al

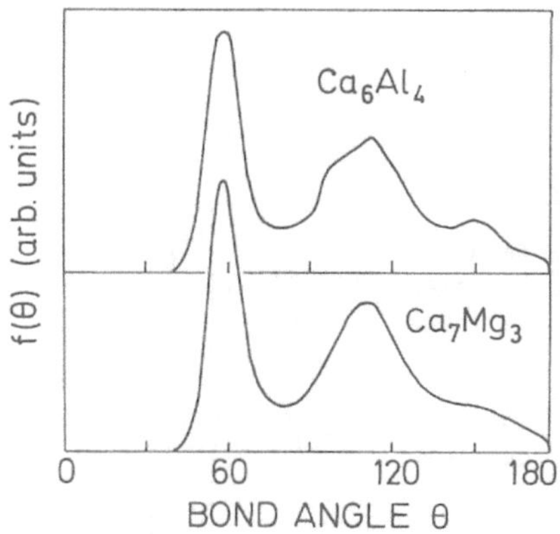

Fig.8: Bond-angle distributions in Ca-based metallic glasses. In Ca-Mg one finds a predominantly icosahedral short-range order. The additional structure found in the Ca-Al glasses indicates a substantially different coordination of the Ca and the Al sites: the Al atoms still have an icosahedral enviroment, but much higher coordination is found around the Ca sites. After [22].

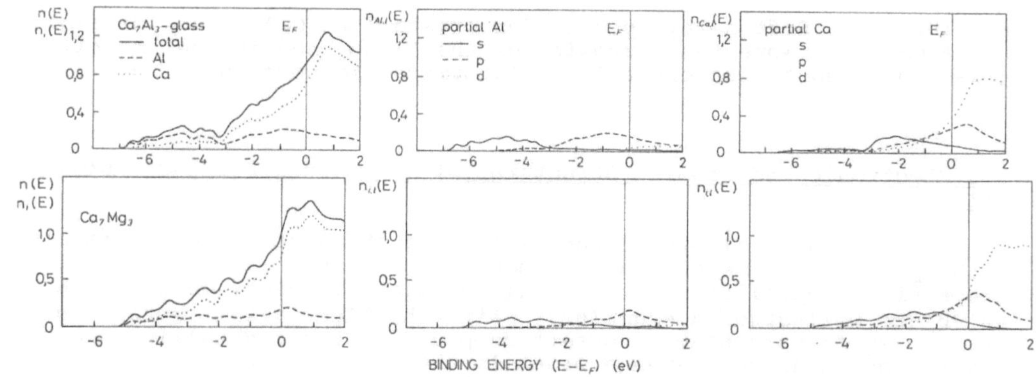

Fig.9: Total and partial electronic DOS for the metallic glasses Ca_7Al_3 and Ca_7Mg_3. After [22].

is established by comparison with the Al-K_β soft-X-ray emission spectra[37,38] (Fig.10).

The results for the electronic structure correlate very well with the electronically driven changes in the interatomic interactions. From the integrated partial densities of states we find that the interatomic charge transfer is small, but there is a substantial intra-atomic electron transfer from s to d states on the Ca sites, and from d to s states on the Al sites, and only a very small such transfer in Ca-Mg. This reflects the compression of the Ca ions on one hand, and on the other hand the increased bonding of the s-electrons on the Al sites leading to the stronger Al-Al potential and the incipient splitting of the band.

The de-hybridization of the conduction band is also very important for the electronic transport properties. Because of the confinement of the electron states to relatively narrow sub-bands, weak localization effects[39] begin to dominate the electronic transport in Ca-Al, but they are still rather weak in Ca-Mg[40]. The localization effects are expected to be even more important in alloys with very strong chemical interactions forming "compounds" in the liquid state.

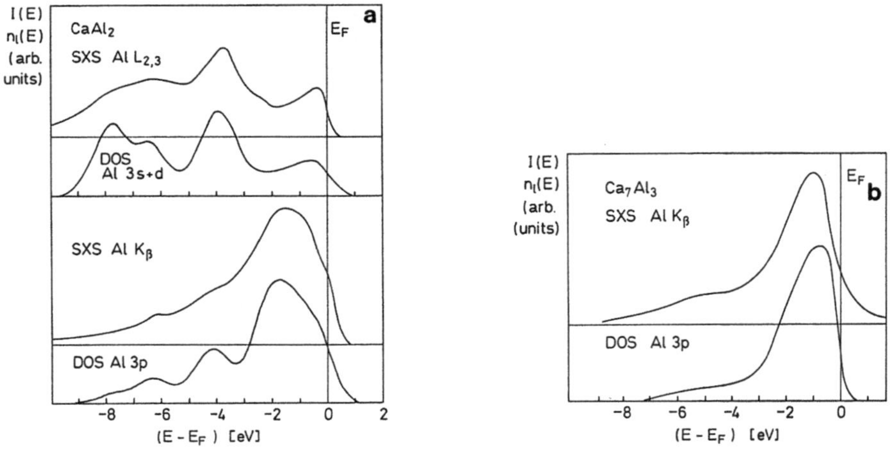

Fig.10(a,b): Soft X-ray spectra for crystalline $CaAl_2$ and for amorphous Ca_7Al_3. Note the narrowing of the Al-3p band in the metallic glass compared to the crystalline intermetallic compound.

3.3 Electronic and Atomic Structure of "Compound-Forming" Liquid Alloys

The liquid alloys of the alkali and the alkaline-earth metals with polyvalent metals have many anomalous and intriguing properties. The alloys of the alkali metals with heavy metals such as Pb are well-studied examples[41]. With the light alkali metals such as Li, Pb forms a stable "octet" compound in the crystalline state, and the chemical short-range order in the liquid alloy is again largest at the "stoichiometric" composition Li_4Pb and may be described simply in terms of a preferred heterocoordination caused by an electronically driven ordering potential[42]. With the heavy alkali metals K, Rb and Cs, Pb forms stable equiatomic compounds characterized by the formation of tetrahedral Pb_4^{4-} clusters ("Zintl ions") isolated in an alkali matrix[43]. Diffraction experiments indicate that these polyanionic clusters are not destroyed upon melting[44]. Computer simulations of the structure of liquid KPb based on the interatomic forces of Fig.3(b) show indeed that the average nearest-neighbour distances in the melt are virtually identical to the nearest-neighbour distances in the crystal (Fig.11). The partial coordination numbers are also found to agree very well[44]. This indicates that the characteristic topological and chemical short-range order survives the solid/liquid transition.

The differences in the atomic order are also reflected in the electronic structure. In the crystalline "octet" compounds one finds a narrow Pb-6s band separated by a gap of about 5 eV from a Pb-6p conduction band[45]. In the equiatomic KPb compound the electronic eigenstates are dominated by the symmetry of the Pb cluster, they may be classified according to the symmetrized eigenstates of the tetrahedral Pb_4^{4-} cluster[45,46]. The Fermi level falls into an energy gap; this explains the semiconducting character of the alloy (Fig.12). The electronic structure calculations for the liquid alloys are rather difficult. The K-K nearest-neighbour distances are rather spread over a broad interval, and this causes considerable difficulties in achieving a well-converged result. Preliminary results for liquid KPb indicate that the main features of the electronic DOS are con-

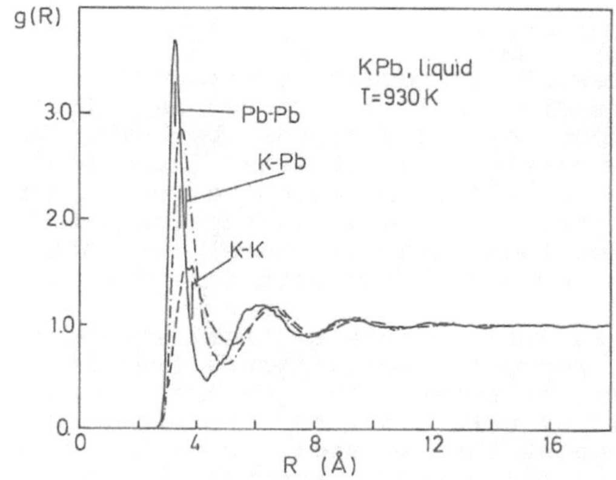

Fig.11.: Partial pair correlation functions for liquid KPb. The vertical lines show the nearest-neighbour distances in the crystalline intermetallic compound KPb.

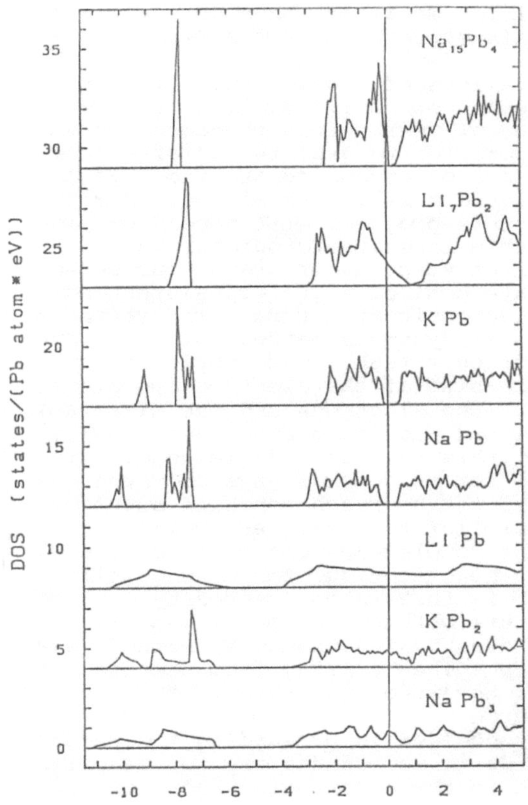

Fig.12: Electronic densities of states for crystalline alkali-lead compounds (Tegze and Hafner, to be published.

served upon melting. This would be the first example of a system where we find a strong similarity between the electronic structure of the crystalline and the liquid states.

4. Conclusions

We have presented an overview of our recent investigations on the interplay of the atomic and the electronic structure in liquid and amorphous metals and alloys. Our results show that by combining pseudopotential methods for the construction of the interatomic forces, molecular dynamics and potential-energy-mapping techniques for the calculation of the atomic structure, and a supercell-LMTO approach to the calculation of the electronic structure, we obtain electronic spectra in quantitative agreement with photoemission and soft-X-ray spectroscopy.

For the pure simple metals we are now able to treat even complex metallic liquids such as Ge, where the solid/liquid transition is accompanied by a semiconductor/metal transition. We are now analyzing the trends in the electronic structure of the pure liquid s-p bonded metals and first results suggest that we shall soon be able to give a full explanation of the characteristic trends in the electronic structures. Combining pseudopotential- and tight-binding methods, reliable interatomic potentials and atomic structures may be calculated for the liquid transition metals[8] - thus we shall soon be able to investigate the electronic structures of the d-bonded liquid metals as well.

For the liquid and amorphous alloys of simple metals important progress has also been made: we can not only explain the weak structure-induced effects in the DOS of nearly-free-electron systems such as Mg-Zn, we are also in a position to explain the often surprising features of the electronic DOS of strong-scattering systems such as the Ca-Al glasses or the liquid alkali-lead alloys. For the liquid transition-metal alloys, future progress will depend on advances in the construction of reliable interatomic potentials.

However, all this is only a first step. As it stands our method falls short of achieving the ultimate goal of a <u>self-consistent</u> calculation of the atomic <u>and</u> the electronic structure. The recently presented density-functional molecular dynamics method of Car and Parrinello[47] aims to solve this difficult task. The coupled equations of motion for the dynamical variables of the electron-ion system (the quantum-mechanical orbitals of the electrons and the classical coordinates of the ions) are formulated and integrated numerically. Even with present-day computers this is a formidable undertaking which appears to be hardly feasible for a complex system such as e.g. K-Pb. The electronic part of the approach contains supercell electronic structure calculations similar to ours, so we think that the experience from our treatment of complex systems will be helpful in implementing this very important new technique for more complicated systems. For liquid Si (which is the system treated by Car and Parrinello[48]) the electronic structure obtained by our method is similar to that calculated using the density-functional MD method, except that it does not show the sharp spikes arising from the use of only the Γ-point in the MD method. Nevertheless, this is the way in which future progress will be achieved.

Altogether, we hope to have succeeded in persuading the reader that the electronic theory of liquids is now on the way to becoming a field of research as respectable as the electronic theory of crystalline solids.

Acknowledgements

The cooperation of Prof. S.S.Jaswal in our work on the metallic glasses is gratefully appreciated.

This work has been supported by the Fonds zur Förderung der wissenschaftlichen Forschung in Österreich (Austrian Science Foundation) under project no. 6191, and by the Hochschuljubiläumsfonds der Österreichischen Nationalbank under project no.2909.

References

1. J.Broughton, W.Krakow, and S.T.Pantelides (Eds.): Computer-Based Microscopic Description of the Structure and Properties of Materials", MRS-Proceedings vol.63 (MRS, Pittsburgh 1986)
2. G.Indlekofer, P.Oelhafen, R.Lapka, and H.J.Güntherodt: Z.Phys. Chemie 157,465(1988); P.Oelhafen, G.Indlekofer, and H.J.Güntherodt: ibid. p.483.
3. G.Indlekofer: Thesis, University of Basel 1988, unpublished.
4. P.Oelhafen: in Amorphous Metals, ed. by H.Matya and P.G.Zielinski (World Scientific, Singapore 1985),p.158.
5. V.Heine and D.Weaire: Solid State Physics 24,247(1971).
6. J.Hafner: From Hamiltonians to Phase Diagrams - The Electronic and Statistical-Mechanical Theory of s,p-Bonded Metals and Alloys, Solid State Sciences vol.70 (Springer, Berlin-Heidelberg 1987).
7. J.M.Wills and W.A.Harrison: Phys.Rev. 28,4363(1983).
8. C.Hausleitner and J.Hafner: J.Phys. F18,1025(1988).
9. A.Arnold, N.Mauser, and J.Hafner: J.Comp.Phys. (submitted).
10. F.H.Stillinger and T.A.Weber: Phys.Rev. B31,5262(1985).
11. J.Hafner, J.Phys. F18,153(1988).
12. J.Hafner and V.Heine: J.Phys. F13,2479(1983); ibid. 16,1429 (1986).
13. J.Hafner and G.Kahl: J.Phys. F14,2259(1984).
14. J.Hafner, to be published.
15. M.A.St.Peters and L.M.Roth: J.Non-Cryst.Solids 75,455(1985).
16. D.Nicholson, A.Chowdhary, and L.Schwartz: J.Non-Cryst.Solids 61+62,1137(1984).
17. M.J.Kelly and D.W.Bullett: J.Phys. C22,2531(1979).
18. B.J.Hickey and G.J.Morgan: J.Phys. C29,6195(1986).
19. T.Fujiwara: in Topological Disorder in Condensed Matter, ed. by F.Yonezawa and T.Ninomyia, Solid State Sciences vol. 46 (Springer, Berlin-Heidelberg 1983), p.111.
20. R.H.Fairlie, W.M.Temmermann, and B.L.Györffy: J.Phys. F12,661 (1982).
21. S.S.Jaswal: J.Non.Cryst.Solids 75,373(1985).
22. J.Hafner and S.S.Jaswal: J.Phys. F18,L1(1988); J.Hafner and S.S. Jaswal, Phys.Rev. B (in print).
23. J.Hafner, S.S.Jaswal, M.Tegze, A.Pflugi, J.Krieg, P.Oelhafen, and H.J.Güntherodt: J.Phys. F (in print).
24. J.Skriver, The LMTO Method, Solid State Sciences vol.41 (Springer, Berlin-Heidelberg 1981).
25. S.K.Bose, S.S.Jaswal, O.K.Andersen, and J.Hafner: Phys.Rev. B (in print).
26. M.Sob, O.Jepsen, and O.K.Andersen: Z.Phys.Chemie 157,515(1988).
27. A.Arnold, N.Mauser, and J.Hafner: J.Phys. F (submitted).
28. R.Biswas and D.Hamann: Phys.Rev.Lett. 55,2001(1985); Phys.Rev. B36,6434(1987).
29. M.D.Kluge, J.D.Ray, and A.Rahman: Phys.Rev. B36,4234(1987).
30. S.K.Bose, K.Winer, and O.K.Andersen: Phys.Rev. B37,6262(1988).
31. G.Kahl and J.Hafner: Sol.State Comm. 49,1125(1984).
32. W.Jank and J.Hafner, Europhys.Lett. (submitted).
33. N.E.Christensen, Phys.Rev.
34. S.R.Nagel and J.Tauc: Phys.Rev.Lett. 35,380(1975).
35. P.Häussler, F.Baumann, J.Krieg, P.Oelhafen, and H.J.Güntherodt: Phys.Rev.Lett. 51,714(1983); P.Häussler, F.Baumann, P.Oelhafen, U.Gubler, and H.J.Güntherodt: Z.Phys.Chemie 157,471 (1988); U.Mizutani and P.Oelhafen, Phys.Rev. B (submitted).
36. U.Mizutani and P.Oelhafen, private communication.
37. S.R.Nagel, U.Gubler, C.F.Hague, J.Krieg, R.Lapka, P.Oelhafen, H.J.Güntherodt, H.J.Evers, A.Weiss, V.L.Moruzzi, and A.R. Williams: Phys.Rev.Lett. 49,575(1982).

38. H.Neddermeyer: in X-Ray Spectra and the Electronic Structure of Matter, ed. by A.Faessler (TU München 1973),p.152.
39. G.J.Morgan, M.A.Howson, and A.Paja: Key Engineering Materials 13-15,377(1987); G.J.Morgan, M.A.Howson, and K.Saub: J.Phys. 15,2157(1985).
40. J.Hafner and M.Tegze: to be published.
41. S.Tamaki: Z.Phys.Chemie 156,537(1988).
42. A.Pasturel, J.Hafner, and P.Hicter: Phys.Rev. B34,8357(1985).
43. W.van der Lugt and W.Geerstma: Can.J.Phys. 65,326(1987).
44. H.T.J.Reijers, W. van der Lugt, and C. van Dijk: Physica B&C 144,404(1987); M.L.Saboungi, R.Blomquist, K.J.Volin, and D.L.Price: J.Chem.Phys.
45. M.Tegze and J.Hafner: to be published.
46. F.Springelkamp, R.A. de Groot, W.Geerstma, W. van der Lugt, and F.M.Mueller: Phys.Rev. B32,3219(1985); W.Geertsma, J.Phys. C28,2461(1985); Z.Phys.Chemie 156,581(1988).
47. R.Car and M.Parrinello: Phys.Rev.Lett. 55,2471(1985).
48. R.Car and M.Parrinello: Phys.Rev.Lett. 60,204(1988).

Charge Correlations in Molecular Fluids and Impurity Absorption Spectra

S. Baer

Department of Physical Chemistry, The Hebrew University,
91904 Jerusalem, Israel

A general quantum statistical perturbation expansion for molecular fluids, with terms ordered in increasing order of molecular susceptibilities, reproduces results of recent theories on absorption spectra in liquids when taken to lowest order, i.e. including only linear polarizability terms.

Inclusion of higher-order, cubic polarizability terms produces solvent effects on impurity electronic absorption spectra. The obtained impurity line shift is a functional of the dipole pair correlations of the pure solvent and of the two-frequency dependent cubic polarizability tensor $\gamma(\omega,\xi,-\xi)$ of the impurity molecule. A simple form of γ based on a 3-level isotropic molecular model, together with a linear polarizability approximation for the solvent dipole correlations leads to a blue shift of the impurity lowest excitation frequency. The magnitude of the shift depends on the position of the impurity line relative to the excitonic band spectrum of the pure solvent and on molecular parameters of the impurity obtainable from the Kerr effect and other third-order, two field frequency optical processes.

1. Charge pair correlation functions

The correlation function [1,2] between charges at two points $(\mathbf{x}_1,\tau_1)$, $(\mathbf{x}_2,\tau_2)$, in space and variable reciprocal temperature ("imaginary time"), associated with molecules positioned at $\mathbf{R}_1$ and $\mathbf{R}_2$, is defined by

$$F(1,2) = \langle\rho(\mathbf{x}_1,\tau_1)\rho(\mathbf{x}_2,\tau_2)\rangle_{\mathbf{R}_1,\mathbf{R}_2} \tag{1}$$

where

$$\rho(\mathbf{x},\tau) = \sum_{a=1}^{N} \rho_a(\mathbf{x},\tau)\ ,$$

the total charge density operator, is defined in terms of molecular charge densities [3]. $F(1,2)$ is decomposable into

$$F(1,2) = F_s(1,2) + F_d(1,2) \tag{2}$$

where F_s is the self part contribution to (1), with the molecular charge densities referring to the same molecule and F_d is the distinct part, with charge densities referring to different molecules. The corresponding dipole pair correlation tensors are denoted respectively by $\underline{F}(1,2)$, $\underline{F}_s(1,2)$ and $\underline{F}_d(1,2)$. They are functions of $\mathbf{R}_{12} = \mathbf{R}_1 - \mathbf{R}_2$ and periodic functions of $\tau = \tau_1 - \tau_2$, with period β. We denote the coefficients of their Fourier series expansions by $X(\mathbf{R}_{12},i\omega)$ etc., were ω belongs to a set

$$\omega = 2\pi n/\hbar\beta,\ \ n = 0, \pm1, \pm2, \ldots \tag{3}$$

$X(\mathbf{R}_{12},i\omega)$ is the mutual susceptibility of two points $\mathbf{R}_1$, $\mathbf{R}_2$ in the liquid.

A Dyson expansion [4] of the statistical operator $\exp(-\beta H)$ leads to a diagrammatic expansion of the pair correlation in terms of molecular charge correlations $f^{(k)}(\tau_1, \ldots, \tau_k)$ (or their Fourier coefficients $\chi^{(k-1)}(i\omega_1, \ldots, i\omega_{k-1})$, the molecular susceptibilities at imaginary frequencies) and Coulomb bonds.

Also defined is the *direct* charge correlation function $C(1,2)$, given by the functional relation [2]

$$F_d(1,2) = \iint F_s(1,1')\, C(1',2')\, F_s(2',2)\; d1'd2'$$

$$+ \iint F_d(1,3)\, C(3,2')\, F_s(2',2)\; d3d2' \tag{4}$$

where $d1' \equiv d^3\mathbf{x}_{1'} d^3\mathbf{R}_{1'} d\tau_{1'}$ etc., and the direct charge correlation tensor $C(\bar{1},\bar{2})$ is related to C(1,2) by

$$C(1,2) = (\mathbf{x}_1 - \mathbf{R}_1)\, C(\bar{1},\bar{2})\, (\mathbf{x}_2 - \mathbf{R}_2)\,. \tag{5}$$

Again, we denote the coefficients of the Fourier series expansion of $C(\bar{1},\bar{2})$ by $C(\mathbf{R}_{12}, i\omega)$.

2. Pure liquid susceptibilities and absorption spectra

$X(\omega)$, the bulk liquid susceptibility tensor per unit volume at frequency ω, is obtained by analytic continuation [5] from its values on a set (3) of imaginary frequencies, where it is given by

$$X(i\omega) = \tilde{X}(\mathbf{o}, i\omega)\,. \tag{6}$$

Here $\tilde{X}(\mathbf{k}, i\omega)$ is the space Fourier transform of $X(\mathbf{R}, i\omega)$, which is in turn a Fourier series coefficient of the imaginary time dependent $F(1,2)$.

The absorption spectrum profile is given by the imaginary part of $X(\omega + i0)$, averaged over all directions of $\mathbf{R}$. This averaged susceptibility is given by

$$\bar{X}(\omega) = \frac{1}{3} \sum_{i=1}^{3} X_{ii}(\omega)\,. \tag{7}$$

Approximate expressions for $\bar{X}(\omega)$ are obtained by partial summation of the diagram expansion of $F(1,2)$ or $X(\mathbf{R}, i\omega)$. Mean field theories take as the main contribution to F the sum of all long range chain diagrams. Thus

$F_s =$ 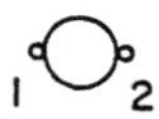 , *i.e.* $X_s(i\omega) = n\chi(i\omega)$ (8)

$F_d =$ 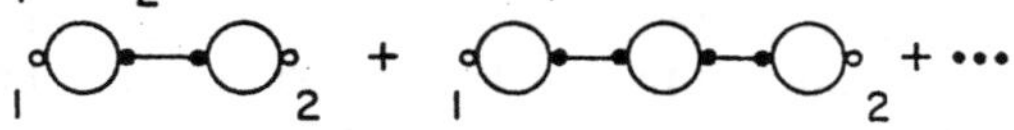 (9)

where the point rooted circles are $f^{(2)}$ factors and the lines are Coulomb bonds.

The effect of the repulsive interactions can be introduced in several approximate ways by redefining the Coulomb bond for short distances $R < \sigma$, where σ is a repulsive core diameter. By (4) the redefined Coulomb bond is equated to $C(\bar{1},\bar{2})$, hence (8) and (9) can be rewritten in Fourier transformed form, combined to give

$$\tilde{X}(\mathbf{k},i\omega) = X_s(i\omega)\,[1-X_s(i\omega)\tilde{C}(\mathbf{k},i\omega)]^{-1}\,, \tag{10}$$

where

$$C(\mathbf{R},i\omega) = \nabla\nabla R^{-1}, \quad R > \sigma\,. \tag{11}$$

and the following approximations are given for $C(\mathbf{R},i\omega)$ at short distances:

(1) The "random phase" approximation (RPA)[2,6]:

$$C(\mathbf{R},i\omega) = 0, \quad R < \sigma\,; \tag{12}$$

(2) The mean spherical approximation (MSA) [7,8]:
$C(\mathbf{R},i\omega)$ is solved from (4) with the condition

$$F_d(\bar{1},\bar{2}) = 0, \quad R < \sigma\,. \tag{13}$$

Adopting a simple Drude model for the molecular susceptibility,

$$\chi(\omega) = \alpha(\omega) = \alpha_o/[1-(\omega/\omega_o)^2]\,, \qquad \omega_o \gg (\hbar\beta)^{-1} \tag{14}$$

where $\alpha(\omega)$, α_o are respectively dynamic and static molecular polarizabilities, and ω_o is a characteristic excitation frequency, one obtains the following absorption spectrum of the liquid with the foregoing approximation schemes:

(1) RPA:
$X(\omega)$ has two sharp absorption lines [1] which can be identified as transverse (ω_T) and longitudinal (ω_L) excitonic lines [9]. They are shifted with regard to the isolated molecule absorption line ω_o and given by

$$\omega_T = \omega_o(1-\frac{4\pi}{3}n\alpha_o)^{\frac{1}{2}}, \quad \omega_L = \omega_o(1+\frac{8\pi}{3}n\alpha_o)^{\frac{1}{2}}\,. \tag{15}$$

(2) MSA:
$X(\omega+i0)$ has a nonzero imaginary part in a frequency range [10,11]

$$\omega_- < \omega < \omega_+ \tag{16}$$

which constitutes an excitonic absorption band.

For comparison, Fig. 1 shows a plot of the full band spectrum as given in [10], together with the pair of lines (15), both calculated from estimated molecular parameters and density of liquid argon. Numerically evaluated band edge values ω_- have been compared with experimental results on photoconduction thresholds [12] in a series of fluid xenon densities, showing a qualitative agreement between theory and experiment.

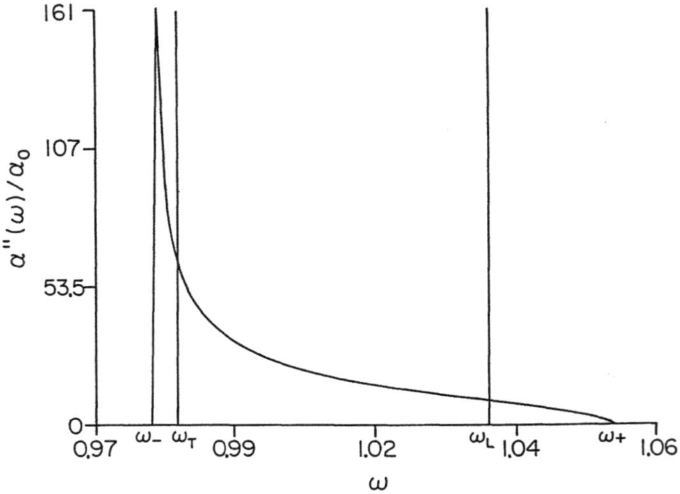

Fig. 1. MSA exciton absorption band [10] and RPA exciton frequencies (scaled by ω_0). $n\sigma^3 = 0.8, \quad \alpha_0/\sigma^3 = 0.01.$

3. Impurity susceptibilities

To obtain an impurity susceptibility, one has to consider the self-correlations F_s of an impurity molecule. From (8) we see that in this approximation scheme F_s is given only to zeroth order, i.e. the dissolved impurity has the same spectrum as the isolated molecule. In the next approximation one can collect all terms including up to 4th order molecular correlations. Summing these terms one obtains

$F_s =$ 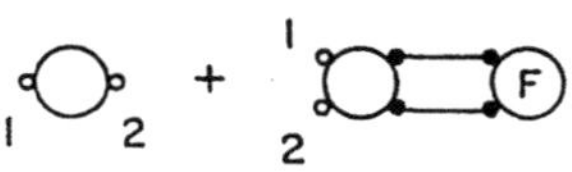 (17)

where, as in the foregoing, the two unlabelled circles represent $f^{(2)}$ and $f^{(4)}$ respectively (each multiplied by the density factor n) and the F labelled circle represents the full charge pair correlation function (tensor) F given by (2) with a series expansion restricted to terms including at most $f^{(4)}$ factors.

Taking the Fourier series expansion of (17), and introducing explicitly tensor component indices, we obtain

$$X_{sii}(i\omega) = n\chi_{ii}(i\omega)$$
$$+ \frac{n}{\beta}\sum_{\omega'}\sum_{j=1}^{3}\chi^{(3)}_{iijj}(i\omega,i\omega',-i\omega')\,\frac{1}{(2\pi)^3}\int d^3\mathbf{k}[\tilde{C}_{jj}(\mathbf{k},i\omega')]^2\tilde{X}_{jj}(\mathbf{k},i\omega')\,. \tag{18}$$

$\tilde{C}(\mathbf{k},i\omega)$ is here the Fourier transformed Fourier coefficient of the redefined Coulomb bond, as given by (11), and $\chi^{(3)}$ is the cubic molecular susceptibility tensor. Because of isotropy the only nonvanishing components of $\chi^{(3)}$ are those with doubly repeated indices.

When F_s, n and the χ's refer to the solvent molecule, (18) and (10) can be solved self-consistently for $X_s(i\omega)$ and $\tilde{X}(\mathbf{k},i\omega)$, and hence the pure solvent absorption spectrum can be obtained in the higher order, i.e. cubic molecular response, approximation. Here we are interested in the impurity molecule and shall refer to it by the subscript I, thus writing n_I, $f_I^{(k)}$, $\chi_I^{(k-1)}$ and X_I, respectively for the density, molecular charge correlations, and susceptibilities of the isolated impurity molecule and of the solute in a unit volume. Hence, substituting impurity f's and χ's in (17) and (18), respectively, with $n = n_I$, while F and $\tilde{X}$ denote the pair correlations and Fourier transformed susceptibility of the pure solvent, we obtain the respective low concentration impurity self-correlations F_I and susceptibility X_I.

The impurity absorption spectrum can be obtained now from the imaginary part of $X_I(\omega)$. To evaluate it, using (18), we make an analytic continuation from the imaginary frequency $i\omega$ to the real frequency ω. We shall restrict the discussion to spherically symmetric molecules possessing no permanent multipoles, so that χ reduces to the linear polarizability α and $\chi^{(3)}$ reduces to γ, the cubic polarizability of the molecule. Furthermore, to simplify calculations assume that the characteristic excitation frequencies of both solvent and solute molecules are much higher than $(\hbar\beta)^{-1}$. In this case the sum over frequencies ω' in (18) can be reduced to an integral through the correspondence

$$\frac{1}{\beta}\sum_{\omega'} \to \frac{\hbar}{2\pi}\int d\omega' . \tag{19}$$

Writing $i\omega' = \xi$ and $-i\omega' = -\xi$, we make use of the requirement that $\tilde{X}(\mathbf{k},\xi)$ and $\gamma(\omega,\xi,-\xi)$ are analytic functions in both the upper and lower complex ξ plane and deform the integration path from the imaginary axis $(-i\infty, i\infty)$ to the contour C: $(\infty - i0, 0, \infty + i0)$, see Fig. 2. Defining now the complex frequency dependent tensor $K(\xi)$ by

$$K(\xi) = \frac{1}{(2\pi)^3}\int d^3\mathbf{k}\; \tilde{C}(\mathbf{k},\xi)\tilde{X}(\mathbf{k},\xi)\tilde{C}(\mathbf{k},\xi) , \tag{20}$$

we rewrite (18) for the impurity as a sum of two terms

$$X_I(\omega)/n_I \equiv \chi_I(\omega) = \chi_{0I}(\omega) + \chi_{1I}(\omega) \tag{21}$$

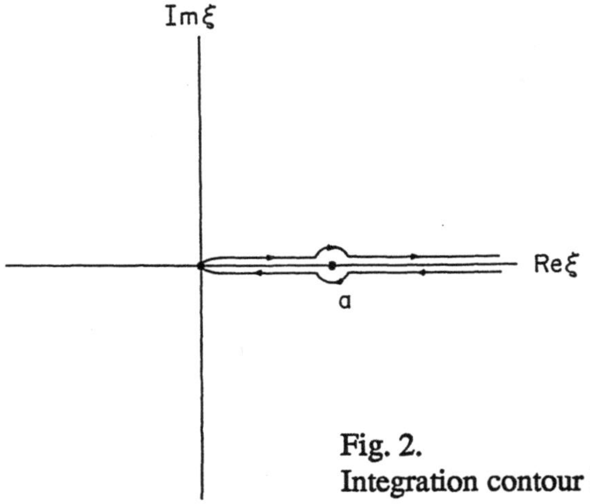

Fig. 2.
Integration contour in the complex ξ plane (see text).

where

$$\chi_{0I}(\omega) = \alpha(\omega)\ , \tag{22}$$

$$\chi_{1I\ ii}(\omega) = \frac{\hbar}{2\pi i}\int_C d\xi \sum_{j=1}^{3} \gamma_{iijj}(\omega,\xi,-\xi)K_{jj}(\xi)\ . \tag{23}$$

We see that the effect of the solvent on the impurity spectrum enters in χ_{1I} through the tensor $K(\xi)$, which is a functional of $\widetilde{X}(\mathbf{k},\xi)$.

Without introducing an explicit expression for $\widetilde{X}(\mathbf{k},\xi)$ in (20), we can make the following general statements about $K(\xi)$: Since by (20) $K(\xi)$ is a linear combination of susceptibility functions, it has the analytic properties of a generalized susceptibility [13], namely it has two analytic branches in the upper and lower halves of the complex ξ plane. Approaching the real ξ axis from either side, these are given by

$$K(\xi \pm i0) = K_{\pm}(\xi) = K'(\xi) \pm iK''(\xi) \tag{24}$$

where $K'(\xi)$ and $K''(\xi)$ are respectively real and imaginary parts of $K(\xi)$ and are respectively even and odd functions of ξ. For an arbitrary complex z we have

$$K(z) = \frac{1}{\pi}\int_{-\infty}^{\infty} \frac{K''(\xi)}{\xi - z} d\xi\ , \tag{25}$$

representing $K(z)$ as a sum of simple poles distributed on the real axis with a density $\pi^{-1}K''(\xi)$.

In the simple molecular model constructed in the following, $\gamma(\omega,\xi,-\xi)$, viewed as a function of ξ, is obtained as a sum of terms which have simple and double poles. In [14] we derive an explicit expression for $K''(\xi)$ in the RPA and give explicit closed forms for (23), which is then evaluated numerically.

4. Model for molecular (hyper)polarizabilities

To obtain explicit expressions for X_I, we introduce a specific isotropic molecular model to evaluate the (hyper)polarizabilities of a molecule. We take the molecule to be in its ground state. The electronic properties of the molecule are represented by a single effective charge e moving in a central field. Thus its states can be classified by angular momentum states $|l,m\rangle$. We suppose that the quantum numbers l specify also energy levels, with excitation frequencies $\omega_{l0} \equiv \omega_l - \omega_0$. The non-zero transition moments are then completely defined by the set of equations [15]

$$\langle l,m-1|\mu_x - i\mu_y|l-1,m\rangle = a_{l,l-1}\sqrt{(l-m+1)(l-m)}\ ,$$

$$\langle l,m|\mu_x + i\mu_y|l-1,m-1\rangle = -a_{l,l-1}\sqrt{(l+m-1)(l+m)}\ ,$$

$$\langle l,m|\mu_z|l-1,m\rangle = a_{l,l-1}\sqrt{l^2-m^2}\ . \tag{26}$$

Since we are concerned only with linear (α) and cubic (γ) polarizabilities, the foregoing model can be restricted to a 3-level model with $l = 0, 1, 2$. It is then completely specified by four parameters defined by $\omega_{10} = \omega_a$, $\omega_{20} = \omega_b$, $a_{10} = a_{01} = a$, $a_{21} = a_{12} = b$.

When applied to the general expression for the linear polarizability, the selection rules (26) lead directly to the Drude model expression

$$\alpha(\omega) = \frac{a^2}{\hbar}\frac{1}{2}\left[\frac{1}{\omega_a+\omega} + \frac{1}{\omega_a-\omega}\right] , \tag{27}$$

written alternatively as

$$\alpha(\omega) = \frac{\alpha_a \omega_a^2}{\omega_a^2-\omega^2}, \quad \text{where } \alpha_a = \alpha(0) = \frac{a^2}{\hbar\omega_a} . \tag{28}$$

The general expression [16,17] for the 3-field frequency cubic polarizability γ, obtained with the selection rules (26), has the following form for the γ tensor components:

$$\gamma_{ihjk}(\omega_1,\omega_2,\omega_3) = I_{-\sigma,1,2,3}$$

$$\frac{1}{\hbar^3}\left\{\sum_{(m)} \frac{<0|\mu_i|1,m_1><1,m_1|\mu_h|2,m_2><2,m_2|\mu_j|1,m_3><1,m_3|\mu_k|0>}{(\omega_a-\omega_\sigma)(\omega_b-\omega_1-\omega_2)(\omega_a-\omega_1)}\right.$$

$$\left. - \sum_{(m)} \frac{<0|\mu_i|1,m_1><1,m_1|\mu_h|0><0|\mu_j|1,m_2><1,m_2|\mu_k|0>}{(\omega_a-\omega_\sigma)(\omega_a-\omega_1)(\omega_a+\omega_2)}\right\} , \tag{29}$$

where $\omega_\sigma = \omega_1+\omega_2+\omega_2$ and $I_{-\sigma,1,2,3}$ is an averaging operation over all permutations of the four frequency tensor index pairs $(-\omega_\sigma,i),(\omega_1,h),(\omega_2,j),(\omega_3,k)$.

Certain simplifications of (29) ensue from the present study. First, we require only the high symmetry components γ_{iiii} and γ_{iijj}, $i \neq j$. Second, we are concerned with the set of frequencies

$$(-\omega_\sigma,\omega_1,\omega_2,\omega_3) = (-\omega,\omega,\xi,-\xi) . \tag{30}$$

The hyperpolarizabilities for this set are related to the optical Kerr effect and other nonlinear optical processes involving two laser fields [18], and relevant parameters of our present model can be related to parameters measured by such processes.

A detailed enumeration of all distinct contributions of the terms in (29), with their relevant poles and residues, is given elsewhere [14]. Analogously to the static polarizability α_a related to transition moments and frequencies by (28), we find that the relevant parameters in γ are two characteristic hyperpolarizabilities defined by

$$\gamma_P = \frac{a^2b^2}{(\hbar\omega_a)^2\hbar\omega_b} , \quad \gamma_Q = \frac{a^4}{(\hbar\omega_a)^3} , \tag{31}$$

or

$$\gamma_Q = \alpha_a^2/\hbar\omega_a\,, \quad \gamma_P/\gamma_Q = b^2\omega_a/a^2\omega_b\,. \tag{32}$$

They are related to static γ elements by

$$\gamma_{iiii}(0,0,0) = 3\gamma_{iijj}(0,0,0) = 4\gamma_P - \gamma_Q\,. \tag{33}$$

Alternatively, this constant is directly related to the molecular (low density) Kerr constant K which is proportional to [19]

$$\gamma_{iiii} - \gamma_{iijj} = \frac{2}{3}(4\gamma_P - \gamma_Q)\,. \tag{34}$$

5. Impurity absorption spectra

Given the foregoing molecular model for the dynamic cubic polarizability $\gamma(\omega,\xi,-\xi)$ of an impurity molecule and adopting the RPA, (10)-(12), the evaluation of $\chi_{1I}(\omega)$ by (23) reduces to the evaluation of a sum of integrals which are given in explicit form and have been evaluated numerically in [14].

Finally we obtain

$$\chi_{1I}(\omega) = \frac{\chi_c(\omega)}{[1-(\omega/\omega_a)^2]^2} \tag{35}$$

where the factor $\chi_c(\omega)$ is regular at $\omega = \omega_a$ but has a singularity at $\omega = \omega_b$. As an example we have plotted $\chi_c(\omega)$ in Fig. 3 as a function of ω/ω_0, scaling it by the factor $\sigma^6/\hbar\omega_0$, where σ is the hard sphere diameter common to both solute and solvent molecule and ω_0 is the excitation frequency in the solvent Drude model polarizability $\alpha(\omega)$. The parameters chosen were

$$\omega_a/\omega_0 = 0.5, \quad \omega_b/\omega_0 = 0.75, \quad b^2/a^2 = 8, \quad \alpha_a\sigma^3 = 0.04$$

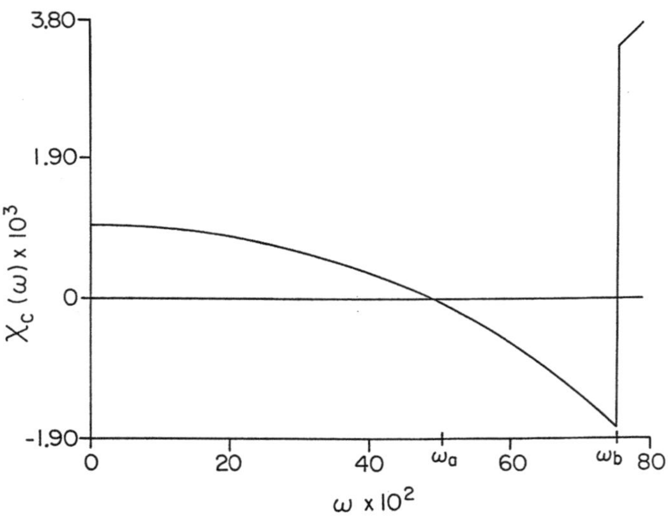

Fig. 3. Plot of $\chi_c(\omega)$ (scaled by $\sigma^6/\hbar\omega_0$) as a function of ω/ω_0.

and

$$y = \alpha_0 n = 0.009.$$

They are meant to correspond roughly to the the situation of a Xe atom dissolved in liquid Ar. The frequencies ω_a, ω_b correspond to the first resonance line and the ionization energy of Xe, ω_0 to the ionization energy of Ar, and α_0, α_a to Ar and Xe polarizabilities. We have estimated b^2/a^2 from (32) and (33), using the value [20,21]

$$\gamma \equiv \gamma_{1111} = 0.065\sigma^6/\hbar\omega_0, \quad \sigma = 0.34\mathrm{nm}, \quad \hbar\omega_0 = 15.56\mathrm{eV}.$$

Besides a singularity at ω_b, a characteristic feature of $\chi_c(\omega)$ is its having a negative value for a range of frequencies including $\omega = \omega_a$. As seen below, this behaviour results in a blue frequency shift of the dissolved impurity.

The truncated, two-term expression (21) for $\chi_I(\omega)$ - the effective dissolved impurity polarizability - presents the qualitatively correct result of polarizability enhancement [22] in the low frequency range. This enhancement is a direct result of the solvent-solute dispersion interactions, producing $\chi_{1I}(\omega) > 0$ for small ω. Yet (21) gives a completely misleading picture of the impurity absorption spectrum. Indeed we have here the same situation as in standard finite order perturbation theory, which produces a perturbed wave function having the same singularities as the unperturbed one, and has to be formulated specifically for the energy in order to obtain the correct perturbed energy values. Here we recast (21) into the elementary Padé approximant form

$$\chi_P = \frac{\chi_{0I}}{1-\chi_{1I}/\chi_{0I}} \tag{36}$$

in order to obtain energy shifts. The approximant $\chi_P(\omega)$ can be interpreted as an approximate summation of the infinite series expansion of χ_I, but it is simpler to note that χ_P is equivalent to an expansion of χ_I^{-1} up to the same order as χ_I in (21).

Noting (35) and (27), the singular frequencies of (36) are the roots of

$$\omega_a^2 - \omega^2 - \omega_a^2\chi_c(\omega)/\alpha_a = 0, \tag{37}$$

from which we can estimate the relative frequency shift

$$\frac{\Delta\omega}{\omega_a} = -\frac{\chi_c(\omega_a)}{2\alpha_a}. \tag{38}$$

Checking (34) and (35), one can see that the contribution of the positive (P) terms to $\chi_c(\omega)$ is zero when $\omega = \omega_a$. Thus we have always

$$\chi_c(\omega_a) < 0 \tag{39}$$

and by (38) obtain a blue frequency shift. When compared with experimental [23] shifts of Xe lines in liquid rare gases at the given density, it is found to be too small by at least an order of magnitude. This could be possibly corrected by a proper modification of the molecular model for γ, in particular discarding the relation (30) between α and γ_Q and taking higher values for γ_Q. Another possible correction is the use of an improved model, such as the MSA, for the solvent dipole correlations, although (20) and (23) suggest that because of averaging the effect of the excitonic band on the solute is still well represented by the crude RPA two-line spectrum.

In summary, a blue shift has been obtained from the dispersion interactions between the impurity and the solvent molecules without having to consider a special effect of repulsive interactions [24]. Yet it is likely that a full quantitative agreement with experiment will require a refined theoretical treatment, including e.g. electron exchange terms [25] in the expansion, thus going beyond the crude treatment of the repulsive interaction as a classical exclusion.

A conspicuous limitation of the truncated perturbation expansion is its failure to produce line broadening. In this respect it has not improved on linear approximation, mean field theories. A correction of this defect should be found by a self-consistent modification of the present treatment without introducing phenomenological parameters [26] into the theory.

References

[1] Baer, S., 1973, *Chem. Phys.* **1**, 55.

[2] Baer, S., 1974, *J. Chem. Phys.* **60**, 435.

[3] Longuet-Higgins, H.C., 1956, *Proc. Roy. Soc. (London)* **A235**, 537; 1965, *Disc. Faraday Soc.* **40**, 7.

[4] Bloch, C., 1965, in *Studies in Statistical Mechanics*, Vol. III, edited by J. de Boer and G.E. Uhlenbeck (North-Holland, Amsterdam), Pt. A.

[5] Baym, G., and Mermin, N.D., 1961, *J. Math. Phys.* **2**, 232.

[6] Baer, S. and Ben-Shaul, A. 1972, *J. Chem. Phys.***56**,1238.

[7] Hoye, J.S., and Stell, G., 1981, *J. Chem. Phys.* **75**, 5123.

[8] Thompson, M.J., Schweizer, K.S., and Chandler, D., 1982, *J. Chem. Phys.* **76**, 1128.

[9] Christiaens, W., and Phariseau, P., 1970, *Physica* **46**, 569.

[10] Chandler, D. Schweizer K.S., and Wolynes, P.G., 1982, *Phys. Rev. Lett.* **49**, 1100.

[11] Hoye, J.S., and Olaussen, K., 1982, *J. Chem. Phys.* **77**, 2583.

[12] Steinberger, I.T., and Baer, S., 1987, *Phys. Rev. B***36**, 1358.

[13] See e.g. Landau, L., and Lifshitz, E.M., 1958, *Statistical Physics* (Pergamon, New York) Sec. 122.

[14] Baer, S., 1988, *Mol. Phys.*, to appear.

[15] See e.g. Landau, L., and Lifshitz, E.M., 1959, *Quantum Mechanics* (Pergamon, New York) Sec. 27.

[16] Orr, B.J., and Ward, J.F., 1971, *Mol. Phys.* **20**, 513.

[17] Prior, Y., 1984, *IEEE J. Quantum Electron.* **20**, 37.

[18] Hellwarth, R.W., 1977, *Prog. Quantum Electron.* **5**, 1.

[19] Buckingham, A.D., and Pople, J.A., 1955, *Proc. Phys. Soc. A***68**, 905.

[20] Buckingham, A.D., and Dunmur, D.A., 1968, *Trans. Farad. Soc.* **64**, 1176.

[21] Shelton, D.P., 1986, *J. Chem. Phys.* **84**, 404.

[22] Pratt, L.R., 1980, *Mol. Phys.* **40**, 347.

[23] Messing, I., Raz, B., and Jortner, J., 1977, *J. Chem. Phys.* **66**, 2239, 4577.

[24] Saxton, M.J., and Deutsch, J.M., 1974, *J. Chem. Phys.* **60**, 2800.

[25] Baer, S., 1968, *J. Chem. Phys.* **49**, 3632.

[26] Schweizer, K.S., 1986, *J. Chem. Phys.* **85**, 4638.

Pair Potential in Liquid Bismuth from Structure Data

M. Dzugutov

The Royal Institute of Technology, Institute of Reactorphysics,
Drottning Kristinas Väg 47, S-11428 Stockholm, Sweden

Pair potential was derived from the structure factor of liquid Bi measured by neutron scattering at 578 K. The derivation was performed in an iterative predictor-corrector procedure using the modified hyper-netted chain equation as predictor and molecular dynamic simulation as corrector. The MD system included 16384 particles. The result demonstrates that the observed anomalies of the structure may be successfully interpreted using a classical system with pair interaction.

The structure and dynamical properties of liquid semimetals and semiconductors were the subject of extensive studies during recent years. A large amount of experimental information collected about these liquids still needs to be interpreted. The most interesting feature of their static structure factor, revealed by both neutron scattering and X-ray diffraction measurements [1], is the more or less pronounced shoulder on the high Q side of the main peak. Some distortions of the following peaks have also been observed. The corresponding radial distribution functions exhibit in some cases an additional peak between the first and the second main peaks. The recent accurate neutron-scattering measurements of the structure of liquid Bi [2] clearly demonstrate these features.

Some attempts to explain these anomalies were based on the assumption that two kinds of structure are co-existing in liquid Bi [2]. From the analysis of the inelastic neutron scattering experiments, Dahlborg and Olsson [3] also came to the conclusion that the molecular dynamics in liquid bismuth cannot be described in terms applicable to classical liquids. In any case, it is clear that the long-range forces in such a system must be different compared to simple liquid metals. Therefore, it is interesting to examine to what extent the anomalies of the structure in question could be interpreted by classical simulation based on the concept of pair interaction. The earlier attempts to derive pair potential in liquid bismuth from the structure data [2] were not convincing.

In this study the Molecular Dynamics (MD) technique was employed to derive the pair potential by fitting it to the measured S(Q) [2]. The large MD system consisting of 16384 particles was used in order to reach the long-wavelength area of Q. The simulation was carried out at the temperature 578 K and the number density $n=0.0289$ Å^{-3}. That corresponds to the conditions at which the experimental S(Q) we attempted to simulate was measured [2]. The Verlet's algorithm with the time-step 0.02 ps provided excellent stability of the system's full energy. All the calculations were carried out on a VAX 8200.

The parametrised pair potential was chosen of the same form as that successfully used earlier to simulate the structure of liquid lead [4]:

$$V(r)=V_1(r)+V_2(r)+V_3(r) \tag{1}$$

$$V_1(r)=a_1(b_1/r)^N \exp([(r-c_1)^{-1}], \quad r<c_1$$

$$V_1(r)=0, \quad r>c_1$$

$$V_2(r)=a_2(b_2/r-c_2)\exp[(r-b_2)^{-1}], \quad r<b_2$$

$$V_2(r)=0, \quad r>b_2$$

$$V_3(r)=a_3 r^{-3}\exp(-b_3 r)\cos(2K_F r+c_3)$$

The first and the third terms represent, respectively, the usual strong repulsion and Friedel oscillations, K_F denotes the radius of the Fermi sphere (for liquid bismuth at the density we studied $K_F= 3.27$ Å^{-1}). The term V_2 represents the long-range repulsion, presumably characteristic for non-simple metals [4,5].

The radial distribution function of the studied system was calculated in the trial MD runs using an assumed form of the potential. The S(Q) was obtained by Fourier transformation of the RDF. Then the potential's parameters were varied in order to achieve fit to the experimental S(Q). After initial "manual" adjustment of the potential, the minimization procedure using the Modified Hyper-Netted Chain equation [6] was employed. The final result of the fitting of S(Q) is shown in Fig. 1. The agreement between the two sets of S(Q) data looks quite satisfactory, including the important small Q area which is shown separately in Fig. 1b. The earlier mentioned anomalies of the S(Q) are well reproduced. The corresponding final version of the effective pair potential is presented in Fig. 2. Table 1 contains the resulting set of values of the potential's parameters. The radial distribution function of the model liquid (Fig. 3) demonstrates an additional pronounced maximum between the first and the second main peaks as well as some other distortions. It may be shown that this additional maximum is related to the shoulder on the main peak of S(Q).

Finally, it may be concluded that the observed anomalies of the bismuth's structure could be well reproduced by a simple system using pair potential. However, minor

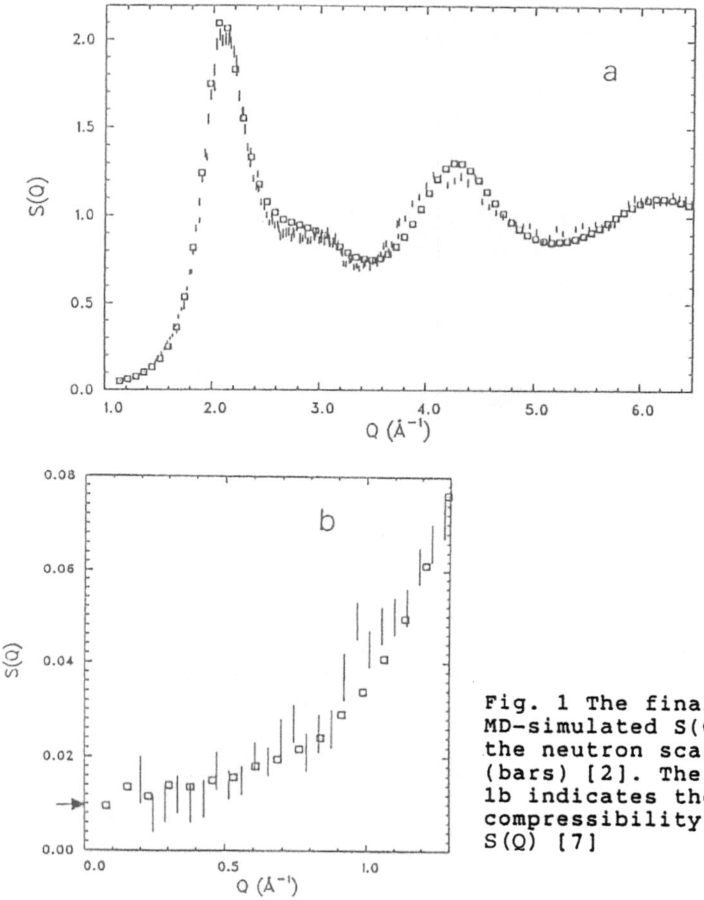

Fig. 1 The final fit of the MD-simulated S(Q) (boxes) to the neutron scattering data (bars) [2]. The arrow in Fig. 1b indicates the experimental compressibility value of S(Q) [7]

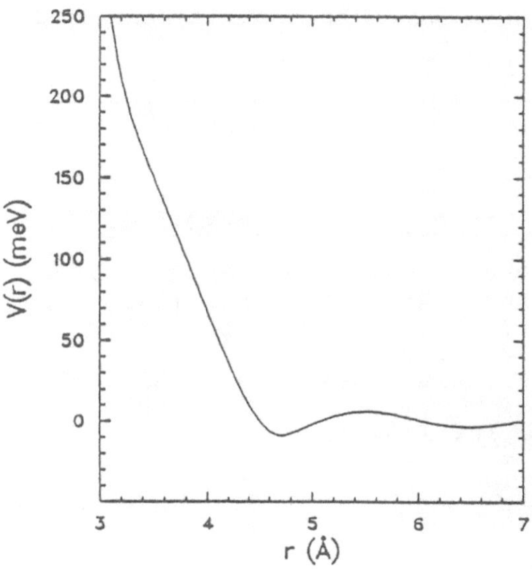

Figure 2. The pair potential in liquid bismuth derived from the measured S(Q) at 578 K

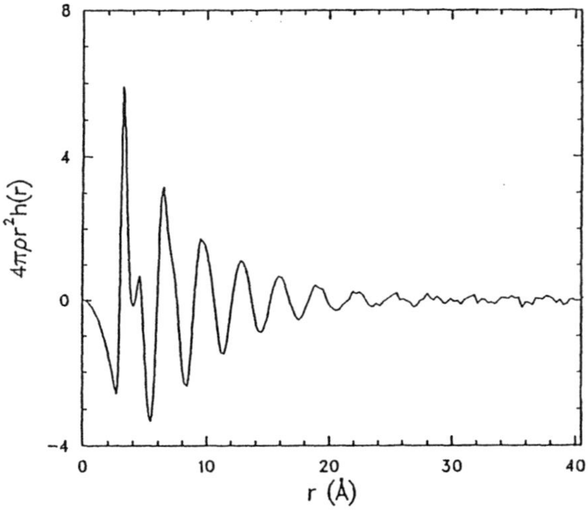

Figure 3. Radial distribution function of liquid bismuth at 578 K calculated by Molecular Dynamics

discrepancies between the experimental data and the simulated S(Q) leave room for suggestion that many-body forces might cause some structural effects in liquid bismuth.

Table 1. The final values of the potential's parameters

N	a_1	b_1	c_1	a_2	b_2	c_2	a_3	b_3	c_3
	meV	Å	Å	meV	Å		meV·Å³	Å⁻¹	rad
15	73.0	3.284	4.9	282.0	4.9	0.59	1514.7	0.2	0.07

References

1. Y. Waseda: The Structure of Non-Crystalline Materials (McGraw-Hill, New York 1980)
2. U. Dahlborg, M. Davidovic: Phys. Chem. Liq. 15, 243 (1986)
3. U. Dahlborg, L. G. Olsson: Phys. Rev. 25, 2712, (1982)
4. M. Dzugutov, K.-E. Larsson, I. Ebbsjö: To be published
5. R. Taylor: J. Phys. 46, C4-309 (1985)
6. L. Reatto, D. Levesque, J. J. Weis: Phys. Rev. A 33, 3451 (1986)
7. O. J. Kleppa: J. Chem. Phys. 18, 1331 (1950)

Extracting the Pair Correlation Function from Structure Factor Data by Monte Carlo Simulation

A.K. Soper

Neutron Division, Rutherford Appleton Laboratory, Chilton, Didcot, Oxon, OX11 0QX, UK

1. Abstract

Measurements of the liquid structure factor, S(Q), by radiation diffraction experiments have now been performed for many years on a broad spectrum of liquids and over a range of state conditions. The analysis of data from these measurements necessarily involves several steps to correct for background, container scattering, multiple scattering and attenuation processes, all of which are regarded as reasonably well understood. However, obtaining the pair correlation function, g(r), by Fourier transform of S(Q) data has frequently proved to be unreliable because of the inherent truncation of the data, and because of statistical and systematic effects introduced in the process of obtaining and analysing the data. Recently there have been several attempts to perform the transform operation using Maximum Entropy methods, although it is now demonstrable that these approaches can also give unreliable results. A new approach to this problem is described here which uses a one-dimensional Monte Carlo simulation to calculate g(r): the fluctuations in g(r) are constrained to the correct, exponential decay with increasing r values. As a consequence, many of the problems associated with previous methods are overcome. Examples are given of simulations of the hard sphere structure factor and real diffraction data.

2. Introduction

The analysis of diffraction data from liquid and amorphous materials to obtain the structure factor and pair correlation function involves a straightforward but error-prone sequence of operations:

(a) correction of raw data for detector and diffractometer backgrounds;

(b) normalization of detector counts to remove detector efficiency and solid angle effects;

(c) correction of normalized counts for multiple scattering effects;

(d) correction of normalized counts for container scattering;

(e) correction for self-attenuation in the sample;

(f) estimation of g(r) by Fourier inversion.

There may be other effects, specific to the type of radiation being used, such as recoil corrections for neutrons and Compton corrections for X-rays. Although the origin of these corrections is well understood there are limits to how accurately they can be estimated and applied. In addition real diffraction data contain two intrinsic qualities which cannot be avoided however accurately the corrections are made: the truncation in the range of Q over which the data are available and statistical noise.

The final Fourier transform stage of this procedure is particularly vulnerable to the limitations of the data because each type of error has its own characteristic effect on g(r): slowly varying effects in S(Q) produce unphysical peaks and valleys in g(r), particularly in the small r region where g(r) is expected to be zero; noise in S(Q) produces noise in g(r); and truncation produces a characteristic 'ringing' in g(r) which is particularly difficult to avoid in general. Therefore it is now generally felt that direct Fourier transform of diffraction data, unless it is exceptionally complete and accurate, is unreliable and should be avoided.

Recently several attempts to estimate g(r) using the maximum entropy (ME) method have been proposed [1,2,3,4]. The underlying principle of these approaches is to regard the quantity $N(r)=4\pi\rho r^2 g(r)$ as a probability distribution (since N(r) is normalizable) and then maximize the entropy $H=-\Sigma N(r)\ln[N(r)/P(r)]$, subject to the constraint that the structure factor calculated from N(r) lies through the data points to a specified precision. Here ρ is the number density of atoms in the liquid, and P(r) is a <u>prior</u> distribution which must be defined before the analysis can proceed. In fact I have recently demonstrated that the result of ME analysis can depend quite markedly on the form of the assumed prior [5]: this is unsatisfactory for real datasets with an unknown prior.

The purpose of the present article then is to propose an alternative to maximum entropy analysis which avoids the main objection, i.e. dependency on the assumed prior. A one-dimensional Monte Carlo simulation is used because it has the important benefit that alternative constraints can be applied to the calculated distributions without modifying the basic algorithm in any way, and because error bars on g(r), which in the past have been glaringly absent in the literature, can be estimated. Full details of this procedure are available elsewhere [5]. Here the main results are highlighted.

There have been previous attempts to analyse diffraction data from amorphous [6] and crystalline systems [7] using Monte Carlo analysis. The present approach differs from the earlier work [6] in that the structure data S(Q) are modelled directly rather than g(r), and being one-dimensional it is quite rapid on a modern mainframe computer, occupying typically some minutes of cpu time on a VAX 8650.

3. Monte Carlo analysis of S(Q)

Monte Carlo techniques are widely used to study the static properties of fluids. Typically, a potential of interaction between atoms or molecules is assumed and the molecules are allowed to proceed on a random walk. At each move the energy of the system is calculated and the move accepted with probability $\exp(-\Delta U/kT)$, where ΔU is the change in energy with the move and T is the absolute temperature. In the present application there is no assumed potential of interaction. Instead the distributions N(r), P(r) are divided into a series of bins, N_j, P_j, $j=1,J$. A move consists of changing the number of atoms at each bin in turn by a random amount in the range $\pm\Delta\varepsilon_j$. The only limitation placed on the number moved is that at the end of each move

$N_j > 0$ for all j. This causes the structure factor, M(Q), associated with the N_j, to change by an amount $\Delta M(Q)$, where

$$\Delta M(Q) = \Delta N_j \left[\sin(Qr_j)/Qr_j \right] . \qquad (1)$$

The fit to the data is measured by an "R-factor", R_f, where

$$R_f = \sum (D(Q) - M(Q))^2 / \sum (D(Q))^2 . \qquad (2)$$

Here D(Q) represents the data and the summations occur over the range of the measured data. Each move is then accepted with probability $\exp(-\lambda_1 \Delta R_f)$, where the multiplier λ_1 determines how close the model distribution is to fit the data and ΔR_f is the change in R-factor. As the value of λ_1 is made larger so the model misfits the data with a smaller variance.

The basic algorithm described above converges extremely rapidly but is not useful as it stands because artificial structure which arises from the discrete nature and limited range in Q of the supplied data is generated. Therefore an additional constraint on the N_j which serves to avoid such improbable distributions is applied. Entropy, H, (defined above) is one such constraint which could serve the purpose, although this form does not place a sufficient restriction on the fluctuations in g(r) to avoid truncation effects. Quite general considerations [5] indicate that the real fluctuations in g(r) (as opposed to those induced by noise and truncation) should fall at least as fast as $\exp(-\alpha r)/r$, where α is the inverse correlation length for the liquid. This exponential behaviour derives from the fact that the peak at say $Q=Q_o$ in S(Q) for a fluid, including the peak at Q=0 if it exists, is by definition always broader than $\delta(Q-Q_o)$. Only at the critical point does α become zero corresponding to an infinite range of correlations and S(Q) diverging at Q=0. It should be noted that in theory [8] g(r) should decay as u(r) at large r values, where u(r) is the effective pair potential, rather than the exponential decline proposed here. In fact the exponential decline is only formally correct for the hard sphere fluid [9]. However, in practice the structure is dominated by hard core interactions, and at r values large enough that these have damped out in g(r) the magnitude of the potential term, even for a screened Coulomb potential, is negligibly small. The simplest constraint which builds in this fluctuation constraint is derived from the first non-zero term in the expansion of H in powers of (N_j-P_j). This constraint, labelled H', is then defined as

$$H' = \sum y_j (N_j-P_j)^2/P_j, \quad \text{with } y_j=\exp(\alpha r_j). \qquad (3)$$

The weight function serves to drive the large r fluctuations to be as small as possible. The fluctuation constraint is then combined with R_f to produce an overall measure of fit, R, which is to be minimized:

$$R = \lambda_1 R_f + \lambda_2 H' , \qquad (4)$$

where λ_2 is an additional parameter which controls the amplitude of the fluctuations. The change in R, ΔR, is used in place of ΔR_f to determine the acceptance probability of each move.

The fluctuation constraint works well for perfect datasets, such as the hard-sphere structure factor. However, most measured datasets have a significant noise component: R_f is biased by the noise and a noise-like structure in g(r) can result. The fluctuation constraint, which attempts to drive the distribution back to the uniform, prior distribution, is insensitive to this noise, and the only way to remove it from g(r) is to worsen the fit to the data. Therefore we have introduced an alternative constraint which attempts to make the length of the line used to draw the function D(r)=N(r)-P(r) as short as possible. For the discrete case used in the calculation, and with D(r) represented by D_j, the square of the modulus of this length is given by

$$\sum (D_j - D_{j-1})^2 \quad + \quad \text{constant} . \tag{5}$$

However D_j is subject to the same fluctuation constraint as N_j so a practical quantity which can be used in the Monte Carlo algorithm is

$$L = \quad \sum y_j \{ (D_j - D_{j-1})^2 + (D_{j+1} - D_j)^2 \} . \tag{6}$$

The weights y_j are the same as before, although it is often necessary to reduce their magnitude in regions where the data imply pronounced structure in g(r). Details of how this is done will appear in the full report [5]. Inspection of (6) will show that it can alternatively be regarded as a constraint on the gradients of D(r). The constraint L is used in place of H′ in (4) - otherwise the calculation proceeds as before. It should be noted that although the standard definition of entropy for the g(r) problem has been abandoned here, there is no fundamental inconsistency because the derivation of the Boltzmann-like probability distribution used to accept or reject Monte Carlo moves itself stems from ME formalism. However with these alternative definitions for the constraint on fluctuations, N(r) is no longer treated as a probability distribution.

4. Results and discussion

The first example of the use of this algorithm concerns the Percus-Yevick hard sphere structure factor for a density of $\rho\sigma^3 = 0.5$ and $\sigma=1$ with σ the hard sphere diameter. This is the example used by Root et al. [2]. The structure factor was deliberately truncated at $Q = 15Å^{-1}$. Figure 1 shows the simulated pair correlation function and structure factor, with $R_f = 1\%$. It is seen that the correlation function is almost free of truncation ringing. Moreover there is no discontinuity in S(Q), at $Q=Q_{max}$, unlike the ME solution of Root et. al., and the extension of S(Q) beyond the region used for the simulation lies remarkably close to the correct S(Q) (not shown here). These results have been achieved without the artifice of adding a "tail" to S(Q) [2].

The second example concerns the oxygen-oxygen structure factor for liquid water obtained in a neutron diffraction isotope experiment [10]. In the previous ME analysis of these data it was necessary to fit a Gaussian to the first peak in order to get a good fit to the data. No such fitting is needed in the MC analysis, Fig. 2, which largely confirms the previous estimate of g(r). Note that a shoulder in g(r) at 3.8Å, present in other data [11], is not present in the MC simulation, so confirming that it was an artifact of truncation. In this case there were no measured data below $Q=1Å^{-1}$, and so the distribution has been forced to go smoothly to the compressibility limit, -0.93. Also a small peak is seen on the low r side of the

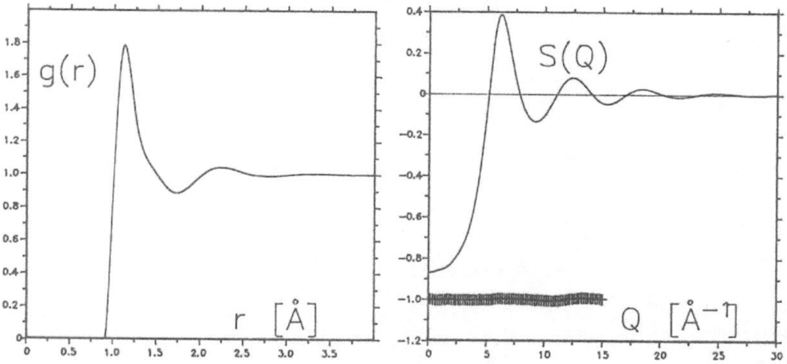

Figure 1. Percus-Yevick hard sphere simulation with Q_{max} = 15Å^{-1}. The calculated g(r) is shown on the left and the simulated S(Q) is shown on the right: the crosses (+) show the residual from the fit.

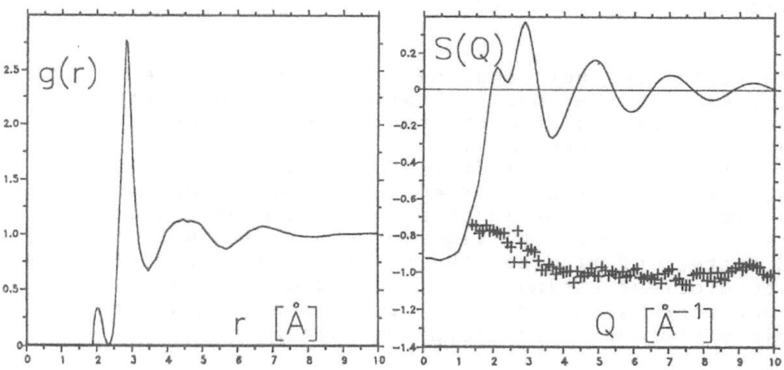

Figure 2. MC simulation of the oxygen-oxygen partial structure factor in liquid water. Same notation as Fig. 1.

main peak at 2.85Å: this is believed to be an artifact which arises from the large Q-dependent background under the original data.

The final example concerns time-of-flight diffraction data on gaseous deuterium near the critical point, Fig. 3 [12]. This result demonstrates the two main advantages of the Monte Carlo algorithm compared to earlier methods: it is able to correctly identify the very sharp intramolecular peak (at 0.74Å^{-1}) at the same time as it finds the broad intermolecular distribution at larger distances, both without any obvious truncation effects. It is also able to determine the uncertainties in g(r) due to the limitations of the data.

5. Conclusion

It is clear from these examples that the Monte Carlo procedure described has produced a significant improvement in our ability to calculate reliable, truncation free g(r)'s. This progress was achieved by abandoning the formal definition of entropy as a constraint on the allowed distributions and instead restricting the

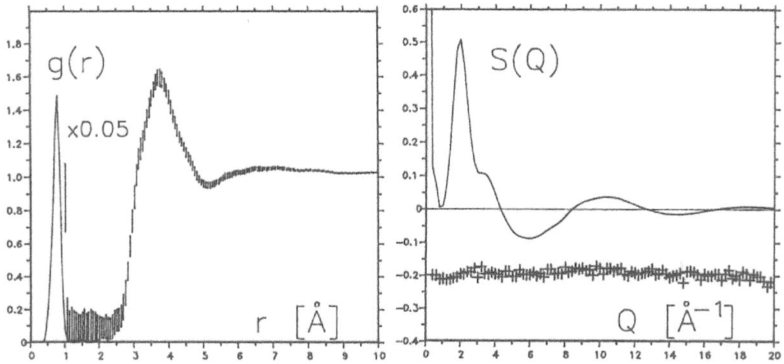

Figure 3. MC simulation of gaseous deuterium at 46.2 K and density of 0.0096 mols. per Å^3. The region below 1 Å has been multiplied by 0.05. Other notation as Fig. 1. In this case the weighting on the fit was lowered so that the distribution would sample a broader range of configurations: the error bars on g(r) show the standard deviation of these configurations about the mean.

fluctuations in g(r) in a manner consistent with the expected r-dependence for these fluctuations. Analysis of noisy data required an alternative but related constraint on the length of the line used to draw N(r) and hence g(r).

The method is not limited to one dimension. Quite recently McGreevy and coworkers [13] have embarked on the more ambitious task of calculating 3-dimensional distributions of particles consistent with a given dataset. Although the time for a simulation is greatly increased (hours instead of minutes) the resulting distributions may eventually lead to a direct estimation of the effective pair potential. This new work is currently in progress and is being applied to a variety of liquid and amorphous materials.

REFERENCES

1. A.K.Soper: Chemical Physics 107, 61 (1986)
2. J.H.Root, P.A.Egelstaff, B.G.Nickel: in Workshop on Neutron Scattering Data Analysis 1986, Chilton, Didcot, England, 13-14 March 1986 (IOP Publishing: Bristol, England) p.61
3. J.A.Potten, G.J.Daniell, B.D.Rainford: in Workshop on Neutron Scattering Data Analysis 1986, Chilton, Didcot, England, 13-14 March 1986 (IOP Publishing: Bristol, England) p.81
4. F.J.Bermejo, F.J.Mompean, J.Santoro, J.C. Dore: Nucl. Inst. Meth. Phys. Res. B22, 581 (1987)
5. A.K.Soper, in preparation. (1988)
6. A.L.Renninger, M.D.Rechtin, B.L.Averbach: J. Non-Cryst. Sol. 16, 1 (1974)
7. S.V.Semenovskaya, K.A.Khachaturyan, A.G.Khachaturyan: Acta Cryst. A41, 268 (1985)
8. J.E.Enderby,T.Gaskell,N.H.March: Proc. Phys. Soc. 85, 217 (1965)
9. P.Perry, G.J.Throop: J. Chem. Phys. 57, 1827 (1972)
10. A.K.Soper, M.G.Phillips: Chem. Phys. 107, 47 (1986)
11. A.H. Narten, W.E.Thiessen, L.Blum: Science 217, 1033 (1982)
12. A.K.Soper, M.Zoppi, R.Magli: submitted to International Conference on Neutron Scattering, Grenoble (July 1988)
13. R.L.McGreevy, L.Pusztai: Clarendon Laboratory Report, Ref:19/88 (Clarendon Laboratory, Oxford, England,1988)

Neutron Diffraction Study of Liquid Chlorine and Liquid Phosphorus Tribromide at 295 K

J.B. van Tricht

Interfacultair Reactor Instituut, Technische Universiteit Delft, NL-2629 JB Delft, The Netherlands

1. INTRODUCTION

Extensive diffraction studies of molecular liquid systems at room temperature have been made. Results of different experiments can be compared as in the case of carbon tetrachloride [1] and bromine. The homonuclear diatomic molecules form a special class, here neutron and X-ray scattering functions can be compared directly.

I compared the chlorine neutron data at room temperature of BOSI et al. [2] and those of SULLIVAN and EGELSTAFF [3]. The origins of differences between the two scattering functions are not at all clear. They might be due to multiple scattering treatment, normalization methods used, or the way inelastic scattering effects were accounted for. Parts in κ-space were omitted because of the scattering of the metal sample cells used, moreover the statistical accuracy was not stated.

Antimony trichloride was studied by TRIOLO and NARTEN [4] because of its solvent properties (not unlike water) and for its associated high dielectric constant. I realized that phosphorus tribromide is a very similar liquid, cf. Table I. Both systems PBr_3 and $SbC\ell_3$ crystallize to the same space group, Pnma [5], however, in PBr_3 the dielectric constant is lower by a factor of ten.

TABLE I. Physical Properties of PBr_3 and $SbC\ell_3$

XY_3		PBr_3	$SbC\ell_3$	Refs.
n	nm^{-3}	6.388	7.037	
T	K	295	353(1)	
T_m	K	233	346	
ρ_{XY}	nm	0.2245	0.2347(2)	11
ρ_{YY}	nm	0.3469	0.3490(20)	11
$b_X \cdot 10^{12}$	cm	0.513	0.564	9
$b_Y \cdot 10^{12}$	cm	0.679	0.9579	9
$r_{s,X} \cdot 10^{24}$	cm^2	3.314	4.2	9
$r_{s,Y} \cdot 10^{24}$	cm^2	5.92	16.63	9
$r_{a,X} \cdot 10^{24}$	cm^2	0.172	5.1	9
$r_{a,Y} \cdot 10^{24}$	cm^2	6.9	33.5	9

Springer Proceedings in Physics, Vol. 40
Editors: M. Davidović and A.K. Soper

Static and Dynamic Properties of Liquids

Diffraction data on liquid $C\ell_2$ and PBr_3 were taken at the liquids diffractometer at the Delft 2MW swimming pool reactor. The results obtained for $C\ell_2$ are compared to RISM [6] calculations and to MD calculations made by RODGER et al. [7]. The data for PBr_3 [8] are compared to RISM calculations and to experimental $SbC\ell_3$ [4] data.

2. EXPERIMENTS

The diffraction experiments were carried out in sealed vitreous silica sample cells with the liquids in orthobaric state. Corrections to the measurements were performed as usual [10]. The statistical accuracy of the corrected data is about 0.003. The normalization was achieved using interval normalization over the ultimate part of the scattering interval, from 80 to 130 nm^{-1}. Figures 1 and 2 show the normalized intensities, which are proportional to the differential scattering cross section

$$\frac{d\sigma}{d\Omega} = \langle b^2\rangle \left(1 + F_1(\kappa) + \frac{\langle b\rangle^2}{\langle b^2\rangle} i_M(\kappa) \right).$$

For liquid $C\ell_2$ the intramolecular distance from the experiment is $R_{C\ell-C\ell}$=0.2021 nm. For PBr_3 the experimental values of the intramolecular distances are given in Table I. The angle Br-P-Br=101.2(6)° is in agreement with [11].

The intermolecular scattering functions, $i_M(\kappa)$, were derived from the normalized intensities as the differences between symbols and lines in Fig. 1 divided by $\langle b\rangle^2/\langle b^2\rangle$, 1.386 and 3.871 for $C\ell_2$ and PBr_3 respectively.

3. RESULTS

The $i_M(\kappa)$'s are shown in Fig. 2 where it should be observed that at low κ the shape of the curve was taken as $i_M(0)(1 + a\kappa^2 + b\kappa^4)$ with a and b determined from a fit to a few data points around κ=10 nm^{-1}. The values of $i_M(0)$ follow from thermodynamics: -0.895 for chlorine, as in [2] and [3], for phosphorous tribromide the value -0.80 was estimated.

The best RISM results are shown as well, calculated at the liquid densities, 11.95 nm^{-3} for $C\ell_2$ and 6.39 nm^{-3} for PBr_3. The fused hard sphere molecules themselves were shaped with the intramolecular distance parameters obtained from the experiment cf. Table I. The atom hard sphere diameters used were $\sigma C\ell$=0.34 nm, σP=0.315 nm according to [1] and σBr=0.34 nm. There is a qualitative agreement of experiment and model.

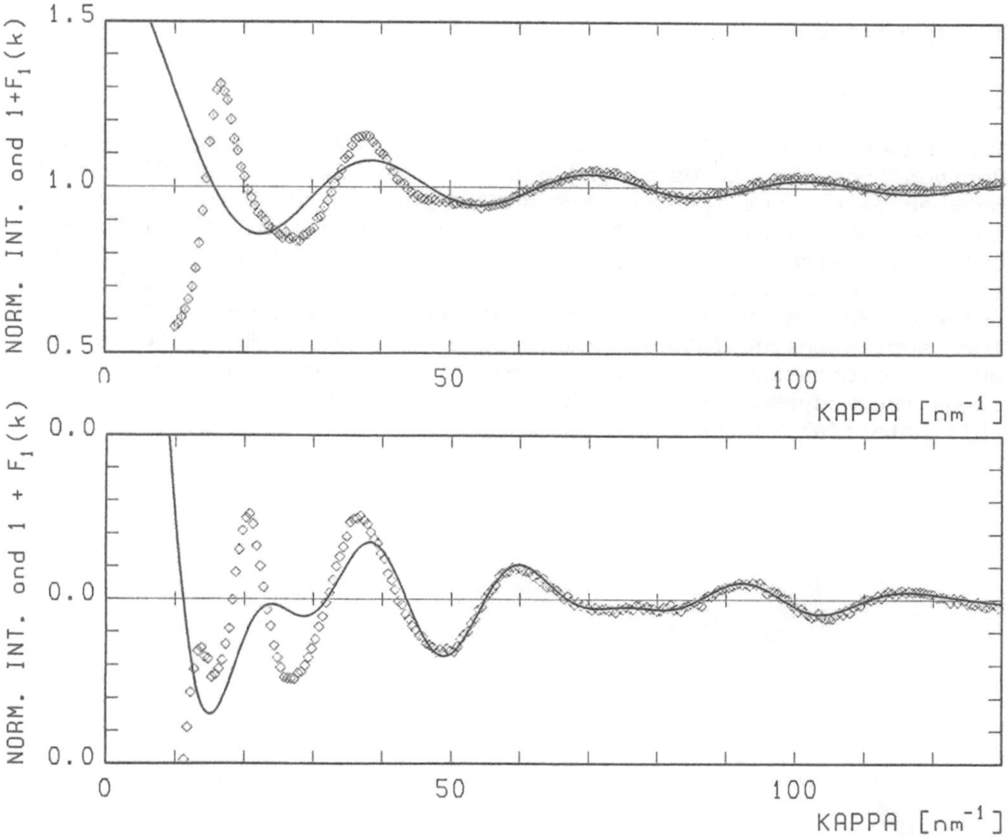

Fig. 1 Normalized intensity (symbols) and the intramolecular form factor (line), of chlorine (top) and phosphorus tribromide (bottom)

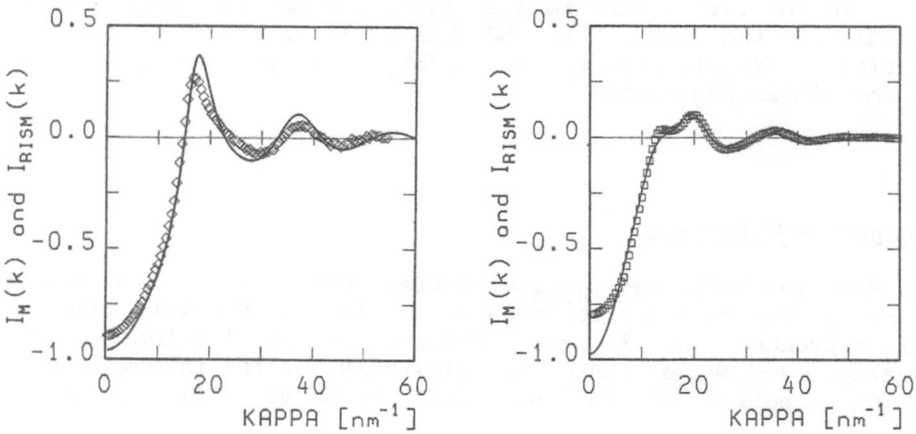

Fig. 2 Chlorine intermolecular scattering function, $i_M(\kappa)$ and $i_{RISM}(\kappa)$ with $\sigma_{C\ell}=0.34$ nm (left). Phosphorus tribromide intermolecular scattering function, $i_M(\kappa)$ and $i_{RISM}(\kappa)$ with $\sigma_P=0.315$ nm and $\sigma_{Br}=0.34$ nm (right)

4. DISCUSSION

4.1 Chlorine

The experimental $i_M(\kappa)$ of this study is shown, together with those derived from the tables in [2] and [3], in Fig. 3. The differences between $i_M(\kappa)$ and $i_{RISM}(\kappa)$ are due in part to the hard sphere atoms approach and in part to intrinsic particularities of the RISM. Molecular dynamics calculations [12] with distributed Lennard-Jones sites yield no significant improvement. A satisfactory correspondence of MD simulations [7] and experiment is obtained, using a much more sophisticated intermolecular potential that takes into account mutual orientations of molecules explicitly. More experiments at various thermodynamic states should be performed to test the use of this potential in the fluid state.

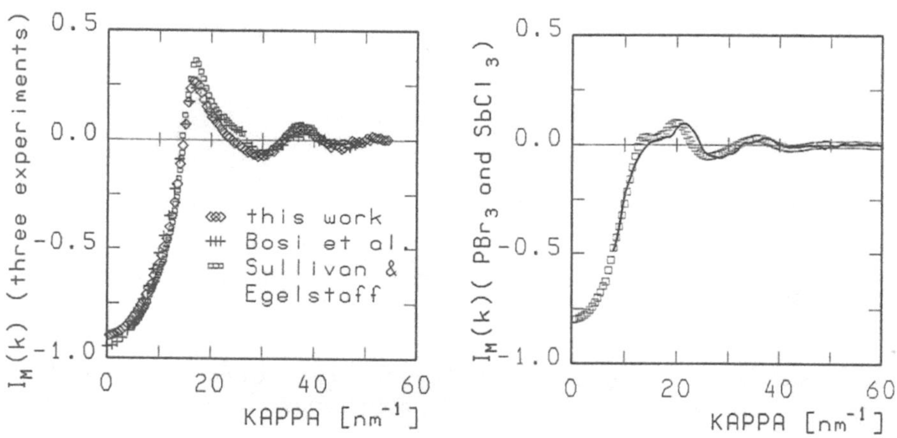

Fig. 3 To the left, experimental $i_M(\kappa)$'s (symbols) according to BOSI et al. [2] and SULLIVAN and EGELSTAFF [3] for liquid chlorine. To the right, experimental $i_M(\kappa)$ for PBr_3 (symbols) and $SbC\ell_3$ from TRIOLO and NARTEN [4] (line)

4.2 Phosphorus Tribromide

In both PBr_3 and $SbC\ell_3$ the $i_{RISM}(\kappa)$ differs from $i_M(\kappa)$ in two important aspects. The shoulder observed at the left of the main peak of $i_M(\kappa)$ is reproduced only in part. Moreover the oscillations at larger κ-values are out of phase. The similarity of the intermolecular scattering functions of PBr_3 and $SbC\ell_3$ is shown in Fig. 3 to the right.

It seems worthwhile to introduce more advanced models to simulate those liquids that consist of polar molecules, to find out whether a distribution of L-J sites, together with dipolar moments, are sufficient to describe the intermolecular scattering function or whether more sophisticated potentials, as were used for $C\ell_2$ [7], are to be called upon.

5. ACKNOWLEDGEMENTS

The author is indebted to Henk Fredrikze and to Piet van der Ende for their help during the preparation of this paper.

6. REFERENCES

1. D.G. Montague, M.R. Chowdhury, J.C. Dore and J. Reed, Mol. Physics 50, 1 (1983)
2. P. Bosi, F. Cilloco, F. P. Ricci and J. Sacchetti, Mol. Physics 49, 567 (1982)
3. J. D. Sullivan and P. A. Egelstaff, Chem. Physics 89, 167 (1984)
4. R. Triolo and A. H. Narten, J. Chem. Physics 69,3159 (1978)
   E. Johnson, A. H. Narten, W. E. Thiessen and R. Triolo, Trans. Farad. Soc. 66, 287 (1978)
5. R. Enjalbert and J. Galy, C.R. Acad. Sc. Paris 287, 259 (1979)
   Acta Cryst. B35, 546 (1979)
6. D. Chandler, The Liquid State of Matter: Fluids, Simple and Complex, Eds. Montroll and Lebowitz, p.239 (1982)
7. P. M. Rodger, A. J. Stone and D. J. Tildesley, Mol. Physics 63, 173 (1988)
8. J. B. van Tricht, H. W. van Zeijl and P. van der Ende, J. Mol. Liquids, 38, 97 (1988)
9. L. Koester and W.B. Yelon, Summary of Low Energy Neutron Scattering Lengths and Cross Sections, (1982)
10. J.B. van Tricht, H. Fredrikze and H. van der Laan, Mol. Physics 52, 115 (1984)
11. K. Kuchitzu, T. Shibata, A. Yokozeki and C. Matsumura, Inorganic Chemistry 10, 2584 (1971)
12. K. Singer, A. Taylor and J.V.L. Singer, Mol. Physics 33, 1757 (1977)

Static Correlations in Rare Gases from Integral Equations for Realistic Model Interactions

S. Celi[1], *L. Reatto*[2], *and M. Tau*[3]

[1]Università di Pavia, Dipartimento di Fisica "A. Volta", via Bassi 6, I-27100 Pavia, Italy
INFN sezione di Pavia, GNSM-CISM unità di Pavia
[2]Dipartimento di Fisica dell'Università, via Celoria 16, I-20133 Milano, Italy
[3]Dipartimento di Fisica dell'Università, viale delle Scienze, I-43100 Parma, Italy

1. INTRODUCTION

We present a detailed test of the radial distribution function (rdf) $g(r)$ as given by the modified hypernetted chain equation (MHNC) [1] and by a crossover MHNC (CRS–MHNC) equation [2]. The test is performed against recent, very accurate results of simulation for Kr and Xe [3,4] using "realistic" pair interactions obtained by fitting experimental two–body data. Since these are not effective interactions like the Lennard–Jones potential, many–body forces must be taken into account explicitly, and in [5] we have analysed the case when the Axilrod–Teller three–body triple dipole interaction is present. The three–body term is taken into account by a suitable generalization of the two equations [6] and we find that its effect is very well reproduced.

In this paper we examine the accuracy of the integral equations in the case of two–body forces only, since from the computations in [5] we know that the main contribution to error comes from the approximations made at this level. We first study the MHNC equation, considering three different approximate forms for the hard sphere bridge function: the Percus–Yevick (PY) form, the one corresponding to the Verlet–Weis (VW) parametrization [7] of $g(r)$ and a form recently proposed by Labík and Malijevský [8]. We find that use of these forms in the MHNC equation is not equivalent and the VW appears to give the best results. The overall best results, however, are obtained with the crossover MHNC equation, where the hard sphere bridge function is used only at short distance, whereas at larger distance the form associated with the mean spherical approximation is used. For really good results we find it necessary to optimize also the choice of the parameters of the crossover function (fully optimized CRS–MHNC).

In Section 2, a brief account of the formalism is given and in Sections 3 and 4 we report our results and conclusions.

2. MHNC AND FULLY OPTIMIZED CROSSOVER EQUATIONS

The MHNC equation is based on an approximation for the bridge function $E(r)$ which enters the formally exact relation

$$g(r) = \exp[-\beta\phi(r) + h(r) - c(r) + E(r)] \tag{1}$$

where $\beta = (k_B T)^{-1}$, $\phi(r)$ is the two–body potential, $h(r) = g(r) - 1$ and $c(r)$ is the direct correlation function. $E(r)$ is only known as a formal series of graphs of $f(r) = \exp[-\beta\phi] - 1$ or of h bonds. In order to get a closed equation for $g(r)$ from (1) a closure relation is needed

Springer Proceedings in Physics, Vol. 40
Editors: M. Davidović and A.K. Soper
Static and Dynamic Properties of Liquids

to approximate E. In the MHNC approximation the bridge function is assumed equal to that of a fluid of hard spheres of suitable diameter d. This diameter is density and temperature dependent and is determined by a free energy minimization criterion due to Lado [9]:

$$\int d^3r[g(r) - g_{HS}(r;d)]\frac{\partial E_{HS}(r;d)}{\partial d} = 0 \ . \tag{2}$$

In the case of the PY and VW bridge functions the g_{HS} entering (2) is the corresponding rdf. In the case of E_{HS} of Labík and Malijevský the corresponding rdf is not known analytically and should be obtained by solving a separate integral equation, so we have used the VW form.

The theoretical basis for the hard sphere approximation for $E(r)$ is limited to the core region where repulsive forces dominate. At larger distances there is good empirical evidence that the mean spherical approximation (MSA), $c(r) = -\beta\phi(r)$, is accurate and this implies that $E(r)$ is not universal but depends on the specific system since in the MSA (1) gives $E(r) = -h(r) + \ln g(r)$. The crossover MHNC approximation hence assumes the following form for the bridge function:

$$E_{CRS}(r) \equiv [1 - l(r)]E_{HS}(r;d) + l(r)[-h(r) + \ln g(r)] \ , \tag{3}$$

where $l(r)$ is a function which is small in the core region and becomes unity in the region of the attractive forces. The function $g(r)$ in (3) is the self–consistent solution of the integral equation and not the mean spherical approximation for the rdf. The choice of $l(r)$ is rather arbitrary and the form originally chosen [2] turns out to be not completely satisfactory because at medium density $l(r)$ does not vanish fast enough at short distance [10]. As in [6] we use the form:

$$l(r) = \begin{cases} 0 & \text{for} \quad r < R - w \\ \frac{1}{2}\left[1 + \tanh\frac{r - R}{\sqrt{w^2 - (r-R)^2}}\right] & \text{for} \quad R - w < r < R + w \\ 1 & \text{for} \quad r > R + w \ . \end{cases} \tag{4}$$

E_{CRS} depends on three parameters: d, R and w. Again minimization of the free energy determines d via the condition

$$\int d^3r[g(r) - g_{HS}(r;d)][1 - l(r)]\frac{\partial E_{HS}(r;d)}{\partial d} = 0 \ . \tag{5}$$

In [2] the parameters R and w were chosen empirically, R being equal to the position of the second zero of $h(r)$ and w being the distance between this zero and the position of the first maximum of $g(r)$. One can invoke the criterion of minimization of the free energy with respect to R and w too, and this gives the two additional equations:

$$\int d^3r\{[g(r) - g_{HS}(r;d)]E_{HS}(r) + \frac{[g(r)^2 - g_{HS}(r)^2]}{2} - \\ - [g(r)\ln g(r) - g_{HS}(r)\ln g_{HS}(r)]\}\frac{\partial l(r)}{\partial R} = 0 \ , \tag{6}$$

$$\int d^3r\{[g(r) - g_{HS}(r;d)]E_{HS}(r) + \frac{[g(r)^2 - g_{HS}(r)^2]}{2} - \\ - [g(r)\ln g(r) - g_{HS}(r)\ln g_{HS}(r)]\}\frac{\partial l(r)}{\partial w} = 0 \ . \tag{7}$$

3. RESULTS AND DISCUSSION

We have solved the MHNC and the CRS–MHNC under the conditions of recent simulations by Levesque and coworkers [3,4]. The numerical method of solution of the integral equation is standard. The first computation is for Xe under triple point conditions $T = 175\,K$ and $\rho = 13.13\,\text{atoms/nm}^3$ with the HFD–B pair potential by Aziz and Slaman [11]. In order to be under the same conditions as the simulation we have truncated $\phi(r)$ at half the simulation box length, $L/2 = 10.078$ Å . With MHNC we have used the three different forms for the hard sphere bridge function; the hard sphere diameter as obtained from Lado's criterion (2) is $d = 0.810\,\sigma$ for the PY form and $d = 0.816\,\sigma$ for the VW form ($\sigma = 3.8924$ Å is the position of the zero of the potential). With the E_{HS} of Labík and Malijevský it is not possible to satisfy Lado's criterion because the integral is always negative since E_{HS} does not increase fast enough inside the hard core, as already noticed [8]. A computation with MHNC using this form of E_{HS} has been recently reported [12], and Lado's criterion appears to be satisfied if the solution of the appropriate integral equation is used for g_{HS}. This rdf, however, near contact is not so good as the WV one and the same is true for the corresponding y_{HS} inside the core; this indicates that the parametrization of E_{HS} of Labík and Malijevský should be improved before it can be used with complete confidence at high density.

The results given by MHNC are in good agreement with simulation and a standard comparison (Fig. 1) shows almost perfect agreement. Actually, the deviations between MHNC and simulation results have a typical value of ± 0.03 (the statistical error of the simulated $g(r)$ is ± 0.005); in order to see the difference between the various approximations more clearly, we also plot in Fig. 1 (left) the difference $\Delta g(r)$ between theory and simulation. The PY $E(r)$ is substantially smaller than the VW $E(r)$ but, as already noticed [1], the errors of PY are compensated to a large extent by a shift in the value of the diameter which satisfies (2). However, at close inspection, the VW form is superior in the region of the first peak of $g(r)$. On the other hand the VW form has a small anomaly around $r = 2d$ which is probably spurious [2].

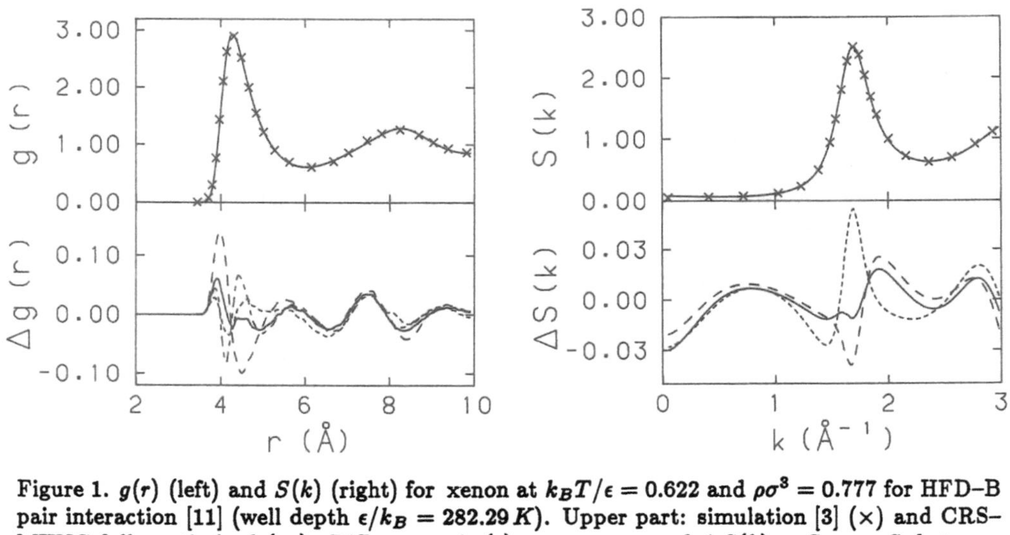

Figure 1. $g(r)$ (left) and $S(k)$ (right) for xenon at $k_BT/\epsilon = 0.622$ and $\rho\sigma^3 = 0.777$ for HFD–B pair interaction [11] (well depth $\epsilon/k_B = 282.29\,K$). Upper part: simulation [3] (×) and CRS–MHNC fully optimized (—). Differences $\Delta g(r) = g_{sim} - g_t$ and $\Delta S(k) = S_{sim} - S_t$ between simulation and theory: CRS–MHNC fully optimized (—), CRS–MHNC (- - -), MHNC with VW E_{HS} (- - -), MHNC with PY E_{HS} (– – –)

The CRS–MHNC with the empirical choice of R and w ($R = 1.062\,\sigma, w = 0.268\,\sigma$, optimized diameter $d = 0.818\,\sigma$) gives a rdf not so good as the MHNC with the VW $E_{HS}(r)$. The fully optimized parameters of CRS–MHNC are $d = 0.819\,\sigma$, $R = 1.240\,\sigma$ and $w = 0.259\,\sigma$. Therefore R is larger than the value given by the empirical rule and in fact it is of the same order as the position of the first minimum of $g(r)$. The rdf is now in better agreement with simulation and the deviations are almost the same as those of MHNC with the VW E_{HS}. Beyond the first peak of $g(r)$ all these equations give similar results, except for the small anomaly at $r = 2d$ given by the VW E_{HS}; this implies that in this range of r the main non-universal part of $E(r)$ is not due to the difference between $h^2(r)$ and $h^2_{HS}(r)$ (notice that $E_{CRS}(r) - E_{HS}(r) \approx \frac{1}{2}(h^2(r) - h^2_{HS}(r)))$ where $l(r) \sim 1$) but is due to the small deviations of $c(r)$ from $-\beta\phi(r)$.

From $g(r)$ as given by simulation it is not possible to obtain $S(k)$ directly via a Fourier transform because $g(r)$ is only known up to a cutoff distance r_C which is usually too small. A common procedure is to extend $g(r)$ beyond r_C with Verlet's algorithm [13]: $g(r)$ is given by simulation for $r < r_C$ and by the Ornstein–Zernike relation for $r > r_C$ assumimg $c(r) = 0$. In Fig. 1 (on the right) we show the difference between the $S(k)$ obtained this way from simulation, and those given by the fully optimized CRS–MHNC and by MHNC with PY and VW E_{HS}. Here CRS–MHNC gives the best result and again the plot of $S(k)$ in the usual format shows perfect agreement, the deviations being at the level of ± 0.02. The percentage error at $k \sim 0$, however, is substantial, due to the very small value of $S(0)$ in the triple point region (from simulation we have $S(0) = 0.049$ and from CRS–MHNC $S(0) = 0.079$). We suspect that this discrepancy at small k is partially due to the method of extension of the rdf. It has already been noticed for the Lennard–Jones potential that this method of extension is not accurate when r_C is of order of $2.5\,\sigma$, as in the present case.

The second computation is for dense krypton gas at $T = 297\,K$ and $\rho = 13.84\,\text{atoms/nm}^3$. As pair potential the HFGKR potential by Aziz [14] truncated at $L/2 = 9.92$ Å is used. In this case (see Fig. 2) the deviations between theory and simulation are smaller than for Xe and are well above the noise level only in the region of the first peak of $g(r)$; the best results are obtained with the fully optimized CRS–MHNC and with MHNC with the VW bridge function.

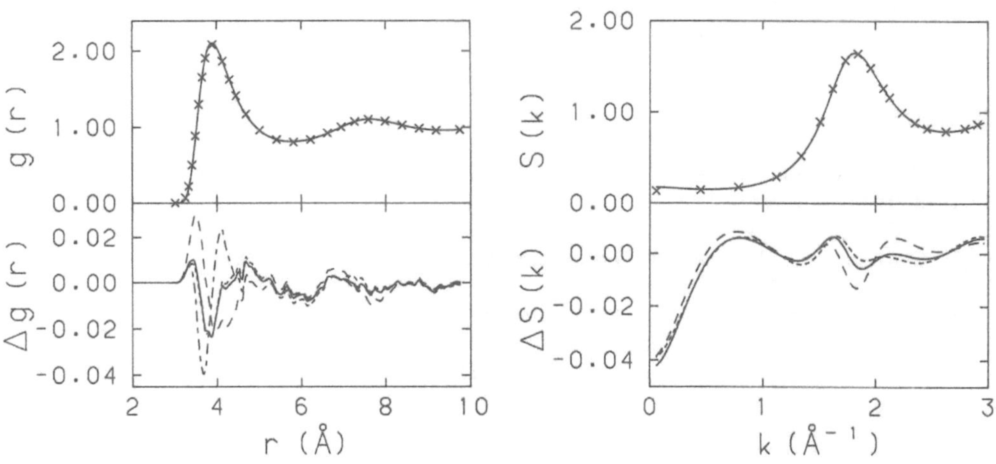

Figure 2. Same as in Fig. 1 for krypton at $k_B T/\epsilon = 1.486$ and $\rho\sigma^3 = 0.635$ for HFGKR pair potential [14] (well depth $\epsilon/k_B = 199.9\,K$). Simulation results from [4]

The parameters of the bridge function are: for MHNC $d = 0.769\,\sigma$ for PY and $d = 0.780\,\sigma$ for VW, for CRS–MHNC $d = 0.781\,\sigma$, $R = 1.249\,\sigma$ and $w = 0.535\,\sigma$ and for the fully optimized CRS–MHNC $d = 0.7806\,\sigma, R = 1.561\,\sigma$ and $w = 0.598\,\sigma$. The zero of the potential for Kr is at $\sigma = 3.579$ Å.

4. CONCLUSIONS

The present computation confirms that both MHNC and CRS–MHNC are fairly accurate equations in the dense fluid regime, the crossover equation being slightly superior. However, their accuracy is not yet good enough to look at the details of the interatomic interactions. For instance, the deviations shown in Figs. 1 and 2 for $g(r)$ are larger than the change in the rdf due to the presence of a three–body interaction, as shown in [5].

If high accuracy is needed, the VW bridge function of hard spheres should be used and not the PY form. The VW E_{HS} is not exact, however, and we cannot exclude that this is responsible in part for the deviations found between theory and simulation. In this respect the attempt of parametrization of E_{HS} made by Labík and Malijevský is useful, but it does not seem to have the required accuracy.

In order to get the best out of the CRS–MHNC equation, one should optimize the choice of the crossover parameters too, but we find that R and w do not appear to depend very much on ρ and T; on the basis of the results reported in Section 3 and of those of other computations we hence conclude that R and w can be approximated by the values given by the following rule: R is equal to the position of the first minimum after the main maximum of $g(r)$, and w is equal to the distance between this minimum and the second zero of $h(r)$.

ACKNOWLEDGMENTS

We thank D. Levesque and J. J. Weis for communicating their results prior to publication. This work has been partially supported by CISM–MPI, by GNSM of CNR and by CNR under the CNR–NSF cooperative science program.

REFERENCES

1. Y. Rosenfeld and N. H. Ashcroft, Phys. Rev. A **20**, 1208 (1979); F. Lado, Phys. Rev. A **8**, 2548 (1973)
2. S. M. Foiles, N. H. Ashcroft and L. Reatto, J. Chem. Phys. **80**, 4441 (1984)
3. D. Levesque, J. J. Weis and J. Vermesse, Phys. Rev. A **37**, 918 (1988)
4. D. Levesque and J. J. Weis, preprint
5. S. Celi, L. Reatto and M. Tau, Phys. Rev. A, in press
6. L. Reatto and M. Tau, J. Chem. Phys. **86**, 6474 (1987)
7. L. Verlet and J. J. Weis, Phys. Rev. A **5**, 939 (1972)
8. A. Malijevský and S. Labík, Mol. Phys. **60**, 663 (1987)
9. F. Lado, Phys. Lett. **89 A**, 196 (1982)
10. P. D. Poll and N. W. Ashcroft, Phys. Rev. A **35**, 866 (1987)
11. R. A. Aziz and M. J. Slaman, Mol. Phys. **57**, 825 (1986)
12. A. Malijevský, R. Pospísil and S. Labík, Mol. Phys. **61**, 533 (1987)
13. L. Verlet, Phys. Rev. **159**, 98 (1967)
14. R. A. Aziz, Mol. Phys. **38**, 177 (1979)

Three-Body Forces and Virial Expansions of S(k) and g(r) in Gaseous Kr

F. Barocchi[1] *and R. Magli*[2]

[1]Dipartimento di Fisica, Università di Firenze,
Largo E. Fermi 2, I-50125 Firenze, Italy
[2]Dipartimento di Energetica, Università di Firenze,
via S. Marta 3, I-50139 Firenze, Italy

We have evaluated the linear coefficient in the virial expansion of the structure factor S(k) and the radial distribution function g(r) in gaseous Kr for several temperatures and with three different models for the irreducible three-body interaction energy. The comparison with the presently available experimental neutron scattering data for S(k) shows large differences.

1 Introduction

In the last few years there has been a wide interest /1-7/ in the study of the role of many-body forces in noble gases. The interest in this field is partly due to a not yet solved problem concerning the structure in which rare gases crystallize and partly to some discrepancies found between recent neutron scattering results and theoretical predictions based on the well known triple dipole three-body interaction potential discussed by Axilrod, Teller and Muto in 1943 /8/.

The necessity of many-body interactions in order to explain static and dynamical properties for noble gases is established /1,2/; in particular, Barker /1/ found that using only a pair interaction experimental values of the third virial coefficient cannot be obtained. In fact, in order to reproduce the experimental data, the introduction of a three-body irreducible interaction is necessary considering the multipolar expansion of the energy up to the triple quadrupole term. Recently, Celi et al. /9/ have pointed out that it is possible to extract with great accuracy the true pair interaction potential from structural data, provided that the three-body interaction is well known.

When many-body forces are included, the problem arises of which analytical form they are to assume. Let us confine our attention to the three-body contribution. Up to a few years ago the Axilrod-Teller-Muto (A-T-M) expression and eventually higher multipolar terms (like triple quadrupole in the case of third virial coefficient of Ar) have been considered satisfactory for the interpretation of thermodynamic properties of rare gases. Recently, some neutron scattering experiments on Kr at room temperature (Teitsma and Egelstaff /3/) and at 220K (Fredrikze /10/) have shown that the A-T-M form is not adequate at all to reproduce the k-behaviour of the structure factor. Moreover, recently Loubeyre /11/ found that by modifying the triple dipole term with a short range contribution which explicitly takes into account exchange effects it is possible to give a good account of thermodynamic properties of noble gases at very high pressure. Therefore the debate about the rigorous representation of three-body non-additive potentials is still open and it is necessary to test the various proposed forms against experimental data.

In order to isolate the three-body contribution to interaction energy from higher orders we consider a virial expansion of the radial distribution function (RDF) g(r) and of the structure factor S(k) with respect to the density and limit our study to the first coefficient depending on non additive three-body energy; this approach applies of course only in particular thermodynamic conditions and not for all r- and k-values, as will be clear from the following.

2 Theory

Let us consider a low density gas composed of N atoms contained in a volume V; if the total interaction energy U(1---N) can be cluster expanded as:

$$U(1---N) = \Sigma_{i>j} U_2(i,j) + \Sigma_{i>j>k} U_3(i,j,k) + --- , \quad (1)$$

where $U_2(i,j)$ is the pair potential, $U_3(i,j,k)$ is the irreducible three-body contribution and so on, we can write the RDF g(r) as a virial expansion with respect to the number density n /12/:

$$g(r_{12}) = g_0(r_{12}) + n^* g_1(r_{12}) + \Theta(n^2) , \quad (2)$$

where:

$$g_0(r_{12}) = \exp[-U_2(1,2)/KT] \quad (3)$$
$$g_1(r_{12}) = g_1^{(2)}(r_{12}) + g_1^{(3)}(r_{12}) \quad (4)$$

and:

$$g_1^{(2)}(r_{12}) = g_0(r_{12}) \int f(r_{13}) f(|\underline{r}_{12}-\underline{r}_{13}|) d\underline{r}_{13} \quad (5)$$

$$g_1^{(3)}(r_{12}) = g_0(r_{12}) \int g_0(r_{13}) g_0(|\underline{r}_{12}-\underline{r}_{13}|) [\exp-(U_3(1,2,3)/KT) - 1] d\underline{r}_{13}$$
$$f(r_{12}) = g_0(r_{12}) - 1 \quad (6)$$

r_{ij} (i,j=1...3) are the sizes of the triangle formed by the atoms 1,2 and 3. In an analogous manner we can expand the structure factor S(k):

$$S(k) = 1 + n \int [g(r_{12})-1] \exp(-\underline{k}.\underline{r}_{12}) d\underline{r}_{12} \quad (7)$$

and in particular the useful function H(k) defined by

$$H(k) = [S(k)-1]/n . \quad (8)$$

From these expansions it is possible to see that both the linear coefficient for g(r) and the analogous coefficient for H(k) are composed of two contributions: the first one, $g_1^{(2)}(r_{12})$ and $H_1^{(2)}(k)$ respectively, depend only on the pair interaction, while the second one, $g_1^{(3)}(r_{12})$ and $H_1^{(3)}(k)$ respectively, depend on the pair and three-body interaction potential. Here we must observe that the virial expansion approach applies for particular thermodynamic conditions and not for all r- and k-values; in particular, far from the critical point, the virial expansions for $g(r_{12})$ and H(k) are convergent provided the two following conditions are fulfilled:

$$n^*\sigma^3 \ll 1 \qquad \text{and} \qquad (n^*\sigma^3)^{1/3} < k^*\sigma/(2^*\pi) \tag{9}$$

where σ is the interatomic distance where $U_2(1,2)$ is zero. The first one is the usual requirement for an expansion parameter, while the second one means that the wavelength probing the system has to be smaller than the average interatomic distance.

3 Results and Discussion

In order to see if the experimental possibilities allow a measurement of the three-body effect, the relative sizes of g_0, $g_1^{(2)}$ and $g_1^{(3)}$ and of the corresponding terms in k-space have to be evaluated and this can be done via eqs.(3),(5),(7) and (8) by a computer calculation once two and three-body potential terms are chosen. We performed the calculation for Kr at seven temperatures between 150 and 297K (T_c=209K); we chose the empirical pair potential of Barker et al. /13/ and for U_3 three different models were used:

1) the "long-range" Axilrod-Teller-Muto triple dipole /8/ (ddd) potential given by:

$$U_3(1,2,3) = \frac{\nu\,[1+3\cos\Phi_1\cos\Phi_2\cos\Phi_3]}{r_{12}^3 r_{23}^3 r_{13}^3} \tag{10}$$

2) the Slater-Kirkwood form recently used by Loubeyre /11/:

$$U_3(1,2,3) = [1+3\cos\Phi 1\cos\Phi 2\cos\Phi 3] - \left\{\frac{\nu}{r_{12}^3 r_{23}^3 r_{13}^3} - A\exp[\alpha(r_{12}+r_{23}+r_{13})]\right\} \tag{11}$$

3) the long range multipolar expansion up to the triple quadupole (qqq):

$$\begin{aligned} U_3(1,2,3) = \; & Z(ddd)_3 - W(ddd)_3 \\ & + Z(ddq)_3 - W(ddq)_3 \\ & + Z(dqq)_3 - W(dqq)_3 \\ & + Z(qqq)_3 - W(qqq)_3 \end{aligned} \tag{12}$$

with the geometrical factors W derived from /14/ and the numerical values from /16/ and /11/. In the previous formulae Φ_i are the angles of the triangle formed by $\underline{r}_i$, i=1,2,3.

In Fig.1 we report $g_1^{(2)}$ and $g_1^{(2)}+g_1^{(3)}$ at T=180K; we can see that the introduction of U_3 strongly cancels the first peak in $g_1^{(2)}$ and its contribution is confined approximately in the range 3.8<r<7 A. The corresponding plot in k-space (Fig.2) shows that the difference between $H_1^{(2)}$ and $H_1^{(2)}+H_1^{(3)}$ is less pronounced and extends from k=o to about k=1 A^{-1}. In Figs.1 and 2 the U_3 we used was the (ddd) one.

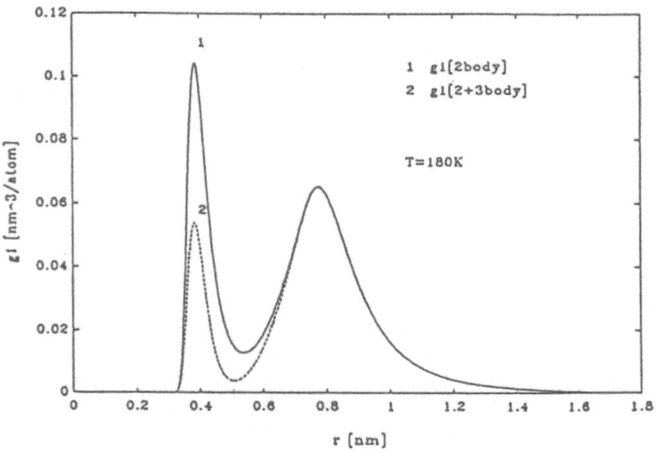

Fig.1
Kr: first virial coefficient of g(r) without (1) and with (2) 3-body potential (ddd) at T=180K

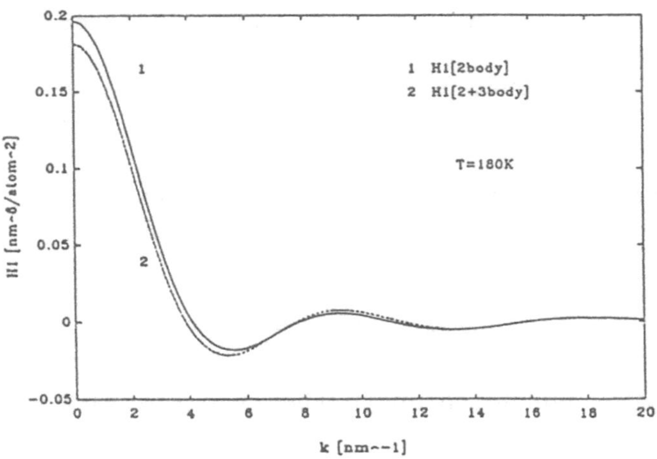

Fig.2
Kr:first virial coefficient of h(k) without (1) and with (2) 3-body potential (ddd) at T=180K

In Fig.3 we report $g_1^{(3)}$ at room temperature for the three U_3 models we used. While the introduction of triple quadrupole slightly increases the negative peak of $g_1^{(3)}$, the Slater-Kirkwood form brings a large change in $g_1^{(3)}$ shifting its hard core and largely modifying the negative height of its peak. In

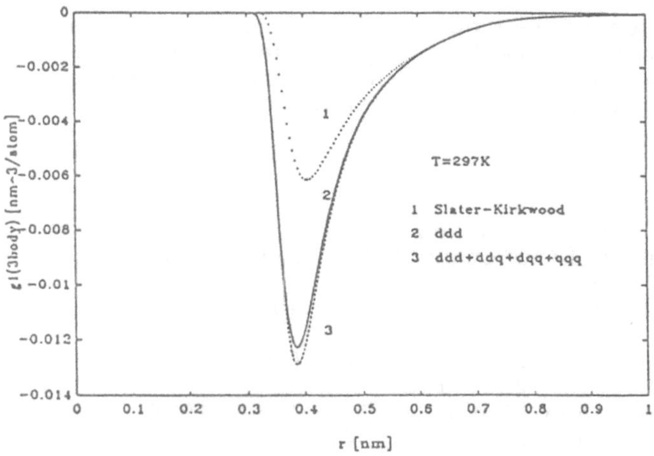

Fig.3

Kr: 3-body contribution to first virial coefficient of g(r) at T=297K for the U_3 models discussed in text

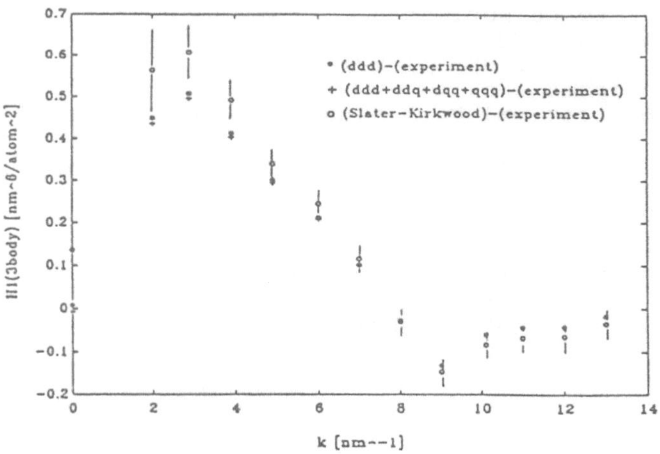

Fig.4

Kr: differences between calculated and measured /3/ $H_1^{(3)}$ at T=297K for the U_3 models discussed in text

Fig.4 the differences between calculated and measured by Teitsma and Egelstaff /3/ $H_1^{(3)}$ are shown. It is evident that while the introduction of triple quadrupole potential makes $H_1^{(3)}$ converge to the experimental value for k=0 derived from compressibility data /16/,the Slater-Kirkwood form used by Loubeyre produces a k=0 value which is in large disagreement with the thermodynamic result; moreover, the overall comparison shows a great discrepancy between calculated and measured $H_1^{(3)}$, in particular for $0<k<0.8$ Å^{-1}.

4 Conclusions

The following conclusions can be drawn: 1) neither multipolar long range potential up to triple quadrupole nor Slater-Kirkwood form for U_3 give a satisfactory account of the present structural data of Kr at low density; 2) new accurate measurements of S(k)

(relative accuracy required of few per thousand) of rare gases are necessary to test definitively the U_3 proposed models for these systems, especially for 0<k<1 A^{-1} and in thermodynamic states for which the virial expansion is sensible.

References

1. J.A.Barker,1976,Rare Gas Solids vol.1
2. P.A.Egelstaff,W.Glaser,D.Litchinsky,E.Schneider,J.B.Suck, Phys.Rev.A 27,1106 (1980)
   P.A.Egelstaff,J.J.Salacuse,W.Schommers,J.Ram,Phys.Rev.A 30,374(1984)
3. A.Teitsma,P.A.Egelstaff, Phys.Rev.A21,367 (1980)
4. J.A.Barker,Phys.Rev.Lett.57,230 (1986)
5. G.C.Aers,M.W.Dharma-Wardana, Phys.Rev.A29,2734 (1984)
6. L.Reatto,M.Tau, J.Chem.Phys.86,6474 (1987)
7. J.J.Salacuse,P.A.Egelstaff, to be published
8. B.M.Axilrod,E.Teller, J.Chem.Phys.11,299 (1943)
   Y.Muto, Proc.Phys.Math.Soc.Japan 17,629 (1943)
9. S.Celi,L.Reatto,M.Tau, to be published
10. H.Fredrikze, Phys.Rev.A36,2272 (1987)
11. P.Loubeyre, Phys.Rev.Lett.58,1857 (1987)
    P.Loubeyre, to be published
12. S.A.Rice,P.Gray in "Statistical Mechanics of Simple Liquids", ed.Wiley 1976
13. J.A.Barker,R.O.Watts,J.K.Lee,T.P.Shafer,Y.T.Lee,J.Chem.Phys. 61,308 (1974)
14. R.J.Bell,J.Phys.B 3,751 (1970)
15. M.B.Doran,I.J.Zucker, J.Phys.C 4,307 (1971)
16. N.J.Trappeniers,T.Wassenaar,G.J.Wolkers, Physica 32, 1503 (1966)
    A.Michels,J.M.Levelt,W.de Graaff, Physica 26,659 (1960)

Interaction-Induced Light Scattering in Dense Gaseous Krypton: Many-Body Effects

F. Barocchi[1], *M. Celli*[1], *M. Zoppi*[2], *and U. Bafile*[2]

[1]Dipartimento di Fisica, Università di Firenze, Largo E. Fermi 2, I-50125 Firenze, Italy

[2]Consiglio Nazionale delle Ricerche, Istituto di Elettronica Quantistica, via Panciatichi 56/30, I-50127 Firenze, Italy

Depolarized Interaction-Induced Light Scattering (DILS) spectra of dense gaseous krypton have been measured in the pressure range 10–270 bars and at room temperature. From these, absolutely calibrated values for the zeroth and second frequency moments have been obtained and compared with Monte Carlo simulation performed in the pairwise-additivity approximation. Two different models for the pair polarizability anisotropy and an updated semi-empirical pair potential have been used in the simulation. The density behaviour of the second moment shows the importance of irreducible many-body contributions to the scattered intensity, and the possibility of using DILS as an experimental probe of many-body properties.

1 Introduction

Static and dynamic microscopic properties of dense simple fluids have been interpreted for a long time in the pairwise-additivity approximation. However, recent work has raised new interest in the problem of many-body interactions, and, in particular, krypton has been widely investigated /1–4/.

Depolarized Interaction-Induced Light Scattering (DILS) is connected to correlation functions of microscopic interaction-induced variables. Therefore, this technique offers, in principle, the possibility of measuring many-body effects and of giving information on higher-order correlations.

In recent years DILS experimental spectra of gaseous and liquid argon have been compared with computer simulation in a wide range of density /5–8/. However, the comparisons were somewhat inconclusive as far as the presence of irreducible three-body effects is concerned, the discrepancy between the measurement and the pair-additivity simulation being of the same order of the experimental uncertainties. Here an analogous comparison is presented for the zeroth and second frequency moments of the DILS spectrum of krypton.

2 The Experiment

DILS spectra of krypton have been measured at T=297 K with pressure and density ranging from 10 to 270 bars and from 0.25 to 8.5 atoms/nm^3, respectively. The experimental set-up is the same as that used for the argon experiment /8/.

In a laboratory reference frame, the direction and polarization of the incoming laser beam are parallel to the y- and x-axis, respectively, while the scattered radiation is collected in the x direction with both polarization components, so that a depolarized 90-degree scattering geometry is defined.

The recorded spectra have been calibrated against the rotational Raman lines S(0) and S(1) of molecular hydrogen, following a procedure described in Ref. 8. At every frequency, the pair and triplet contributions have been extracted with a least-square fit procedure. The resulting two-body spectrum is consistent within the errors with a previous measurement /9/, while the triplet spectrum has been published in Ref. 10. The zeroth and second frequency moments have been evaluated and will be used for comparison with theoretical and computer-simulation values.

3 Theory

The double differential cross section for a system of N classical identical particles in the volume V is, in the depolarized geometry, expressed by /11/

$$\frac{d^2\sigma}{d\Omega d\omega} = 2\, k_0\, k_s^3\, V\, I(\omega)\,, \tag{1}$$

where $\omega = 2\pi\,(\nu_0 - \nu_s)$ is the frequency shift from the laser line, $\vec{k}_0$ and $\vec{k}_s$ are the incident and scattered wave vectors with $k_0 \approx k_s$, and

$$I(\omega) = (1/2\pi)\int dt\, \exp(-i\omega t)\, G(t) \tag{2}$$

with

$$G(t) = (1/V)\left\langle \sum_{P,Q} \alpha_{yx}(P,t)\,\alpha_{yx}(Q,0)\right\rangle . \tag{3}$$

In (3) P and Q refer to all possible clusters of atoms and α_{yx} is the yx component of the irreducible cluster polarizability tensor; the yz component gives an identical contribution which is accounted for by the factor of 2 in (1). In the following the indices yx will be omitted for brevity.

Since the contribution of isolated spherical atoms is zero, in the pair approximation G(t) reduces to a double sum over pairs of atoms which can be rewritten as a density expansion as

$$G(t) = G_{\text{2-BODY}}(t) + G_{\text{3-BODY}}(t) + \ldots \tag{4}$$

where

$$G_{\text{2-BODY}}(t) = (V\rho^2/2)\,\langle\alpha(12,t)\alpha(12,0)\rangle_0^{(2)}\,,$$

$$G_{\text{3-BODY}}(t) = (V\rho^3/2)\,\langle\alpha(12,t)\alpha(12,0)\rangle_1^{(2)} + V^2\,\rho^3\,\langle\alpha(12,t)\alpha(13,0)\rangle_0^{(3)}\,,$$

ρ is the number density, $\langle ...\rangle_0^{(2)}$ and $\langle ...\rangle_0^{(3)}$ are statistical averages over pairs and triplets, while $\langle ...\rangle_1^{(2)}$ is the irreducible part of the pair average due to the presence of a third particle. We will neglect all the terms coming from correlations involving more than three particles.

The double differential cross section can be written as

$$\frac{d^2\sigma}{d\Omega d\omega} = V\ [(\rho^2/2)\ D_{\parallel}(12,\omega) + \rho^3\ D_{\parallel}(123,\omega)] \qquad (5)$$

where the depolarized two- and three-body spectra are defined.

The frequency moments

$$M_n = \int d\omega\, \omega^n\ 2\ k_0^4\ V\ I(\omega) = 2\ k_0^4\ V\ (-i)^n\ (d^n G(t)/dt^n)_{t=0}$$

are then, for n=0,2 and with $G^{(2)} = -\ d^2G/dt^2$:

$$M_0 = 2\ k_0^4\ V\ G(0) = V\ [(\rho^2/2)\ M_0^{(2)} + \rho^3\ M_0^{(3)} + ...]\ , \qquad (6)$$

$$M_2 = 2\ k_0^4\ V\ G^{(2)}(0) = V\ [(\rho^2/2)\ M_2^{(2)} + \rho^3\ M_2^{(3)} + ...]\ . \qquad (7)$$

$M_n^{(j)}$ (j=2,3, n=0,2) can be calculated by performing the ensemble averages for a given model pair polarizability as a function of the interatomic distance and a given pair potential.
On the other hand, M_0 and M_2 can be evaluated at high density by computer simulation in the pair approximation by calculating

$$G(0) = (1/V)\ \langle [\sum_{i>j} \alpha(ij,0)]^2\rangle$$

$$G^{(2)}(0) = (1/V)\ \langle [\sum_{i>j} \dot{\alpha}(ij,0)]^2\rangle$$

where the property of correlation functions $\langle \ddot{A}(t)B(t)\rangle = -\ \langle \dot{A}(t)\dot{B}(t)\rangle$ is used.

4 The Pair Polarizability

By means of a tensor algebra calculation, the correlation functions involving the yx component of the pair polarizability tensor $\overleftrightarrow{\alpha}(\vec{r})$ can be rewritten in terms of the second-rank irreducible spherical component which is proportional to the so-called polarizability anisotropy $\beta(r)$ /11/.

The simplest model for $\beta(r)$, the first-order Dipole-Induced Dipole (DID) model, is given by

$$\beta(r) = 6\alpha_0^2/r^3 \qquad (8)$$

where α_0 is the single-atom scalar polarizability. A general feature of DILS is that (8) doesn't account for the scattered intensity in the wings of the pair spectrum, due to the fact that the DID mechanism accounts for long-range interaction only. Therefore, a semi-empirical three-parameter model

$$\beta(r) = 6\alpha_0^2/r^3 + A/r^6 - B\exp(-r/r_0) \qquad (9)$$

has been introduced /11,12/, which reproduces the measured pair spectra very well over a large intensity range, once the parameters A, B and r_0 are fitted to the low-density (i.e. pair) spectral moments.

If (9) is used in a computer simulation, it is possible to test the validity of the pair-additivity approximation in the dense fluid. The quantities used for the comparison with the experiment are the zeroth and second reduced frequency moments $\overline{M}_n = M_n/2\,k_0{}^4$ as functions of the density.

5 Results

In Table 1 the comparison between experiment and theory is reported in absolute units for $M_n^{(j)}$ (j=2,3, n=0,2). The theoretical values are calculated with both DID and model (9) for j=2, and with DID only for j=3.

Table 1. Two-body and three-body moments: (a) calculation with DID model (8),(b) calculation with model (9), with A=538 $Å^9$, B=580 $Å^3$, r_0=0.48 Å /12/.

	$M_0^{(2)}$ ($Å^9$)	$M_2^{(2)}$ ($10^{25}\,Å^9\,s^{-2}$)	$M_0^{(3)}$ ($10^3\,Å^{12}$)	$M_2^{(3)}$ ($10^{27}\,Å^{12}\,s^{-2}$)
(a)	14.4	6.73	-0.97	-2.39
(b)	14.2	5.23		
expt	12.7±1.7	4.95±0.28	-0.72±0.07	-1.52±0.13

The necessity of shorter-range corrections to (8) is apparent when the two-body second moment is considered, while for the three-body terms the DID model overestimates greatly the magnitude of both moments. However, this discrepancy cannot be attributed to the inadequacy of DID only, as is shown in Figs. 1a and 1b, where the density dependence of $\overline{M}_0/N$ and $\overline{M}_2/N$ is plotted, respectively.In both figures the solid line is the contribution of ρ^2- and ρ^3-terms only in the density expansion of the experimental moments. Also shown are the measured moments and the Monte Carlo 256-atom simulation results with both polarizability models (8) and (9) and an updated semiempirical pair potential /13/.

Deviations from the two-plus-three-body expansion of the experimental moments appear at intermediate density (the critical density of krypton is 6.6 atoms/nm^3). Moreover, Fig. 1a shows that in the whole measured density range the pair approximation accounts rather well for the density behaviour of M_0, with very

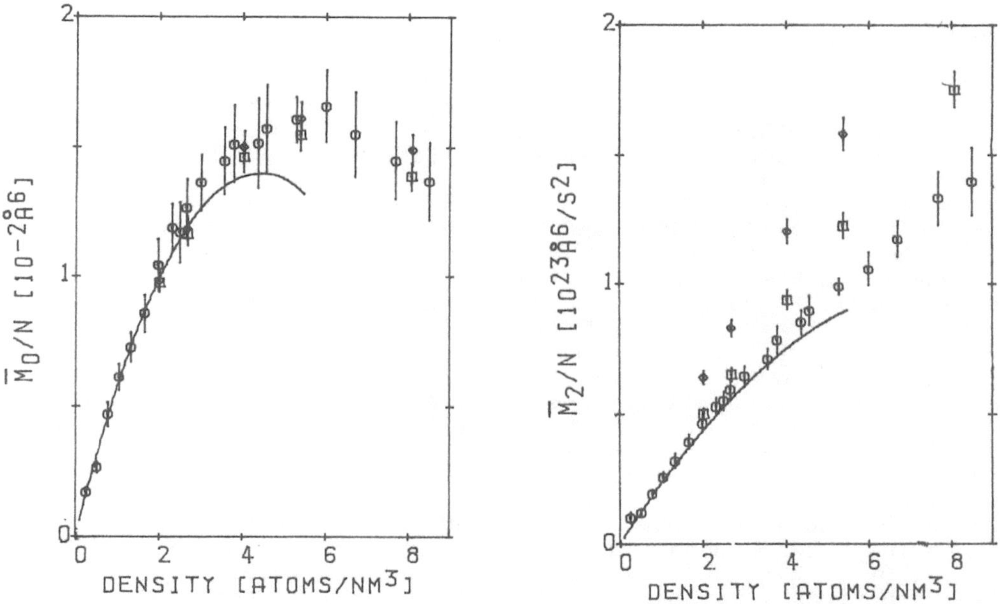

Fig. 1. Density behaviour of (a) $\overline{M}_0/N$ and (b) $\overline{M}_2/N$. Circles are the experiment, diamonds and squares are simulation results with β(r) given by eqs (8) and (9), respectively. See text for the meaning of the solid line.

little differences between models (8) and (9) for (r). On the contrary, in Fig. 1b the large discrepancy between the DID simulation results and the experimental M_2, which is only partially reduced by using the more refined β(r) model (9), shows the onset of irreducible many-body contributions to the polarizability and/or the interaction energy of the N-atom system.

6 Conclusion

From a DILS experiment in simple gaseous systems at various densities it is posssible, in principle, to detect irreducible many-body effects by comparison with a computer simulation performed with a pair interaction potential and a pair polarizability anisotropy function derived from the pair spectrum in the zero-density limit. In order to perform a reliable comparison, absolutely calibrated data and a good model function β(r) fitted to the data are necessary. By applying this method, we have seen that the second frequency moment of DILS in dense gaseous krypton shows evidence of irreducible many-body contibutions /14/.

References

1. A. Teitsma, P.A. Egelstaff: Phys. Rev. A21, 367 (1980).
2. P.A. Egelstaff, W. Glaser, D. Litchinsky, E. Schneider, J.-B. Suck: Phys. Rev. A27, 1106 (1983).
3. P.A. Egelstaff, J.J. Salacuse, W. Schommers, J. Ram: Phys Rev. A30, 374 (1984).
4. H. Fredrikze: Ph.D. Thesis, Technical University of Delft, The Netherlands (1985) and Phys. Rev. A36, 2272 (1987).
5. F. Barocchi, M. Zoppi: Physics Letters, 66A, 99 (1978) and Physics Letters 69A, 187 (1978).
6. M. Zoppi, F. Barocchi, D. Varshneya, M. Neumann, T.A. Litovitz: Can. J. Phys. 59, 1475 (1981).
7. D. Varshneya, S.F. Shirron, T.A. Litovitz, M. Zoppi, F. Barocchi: Phys. Rev. A23, 77 (1981).
8. M. Zoppi, G. Spinelli: Phys. Rev. A33, 939 (1986).
9. M. Zoppi, M. Moraldi, F. Barocchi, R. Magli, U. Bafile: Chem. Phys. Lett. 83, 294 (1981).
10. F. Barocchi, M. Celli, M. Zoppi, Europhys. Lett. 5, 607 (1988).
11. F. Barocchi, M. Zoppi: In Proceedings of the NATO Adv. Res. Workshop "Phenomena Induced by Intermolecular Interactions", ed. by G. Birnbaum (Plenum Press, New York 1985).
12. F. Barocchi, M. Zoppi, U. Bafile, R. Magli: Chem. Phys. Lett. 95, 135 (1983).
13. R. Aziz: Mol. Phys. 38, 177 (1979).
14. F. Barocchi, M. Celli, M. Zoppi: to be published.

Part VI

Summary

Summary of the Workshop on Static and Dynamic Properties of Liquids

U. Dahlborg[1] *and M. Davidović*

[1]Department for Reactorphysics, Royal Institute of Technology, S-10044 Stockholm, Sweden

[2]The "Boris Kidrič" Institute of Nuclear Sciences, Vinča, Belgrade, Yugoslavia

This Summary is the result of the round-table discussion held on the last day of the Workshop, in which all the participants took part. All the topics of the Workshop were covered during the discussion and particular attention was paid to the following subjects:

- liquid-glass transition
- mode-coupling theory
- dynamics of simple liquids
- effective potential.

Some very important contributions at the Workshop dealt with amorphous materials and mode coupling theories, a fact which was also reflected in the round-table discussion. However, in order to make this Summary of more general interest and to make it also readable by the people who did not attend the Workshop, some general background information has been included.

1. The Liquid-Glass Transition and Mode Coupling Theories

When a liquid is cooled below its ordinary melting point rapidly enough to avoid crystallization, the system transforms into a glassy state. During this process, the viscosity increases by several orders of magnitude, which implies that the relaxation time for atomic or molecular processes becomes increasingly longer. In most cases, a glass transition temperature T_g can be identified. It should be mentioned that T_g is conventionally defined as the temperature where the viscosity reaches 10^{13} Poise and thus is not uniquely defined. However, at T_g or close to it the specific heat, C_p, the thermal expansion coefficient, α_p, and the compressibility, κ, in many cases exhibit discontinuities. The magnitude of these responses can be combined to the Prigogine-Defay ratio

$$R = \Delta C_p \, \Delta\kappa / T(\Delta\alpha_p)^2 \, . \tag{1}$$

Springer Proceedings in Physics, Vol. 40
Editors: M. Davidović and A.K. Soper
Static and Dynamic Properties of Liquids

For a system with an ordinary equilibrium transition the ratio is unity while for a glass it was found, as early as the 1950s, to be larger than unity /1/. As mentioned above, the viscosity, η, increases by many orders of magnitude when one goes from the ordinary liquid region to T_g. In a large class of glass-forming systems such a non-Arrhenius behaviour of η could be described up to about 10^7 Poise by the empirical Vogel-Fulcher relation:

$$\eta = A \exp |B/(T-T_0)| \quad . \qquad (2)$$

Both parameters A and B can be adjusted to the experimental data. The transition temperature T_0, defined as the temperature where the viscosity appears to diverge, is usually found to be rather close to T_g. For larger viscosities, an Arrhenius law seems to describe the data better.

In recent years, considerable progress in our understanding of the fundamental nature of the liquid-glass transition (LGT) has been made. Various aspects have been analysed within the mode coupling theories developed independently by Leutheusser /2/ and Bengtzelius et al./3/ some years ago. Further developments and generalizations have been made since then by Götze in a number of papers (see for example /4/) and by the Gothenburg group, both separately and in collaboration /5,6/. The theories predict that an ergodic to non-ergodic phase transition occurs, when the LGT is passed from the liquid side, and they contain many of the features observed in real systems /7/. The behaviour of the system is expected to obey the scaling law

$$\eta(T) = \text{const } (T-T_0)^{-\gamma} \; ; \qquad (3)$$

γ is a non-trivial exponent. Originally, T_0 was identified with the glass transition temperature T_g, but the later work, for example a systematic treatment of existing viscosity data /8/ has indicated that T_0 should instead be considered as a kind of critical temperature for the LGT. The same conclusion could be made from the beautiful neutron scattering data of Mezei and co-workers /9/.

The theories mentioned also predict a particular shape of the density--density correlation function and generalized viscosities. The first experimental evidence for this was provided by Mezei et al., who found that the second relaxation above T_0 could be described by a scaled stretched exponential form $\exp(-t/t_b)^b)$. The exponent b was found to be in fair agreement with theory. The round-table discussion clearly indicated that we today do not know the mechanism behind the first relaxation.

Several suggestions were made, such as the possible existence of extra vibrational modes or some relaxation mechanism involving S(Q). In connection with the first suggestion, the results on silica obtained by Buchenau et al./9,10/ were mentioned. They claim that in order to explain their spectra, in spite of a poor correction for multiple scattering, some low frequency models additional to the ordinary sound modes have to exist in the system.

During the discussion, Sjölander emphasized that the mode coupling theories (MCT) cannot explain everything and that the short time region has to be taken care of separately. Furthermore, in the liquid region a crucial and rigorous test of MCT is still lacking. It was argued by Bosse that many approximations are made in numerical calculations of the MCT and thus that 1% agreement between experiment and theory is not necessary and not a thing to strive for. The emphasis should instead be laid upon the understanding of the underlying physics and on qualitative interpretations of experiments, as for example was demonstrated in the work reported by both Morkel /11/ and Verkerk /12/, earlier and at this Workshop.

At the end of the discussion it was concluded that many more real experiments are needed to understand the physical mechanisms that drive the LGT. Molecular Dynamics (MD) simulations on large systems and long times, which approach the LGT both from the liquid and the solid side, might help but will probably not be conclusive enough to judge whether the theories are really applicable to real systems. This is so because as the LGT is dynamic by nature it will show cooling rate dependencies and hysteresis and in MD simulations the cooling rates are necessarily much larger than those which can be obtained for real systems. It was also stated that the mode theories of today have some shortcomings. Thus, for example, no hopping mechanism which can account for an activated diffusion process is included. It was also shown at this Workshop, by Gaskell, that in a wavevector range where well-defined shear waves are supported by a liquid, the mode coupling theory is unsatisfactory.

2. Dynamics of Simple Liquids

De Graaf summarized our present knowledge about the dynamics in simple liquids and concluded that in order to extract some relevant information today from a neutron inelastic scattering experiment on a liquid it is necessary to have high quality data (good statistics, large enough Q-w range, etc) and to have a full control of the experiment itself. What is even more important is that the experimentalist must be able to accurately correct the data for all the experimental effects. These criteria work hand in hand, and to improve one and not the others to the same degree is meaningless. With the advent of dedicated spectrometers for measurements of dynamic scattering functions S(Q,w) at ILL, ISIS, IPND and Studsvik, it is thus essential to make a much greater effort for development of accurate correction procedures. In order to increase the experimental accuracy, one can anticipate that in the future it will be more and more common to simulate the complete experiment on a computer by Monte Carlo methods before it is actually performed.

The result of an experiment has no value in itself unless there exists also a theory or a model to which the data can be compared and from which the relevant information with regard to the molecular dynamics in the studied system can be extracted. However, no outstanding theory exists today and the theoretical analysis proceeds along several different lines, such as the revised Enskog theory /13/, an eigenmode analysis /14,15/ and an approach based on the generalized hydrodynamics and mode-coupling theories /16,17/. Analysis of data is also commonly made in conjunction with extensive MD simulations. Actually, comparisons of theories have mostly been done against computer simulations. A reason for this is the large flexibility one has in extracting information from these as compared to real experiments from which S(Q,w) is essentially obtained. It is, however, very important to compare the MD results with real experiments to ensure that the simulated system really contains the essential features of a real physical system. In this respect, similar quality criteria have to be imposed on both experimental methods. It is, however, fair to state that the agreement between theory, based on the Zwanzig-Mori formalism and the mode coupling approximations, and experimental and computer simulation data is quite impressive.

At present, it seems thus that we have reached a point where it is appropriate to raise the question whether we, from the physical viewpoint, understand to a satisfactory degree the data we have on simple liquids, or whether we need more data for it. The main opinion expressed at the round-table discussion seemed to be that, at least at present, we do not need much more data of similar accuracy to those we already have. The existing extensive data sets on noble-gas fluids like argon and neon at different thermodynamic states, obtained by the group at Delft /14,15/, and those on liquid metals obtained by the Stockholm group /18/ are accurate enough and also cover a sufficiently large Q-w region to make comparisons to current theories feasible. It was agreed that during the last five years we have substantially increased our knowledge about the microscopic properties of these systems, so much that it can be said now that this field of condensed matter physics has been brought to maturity.

The question was raised whether there was a need for a dedicated spectrometer for neutron Brillouin scattering. At present, the low-Q investigations can be made at very few laboratories and then only with neutrons of energies up to thermal. The main opinion seemed to be that there was such a need and that such a spectrometer would be most useful for studies of collective excitations in amorphous materials, besides, of course, being generally of great value for studies of magnetic excitations. If such a spectrometer is constructed, maybe at ISIS, it is essential that high energy neutrons are used in order to expand the available Q-w space as much as possible and to make it possible to measure collective excitations of the highest possible speed of propagation. It was also pointed out that a neutron Brillouin spectrometer would be very useful for liquids too, for example for tests of mode coupling effects in non-noble gas fluids and in liquid metals in the hydrodynamic regime.

It was argued during the discussion that as our knowledge of simple liquids today is very good, more studies on molecular fluids should be initiated. However, neutron inelastic scattering is rather insensitive to such studies because of the fact that in molecular fluids the time scales for the translational and the rotational motions are similar. Since a neutron is sensitive to both, the two types of motion will mask each other in the measured spectra and, as it was also demonstrated long ago, it will be very difficult to extract any new information from such an experiment. It is important to emphasize that in order to increase our knowledge of the microscopic properties in molecular fluids many different experimental methods, such as light scattering, MD simulations and neutron scattering probably have to be employed simultaneously.

3. Potentials

For a very long time experiments have tried to extract an effective pair potential from the static structure factor S(Q). A number of different theoretical methods are available (hypernetted chain (HNC) equation, etc.) but the conclusion from the first 15 years of work on liquid metals was rather disappointing. With regard to noble-gas fluids, the situation is different since we know that it is possible by the use of the LJ potential to describe both structure and the atomic dynamics with a large degree of precision.

It is thus very satisfactory to note that during the last five years there has been a considerable progress in this field. Both the theoretical development by Hafner and co-workers /19/, the application of the modified HNC to an experimentally determined S(Q) /20,21/, originating from the ideas first put forward by Rosenfeld and Ashcroft /22/, and the phenomenological approaches using the MD simulation method in an iterative way, as developed by Schommers /23/, Dzugutov /24/ and others, have resulted in effective "pair" potentials of similar shape. One reason for this recent development has been considerably increased accuracy of a measured S(Q). It is, however, important to emphasize that the inversion schemes yield effective "pair" potentials which incorporate pair as well as many-body forces and that these contributions cannot be separated from each other. Thus, in order to make comparisons between experimentally derived and theoretically calculated "pair" potentials the theoreticians must be able, within a particular theory or model, to calculate effective 2-body potentials with small 3-body corrections. Information about the short-range order in a disorder system can be obtained from the 3-body term. According to Hafner, this is possible today! This is also a necessary condition to be fulfilled if more accurate MD simulations on molecular systems are to be made, as the inversion schemes are very difficult to use in these cases.

In noble-gas liquids, the 3-body forces are generally small, but it should be realized that the LJ is not a true two-particle potential. In order to determine the contribution from the 3-body forces, experimental methods other than neutron inelastic scattering have to be used. It was suggested by Bratos that one method should be collision-induced infrared absorption of rare gas fluids in which process triplets of atoms make the first contribution. Its interest is due to the fact that a system formed by one or two atoms always has a center of symmetry and cannot contribute to the absorption. The absorption is however weak and the experiment is accordingly very difficult to do.

4. References

1. R.O. Davies and G.O. Jones: Adv. Phys. 2, 370 (1953).
2. E. Leutheusser: Phys. Rev. A29, 2765 (1984).

3. U. Bengtzelius, W. Götze and A. Sjölander, J. Phys. C17, 5915 (1984).
4. W. Götze: Z. Phys. B60, 195 (1984), Phys. Scr. 34, 66 (1986), Z. Phys. Chem. 156, 3 (1988).
5. U. Bengtzelius and L. Sjögren: J. Chem. Phys. 84, 1744 (1986).
6. W. Götze and L. Sjögren: Z. Phys. B65, 415 (1987).
7. F. Mezei, private communication.
8. P. Taborek, R.N. Kleiman and D.J. Bishop, Phys. Rev. B34, 1835 (1986).
9. F. Mezei, W. Knaak and B. Farago: Phys. Rev. Lett. 58, 571, (1987).
10. U. Buchenau, N. Nücker and A.J. Dianoux: Phys. Rev. Lett. 53, 2316 (1984).
11. A.J. Dianoux, U. Buchenau, M. Prager and N. Nucker: Physica B138, 264 (1986).
12. Chr. Morkel, Chr. Gronemeyer, W. Gläser and J. Bosse: Phys. Rev. Lett. 58, 1873 (1987).
13. P. Verker, J.H. Builtjes and I.M. de Schepper: Phys. Rev. A31, 173 (1985).
14. I.M. de Schepper and M.H. Ernst: Physica A98, 189 (1979).
15. A.A. van Well and L.A. de Graaf: Phys. Rev. A32, 2384 (1985), Phys. Rev. A32, 2396 (1985).
16. A.A. van Well, P. Verkerk, L.A. de Graaf, J.-B. Suck and J.R.D. Copley: Phys. Rev. A31, 3391 (1985).
17. O. Söderström, U. Dahlborg and W. Gudowski: J. Phys. F15, L23 (1985).
18. K.E. Larsson and W. Gudowski: Phys. Rev. A33, 1968 (1986).
19. U. Dahlborg, O. Söderström, W. Gudowski, K.-E. Larsson and M. Davidović: J. Phys. F15, 2053 (1985).
20. J. Hafner and V. Heine: J. Phys. F13, 2479 (1983).
21. M.W.C. Dharma-wardana and G.C. Aers: Phys. Rev. B28, 1201 (1983).
22. L. Reatto, D. Levesque and J.J. Weis: Phys. Rev. A33, 3451 (1986).
23. Y. Rosenfeld and N.W. Ashcroft: Phys. Rev. A20, 1208 (1979).
24. W. Schommers: Phys. Rev. A28, 3599 (1983).
25. M. Dzugutov, K.-E. Larsson and I. Ebbsjö: Phys. Rev. A38, (1988), in press.

Index of Contributors

MIX
Papier aus verantwortungsvollen Quellen
Paper from responsible sources
FSC® C105338
FSC
www.fsc.org

If you have any concerns about our products,
you can contact us on
ProductSafety@springernature.com

In case Publisher is established outside the EU,
the EU authorized representative is:
Springer Nature Customer Service Center GmbH
Europaplatz 3, 69115 Heidelberg, Germany

Printed by Libri Plureos GmbH
in Hamburg, Germany